QCD VACUUM STRUCTURE

Proceedings of the Workshop on QCD Vacuum Structure and Its Applications

The American University of Paris
June 1 – 5, 1992

QCD VACUUM STRUCTURE

Editors

HM FRIED
Brown University

B MÜLLER
Duke University

World Scientific
Singapore • New Jersey • London • Hong Kong

Published by

World Scientific Publishing Co. Pte. Ltd.
5 Toh Tuck Link, Singapore 596224
USA office: 27 Warren Street, Suite 401-402, Hackensack, NJ 07601
UK office: 57 Shelton Street, Covent Garden, London WC2H 9HE

British Library Cataloguing-in-Publication Data
A catalogue record for this book is available from the British Library.

QCD VACUUM STRUCTURE
Proceedings of the Workshop on QCD Vacuum Structure and Its Applications

ISBN-13 978-981-02-1280-3
ISBN-10 981-02-1280-1
ISBN-13 978-981-02-1352-7 (pbk)
ISBN-10 981-02-1352-2 (pbk)

PREFACE

Quantum chromodynamics was proposed as the fundamental theory of the strong interactions twenty years ago, and this anniversary has been reason for celebration. Indeed, recent studies of hadronic decays of the Z^0 boson, the carrier of weak neutral interactions, have provided tests of QCD with previously unattained precision. Many aspects of perturbative QCD, including its running coupling constant α_s and the spin of the gluon, are now established and confirmed with considerable accuracy.

This bright picture darkens when one considers QCD phenomena at the energy scale of hadrons and nuclei, or more generally, low-energy excitations of the QCD vacuum. This is so partly because α_s becomes of order unity at hadronic momentum scales, and perturbation theory fails. But this simple explanation is not the whole story. The failure of straightforward perturbation theory is a standard occurrence in many-body physics, and numerous methods have been devised to circumvent it. A basic prerequisite of all such approaches, however, is a thorough understanding of the structure of the ground state of the many-body system, and of the nature of its low-lying excitations.

It is here that our present knowledge is inadequate. After two decades of study, and more than ten years of enormous effort spent on numerical simulations of QCD on space–time lattices, we still do not have a simple, undisputed model of its vacuum state. It is still uncertain which property of the QCD vacuum is responsible for quark confinement, and exactly which mechanism causes the breaking of chiral symmetry. The unprejudiced observer is confounded by the multitude of models that have been proposed for the QCD vacuum state: the dual superconductor, the monopole condensate, the ferromagnetic vacuum, the "spaghetti vacuum," the instanton liquid, the chaotic vacuum, to name only the most widely studied. These all capture certain aspects of QCD dynamics, but none has yet been shown to provide a consistent starting point for the analysis of QCD phenomena at low energies.

When we first proposed to organize a workshop to survey the present status of knowledge about the structure of the QCD vacuum, we expected interest mainly among a small group of experts. To our surprise the number of participants quickly grew to more than sixty. This volume contains the written versions of the lectures and seminars presented at the workshop.* It provides the first, almost-complete overview of the status of this wide-ranging field, and we hope it will serve as a useful reference for experts, as well as a readable introduction for novices.

The workshop would not have been possible without the generosity of the American University of Paris, which provided the conference facilities and organizational support. First and foremost, we express our deepest gratitude to Bill Cipolla, Dean and then Acting President of the University, who enthusiastically welcomed our workshop and suggested that it be held in the University's Grand Salon. We especially thank Karen Archer and Karen Wagstaff, who helped set up the necessary infrastructure; and Susan Bell, who did a marvellous job as workshop secretary and treasurer, and in quickly solving all problems presented to her. We thank Sandra Johnson for handling correspondence and keeping registration files, and Julia Clark for help in editing the proceedings. We also acknowledge the support of the National Science Foundation (Grant No. PHY-9207950), which made possible the participation of six young US scientists.

It is a pleasure to thank S. J. Brodsky, A. Capella, P. Carruthers, N. Cottingham, O. Nachtmann, P. Olesen, H. Reinhardt, and R. Vinh Mau, who, as members of the organizing and advisory committee, helped form the program. And we thank all lecturers who made the workshop successful through the high quality of their presentations. Finally, our gratitude goes to the city of Paris, for providing the incomparable surroundings for this scientific meeting.

H. M. Fried, Brown University
B. Müller, Duke University
November 1992

---

*For completeness we list here the lectures that are not contained in this volume: "Potential Models of Quark Confinement" (A. Martin), "Gauge Invariance of Thermal Processes in QCD" (D. Schiff), "Intermittency and QCD Phase Transition" (R. Peschanski), "Nonperturbative Effects in the Electroweak Vacuum" (P. Olesen), and "Quark Confinement" (V. N. Gribov).

WELCOME

It is with great pleasure that I welcome you to the American University of Paris. We are particularly happy to be hosts of the Workshop on QCD Vacuum Structure for several reasons. As most of you probably know, the American University of Paris is a private, independent four-year liberal arts institution in the tradition of the small American college. Operating as we do in one of the major cultural and scientific crossroads of the world, means that we have immediate access to many distinguished people in many fields.

Unfortunately, we have not yet begun an important program in fundamental science at our University. As strong as our humanities and social science offerings may be, this leaves us with a certain void. (I am trying to avoid "vacuum" since I know of its structure!) For this reason, an important scientific meeting brings, even for a brief and rainy week, a fresh reminder of the vitality and importance of science and scientists.

We hope that your stay with us will be pleasant and rewarding. The success of your Workshop is assured by the excellent organizers, Professors Herb Fried and Berndt Müller, who have, by now, become friends. Let me wish you, then, a happy and productive week and extend my welcome for your next visit.

William F. Cipolla
Dean and Acting President
The American University of Paris

TABLE OF CONTENTS

Part V: Beyond QCD

THOUGHTS ON THE VACUUM: INTRODUCTORY REMARKS

PETER CARRUTHERS
Physics Department, University of Arizona
Tucson, AZ 85721, USA

It is a privilege to expose my ignorance about the important subject of the so-called vacuum to such a distinguished assemblage of experts. As with some other problems in science, the more I think about the vacuum the more confused I get.

As we know, the key breakthrough concerning the vacuum was made by Maxwell, working with an explicit mechanical model of the ether. The vibrations of the ether were identified with the propagating electromagnetic waves, much as sound waves are density waves in a gas or collective oscillations in a crystal lattice.

Einstein's work of genius seemingly made the ether irrelevant. The smooth continuous transformations of the Poincare group became the basis of believing that the vacuum was truly a void. Yet the space- time seemingly is equipped with rulers, on which hang fields whose dynamical behavior is imagined to have no effect on the space time. On the other hand, Einstein found in his general theory that the structure of space-time depends on the matter therein. It seems that something is missing. Of course nowadays the vacuum is ordered to play host to all kinds of symmetry breaking. How does this really happen? If there is texture, how is it compatible with continuity of rotations and Lorentz translations?

In perturbation theory you inject bare particles into a nonexistent bare vacuum. As shown by the vacuum polarization and the Lamb shift, Quantum Electrodynamics gives a spectacular numerical account of vacuum-related effects. But who can be content with the infinite renormalizations that destroy our ability to discuss the basic parameters of field theories? More refined formulations, say the LHZ approach that builds a Hilbert space by applying asymptotic creation operators to an alleged physical vacuum, do not really help very much. The truth must be that since there is no vacuum in the physical universe, our problems might be self inflicted due to a philosophical bias and fundamental error. That is, it is a mistake to identify the ground state with the vacuum. A possibly analogue is the development of the BCS theory of superconductivity. To make sense one must work in a ground state with electron-hole pairs, which is really an entity filled with particles.

Even if we accept the notion that the ground state is not a vacuum,, we have to consider the nature of the medium between the observed particles and the transient excitations that occur in collisions. In addition we learn from condensed matter physics that many phases are possible in various conditions. In addition, how do particles moved through a structured medium without friction? For superfluid helium there is a backflow mechanism; for metals the lattice periodicity gives rise to Bloch's theorem, that is,free electrons. What is the analogous mechanism for the vacuum?

For a long time we have been aware of the reality of vacuum "fluctuations".

2

The mysterious success of perturbative QED has perhaps retarded our progress in this area. My personal confusion on this issue is the following. If the so-called vacuum has a texture, how is this to be reconciled to the apparent validity of the continuous transformations of the Poincare group, which stands like a policeman enforcing continuity at every space-time point in conventional theories? I suspect that the "empty"space has its field energy living in a fractal subspace, due to the nonlinear nature of the field equations. This could help divergence problems if the fluctuations were sufficiently rapid to maintain macroscopic smoothness.

Likewise, when the host medium is excited in a rendezvous of two particles, what textures could we imagine? Clearly, filamentary structures of color flux tubes are possibilities. Domains of chiral condensates are another. A favorite picture, especially for relativistic heavy ion collisions, is the the physical vacuum melts as the quark-gluon plasma is formed. The invasion of this phase could in principle give information on the nature of the usual vacuum state, as could the "bag constant". Current research in condensed matter physics suggests an enormous variety of other possibilities. No doubt many of you have fantasized about such things.

High multiplicity multiparticle events may provide clues, since one can reasonably imagine that they arise from configurations with very high energy density. Likewise jets, penetrating a vacuum hostile to color, should be considered as probes of the vacuum. After all, the conversion of partons to hadrons might depend as much on the friction of the vacuum to color currents as on the nonlinear couplings of QCD. In fact the decceleration of a classical (high momentum) current produces coherent states if we ignore nonabelian charges. The resulting Poissonian distribution closely resembles the two-jet multiplicity distributions seen in electron-positron annihilations. Finally we note that, event by event, one should expect highly irregular, perhaps fractal space time geometry. As the soft and hard jets penetrate the nearby vacuum, there should be boundary effects whose character I cannot describe today.

Hadronization is schematically similar to the behavior of a Type II superconductor in two dimensions. This medium (the vacuum) can play host to flux tubes, whose equilibrium configuration is a hexagonal lattice. At $T = 0$ we inject a cylinder of magnetic flux; which is unstable with regard to the final state lattice. By detecting the lattice and perhaps the time dependence of the process, we might be led to an understanding of the ground state. Unfortunately, in high energy collisions we are deprived of the capability to track the time dependence of the reaction, with the possible exception of the so-called Bose-Einstein effect.

To summarize the biases of an outsider, I restate the challenge: there is no such thing as a "vacuum". It is nearly impossible to analyze those apparently empty parts of space time without paying attention to the existing matter in the universe. Remember that it took thirty years to get from the Fermi sea to the BCS ground state.

This is not an easy problem!

RENORMALIZATION AND CONFINEMENT

JANOS POLONYI [†]

Department of Physics, MIT, Cambridge, MA 02139, USA

ABSTRACT

The numerical results of lattice gauge theory indicate that the configurations contributing to the path integral in the QCD vacuum contain Dirac delta singularities. An effective theory is introduced to describe the effects of these singularities. The physical content of the effective theory is briefly discussed.

1. Introduction

The confinement of quarks remains a puzzle in high energy physics and blocks the way of the systematical understanding of the strong interactions. The phenomenological models suggest several realistic details of the long distance features of the strong interactions but their relation with the fundamental theory, QCD, is lacking. The experimental guidance to establish this connection is missing as well because the confinement of quarks is an essentially negative result. The absence of the isolated quark states in the Rosenfeld table and the studies of the decay pattern of highly excited hadrons offers no hint as of the nature and the origin of the confining forces are concerned.

There are two fundamentally different frameworks to explain the absence of charged states in the asymptotic sector of a theory. In the more economical scenario the charges are permanently screened. This can be illustrated on a hand-waving level by the analogy of the positronium in the following manner. Suppose that we modify the coupling charge, e, in QED that the the ionization energy,

$$\mathcal{E}_{\text{ion}} \approx \frac{e^4}{m},\tag{1.1}$$

becomes comparable or larger than twice the rest mass of the electron, $2m$. With the actual values of e and m this is the case for $e = O(1)$ if the perturbative result (1.1) remains valid in this case. Such strong coupling QED, if consistent, does confine

---

[†] On leave from L. Eötvös University and CRIP, Budapest, Hungary

4

the electrons. In fact, the attempt of removing an electron from the positronium would require more energy than the creation of an e^+e^- pair. Thus the splitting of the positronium into the electron and the positron is followed by the conversion of a virtual e^+e^- pair into a real one and the screening of the separated charges. This might be called soft confinement mechanism since it involves no long range forces. It can certainly be found in the form of the quark-anti quark vacuum polarizations in QCD [1].

The soft screening of the isolated color charges is not sufficient to explain the emergence of the Regge trajectories which suggest the presence of a linear potential, $V(r) \approx \sigma r$, acting between the quarks. The unbounded potential of the static color charges, the hard confinement mechanism, is competing with the soft one in the real world. Their balance is measured by

$$\mathcal{R} = \frac{r_\pi \sigma}{m_\pi}, \tag{1.2}$$

the ratio of the size of the lightest meson and the distance which corresponds to the increase of the static quark potential by the meson mass. (1.2) is close to one and makes the experimental isolation of the confinement mechanism too complicated. From the theoretical point of view one should study the hard mechanism first since it is supposed to be realized in a simpler system which consists of gluons only. The mechanism which pushes the running coupling constant, g, into the region $g = O(1)$ already involves the long range features of the gluonic interactions and the soft screening process should be studied after clarifying the non-perturbative aspects of the pure gluon system.

Here we suggest an approach which might be able to explain the the emergence of the linear potential in QCD [2]. We are far from being able to prove the existence of the string tension, σ, for pure gluons. Nevertheless the merit of our construction is that it is an attempt to incorporate the lesson of the numerical studies of lattice gauge theory into an effective field theory in the continuum.

We start by emphasizing the importance of the exact gauge invariance for the understanding of the origin of the string tension. Suppose that we try to separate a quark-anti quark pair in a gluonic vacuum whose gauge invariance is only satisfied in an approximative manner. Then we have gauge-dependent components in the vacuum state which are not under our control. The gauge dependence signals the presence of charges in these components which may screen the quarks by the soft mechanism and the string tension is lost. In order to derive an unbounded potential we can not afford any leakage in Gauss' law.

A more explicit relation between confinement and gauge invariance is revealed by the studies of QCD at finite temperature. It was found that the confining forces prevail as long as the center symmetry, a discrete part of the gauge symmetry, is preserved [3]. Our starting point is that the presence of the symmetry under the global gauge transformations goes together with the existence of the string tension.

The center symmetry has several distinguished features [4], such as

(i) It is an 'invisible' symmetry, like the rotation of a a particle moving along the circle by 2π. It is the symmetry under the fundamental group of the Yang-Mills system.

(ii) It tends to break dynamically at high energy, in contrast with the usual spontaneous symmetry breaking mechanism.

(iii) Its presence leads to Dirac delta type singularities,

$$A_0^a(x) \approx \delta(x^0)a_0^a(x), \tag{1.3}$$

in the gluon field. These singularities in turn alter the structure of the gluonic vacuum in a substantial manner.

We shall demonstrate the last claim here. We start by elaborating the notion of the center symmetry in Chapter 2 and continue by the discussion of the role of the singular configurations in Quantum Field Theories in Chapter 3. A difference between the continuum and the lattice regulated QCD is pointed out in Chapter 4. The renormalization of QCD is discussed briefly in Chapter 5 and a possible realization of this scenario based on the instantons of size comparable with the cut-off is presented in Chapter 6. The physical content of these singularities is summarized by the effective theories presented in Chapter 7. Chapter 8 is reserved for the concluding remarks.

2. Center Symmetry

Lattice Yang-Mills models posses a discrete symmetry which consists of multiplying the time directed link variables, $U_0(x)$, of a chosen equal-time hyperspace $x^0 = t$ by an element of the gauge group, z, which commute with the whole group, $[z, G] = 0$,

$$U_0(x) \rightarrow zU_0(x). \tag{2.1}$$

In fact, the plaquettes remain invariant under (2.1) due to $z^\dagger z = 1$ and the domain of integration of the path integral is unchanged because $z \in G$. The subgroup C of the group G which commutes with G, $[C, G] = 0$ is called the center of G. It consists of the identity matrix multiplied by the N-th roots of one for G=SU(N).

The quarks have vanishing propagators when the center symmetry is realized [4]. To see this write the quark propagator, $G(x,y)$, with $x^0 < t < y^0$ as the sum over trajectories γ with end points x and y by the help of Schwinger's proper time method,

$$G(x,y) = \int D[\gamma]\Gamma[\gamma] < Pe^{\int_x^y dz^\mu A_\mu(z)} >, \tag{2.2}$$

6

where $\Gamma[\gamma]$ is a path dependent matrix for the spinor components, P stands for path ordering,

$$A_\mu = gA_\mu^a \frac{\lambda^a}{2i}, \tag{2.3}$$

and $< \mathcal{O}[A_\mu] >$ denotes the expectation value of the operator $\mathcal{O}[A_\mu]$ in the gluonic vacuum. $Pe^{\int_x^y dz^\mu A_\mu(z)}$ is the product of the link variables along the trajectory γ, $\prod U_\nu(z)$, in lattice regularization. This product contains the complex adjoint of the link variable, $U_\mu^\dagger(z)$, when the direction of the trajectory γ at the location z is $-\hat{\mu}$ i.e. the negative direction along the μ-th coordinate axis. The trajectory γ contains one more time directed link, $U_0(x)$, at the chosen hyperspace than a complex conjugate, $U_0^\dagger(x)$, since $x^0 < t < y^0$. Thus $Pe^{\int_x^y dz^\mu A_\mu(z)}$ transforms as

$$\prod U_\nu(z) \rightarrow z \prod U_\nu(z) \tag{2.4}$$

under (2.1). Since the sum of the n-th roots of one is vanishing,

$$\sum_{j=1}^{N} e^{ij\frac{2\pi}{N}}, \tag{2.5}$$

$G(x,y) = 0$ for the center symmetrical vacuum. Note that the argument remains valid for any gauge fixing at the points x and y.

The continuum Yang-Mills theory is formally invariant under the center transformations, too. Consider the matrix element of the time evolution operator, $< \mathbf{A}_f(x)|e^{-itH}|\mathbf{A}_i(x) >$, in temporal gauge hamiltonian formalism [5]. The canonical variables are $g\mathbf{A}^a(\mathbf{x})$ and

$$\mathbf{E}^a(\mathbf{x}) = \frac{1}{i}\frac{\delta}{\delta g\mathbf{A}^a(\mathbf{x})}, \tag{2.6}$$

with

$$[A_i^a(\mathbf{x}), E_j^b(\mathbf{y})] = i\delta^{ab}\delta_{ij}\delta(\mathbf{x} - \mathbf{y}). \tag{2.7}$$

The gauge condition $A_0(x) = 0$ leaves the static gauge transformations,

$$A_\mu(x) \rightarrow A_\mu^h = h(\partial_\mu + A_\mu)h^\dagger, \tag{2.8}$$

with $\partial_0 h = 0$ symmetry of the system. The infinitesimal form of (2.8) is

$$\delta\mathbf{A} = \mathbf{D}\epsilon, \tag{2.9}$$

with $h = 1 + \epsilon + O(\epsilon^2)$ and

$$\mathbf{D}\epsilon = \partial + [\mathbf{A}, \epsilon]. \tag{2.10}$$

The vacuum sector what we are mainly interested in is made of gauge invariant states. The matrix element $< \mathbf{A}_f(x)|e^{-itH}|\mathbf{A}_i(x) >$ is not gauge invariant as it stands,

$$< \mathbf{A}_f^h(x)|e^{-itH}|\mathbf{A}_i(x) > \neq < \mathbf{A}_f(x)|e^{-itH}|\mathbf{A}_i(x) >, \tag{2.11}$$

but is invariant under global center transformations,

$$< A_j^z(\mathbf{x})|e^{-itH}|A_i(\mathbf{x}) >=< A_j(\mathbf{x})|e^{-itH}|A_i(\mathbf{x}) >, \tag{2.12}$$

with $z(\mathbf{x}) = z_{\text{glob}}$, $[z, G] = 0$. In fact, the gauge field, $A(\mathbf{x})$, remains unchanged under such gauge transformations according to (2.8) and (2.12) is trivial.

What is non-trivial is whether this formal symmetry is respected in certain matrix elements of the vacuum sector,

$$< A_j^z(\mathbf{x})|e^{-itH}|A_i(\mathbf{x}) >_0=< A_j(\mathbf{x})|e^{-itH}|A_i(\mathbf{x}) >_0, \tag{2.13}$$

where

$$|A >_0= \mathcal{P}_0|A >= \int D[h(\mathbf{x})]|A^h > . \tag{2.14}$$

Here $\mathcal{P}_0$ denotes the projection operator into the vacuum sector, which is the averaging over the symmetry group,

$$\mathcal{P}_0 = \int D[h(\mathbf{x})]\hat{h}. \tag{2.15}$$

Note that the averaging over the symmetry group must be made by the help of the invariant Haar-measure,

$$\int D[h]f(h) = \int D[h]f(hh'). \tag{2.16}$$

In fact, it is easy to verify that (2.13) holds when the integral measure satisfies (2.16).

The physical matrix elements of the vacuum sector,

$$< A_j|\mathcal{P}e^{-itH}|A_i >= \int D[h] < A_j|\hat{h}e^{-itH}|A_i >= \int D[h] < A_j^h|e^{-itH}|A_i >, \tag{2.17}$$

contain the functional integral over the infinite dimensional symmetry group $\mathcal{G}$ which is the direct product of the local gauge group at each space point, $G_{\mathbf{x}}$,

$$\mathcal{G} = \otimes \prod_{\mathbf{x}} G_{\mathbf{x}}. \tag{2.18}$$

The phenomenon of phase transitions corresponds to the non-uniqueness of the integral measure in an infinite dimensional space. It may happen that the domain of integration is reduced in (2.17) in a manner which is not detectable by local, physical conditions. In our case, depending on the h-dependence of the integrand of (2.17), the integration may be constrained to

$$\mathcal{G}_{\text{dec}} = \mathcal{G}/C_{\text{glob}}, \tag{2.19}$$

where C_{glob} denotes the global center. This turns out to be the case at high temperature where the symmetry under global gauge transformations is dynamically broken

8

and the system becomes deconfined. For low energy or long time the matrix elements exhibit the invariance under the full symmetry group, $\mathcal{G}$, according to the numerical results in lattice gauge theory. Note that we are talking about the phase of the matrix elements rather than the theory itself. Different matrix elements of the same theory might be in different phase.

In order to connect this formal symmetry with the more conventional observables we need another ingredient. First we write the matrix element (2.17) in the form,

$$\int D[h] < \mathbf{A}_f |\hat{h} e^{-itH}|\mathbf{A}_i > = \int D[\alpha^a(\mathbf{x})] < \mathbf{A}_f |e^{-itH+i\int d^3 x \alpha^a(\mathbf{x})\mathbf{DE}^a(\mathbf{x})}|\mathbf{A}_i >, \tag{2.20}$$

by the help of the parametrization $h = e^{i\alpha^a \lambda^a}$. The commutation relations, (2.7), can be used to prove that

$$\mathbf{DE}(\mathbf{x}) = \partial \mathbf{E}(\mathbf{x}) + [\mathbf{A}(\mathbf{x}), \mathbf{E}(\mathbf{x})], \tag{2.21}$$

with $\mathbf{E} = E^a \frac{\lambda^a}{2i}$ are the generators of the gauge transformations in the Hilbert space and they replace the Gell-Mann matrices λ^a in $\hat{h}$. The gauge invariance of the dynamics, $[H, \mathbf{DE}] = 0$, allows to collect $\hat{h}$ and the time evolution operator into one exponential. The next step is to obtain the path integral representation for the matrix element in (2.20). After performing the standard manipulations one finds

$$< \mathbf{A}_f |e^{-itH+i\int d^3 x \alpha^a(\mathbf{x})\mathbf{DE}^a(\mathbf{x}))}|\mathbf{A}_i > = \int D[\mathbf{A}(\mathbf{x},t)]e^{iS_{\text{YM}}[A,A_0=\frac{\alpha}{t}]}, \tag{2.22}$$

where

$$S_{\text{YM}}[\mathbf{A}, A_0] = \frac{1}{2g^2} \int d^4 x \, \text{tr}(\partial_\mu A_\nu(x) - \partial_n u A_\mu(x) + [A_\mu(x), A_n u(x)])^2. \tag{2.23}$$

In other words, the angle parameter of the gauge transformation which reinforces Gauss' law appears as a static A_0 field in the path integral expressions. Thus the complete matrix element, (2.17), reads as

$$\int D[h] < \mathbf{A}_f^h |e^{-itH}|\mathbf{A}_i > = \int D_H[tA_0(\mathbf{x})]D[\mathbf{A}(\mathbf{x},t)]e^{iS[A,A_0]}, \tag{2.24}$$

where the integration is performed by the invariant Haar-measure with the angle parameters $tA_0^a(\mathbf{x})$. In the first equation the field $tA_0^a(\mathbf{x})$ is introduced as the angle variable of the static gauge transformation.

We are accustomed to deal with path integral with flat integral measure. It is not difficult to render (2.24) into such form. We write the Haar-measure for the group element e^{tA_0} as $d_f tA_0 \rho(tA_0)$ where $d_f tA_0 = \prod_a tdA_0^a$ is the flat integral measure. We shall consider here only the case of SU(2) gauge theory for simplicity, but the generalization for larger groups is straightforward. For the SU(2) matrix $e^{\alpha^a \sigma^a/2}$ the Haar-measure is

$$d_H \alpha = d\omega d\alpha \sin^2 \alpha = \omega d\alpha \alpha^2 \frac{\sin^2 \alpha}{\alpha^2} = d_f \alpha \frac{\sin^2 \alpha}{\alpha^2}, \tag{2.25}$$

where the parametrization $\alpha^a = \omega^a \alpha$, $\omega^2 = 1$ is used. Thus (2.24) can be written as

$$\int D[h] < \mathbf{A}_f^h | e^{-itH} | \mathbf{A}_i > = \int D_H[A_0(\mathbf{x})] D[\mathbf{A}(\mathbf{x},t)] e^{i \int d^4 x \{ L_{\mathrm{YM}}[\mathbf{A},A_0] + \frac{1}{ta^3} \ln \rho(tA_0)\}}, \qquad (2.26)$$

where

$$S_{\mathrm{YM}} = \int d^4 x L_{\mathrm{YM}} \qquad (2.27)$$

is the usual Yang-Mills lagrangian which is quartic in the gauge field and the functional integration is according to the flat measure. There is a single dimensionless correction factor, ρ, for each degree of freedom in $A_0(\mathbf{x})$, so the last term in the bare action should be normalized by t times the volume element of a single degree of freedom, a^3. The non-flat measure can only be treated properly by regularizations which are local is space and the cut-off, the lattice spacing a, enters in the measure correction term explicitly.

The expression (2.24) is like in the text books except that the measure is non-flat and A_0 is static. We can remove this latter restriction by recalling that $A_0(\mathbf{x})$ is the angle parameter of the integral variable in the projection operator $\mathcal{P}$. It is static in (2.24) because $\mathcal{P}$ was inserted once in the matrix element (2.17). By inserting $\mathcal{P}$ at each time slice in the derivation of the path integral formula one is left with a time dependent A_0 field,

$$< \mathbf{A}_f | e^{-itH} | \mathbf{A}_i >_0 = \int D_H[\Delta t A_0(\mathbf{x},t)] D[\mathbf{A}(\mathbf{x},t)] e^{iS[\mathbf{A},A_0]}, \qquad (2.28)$$

where Δt is the infinitesimal time interval of the short time matrix element appearing in the derivation of the path integral expression and the variable $\Delta t A_0^a(x)$ is integrated over by the invariant measure. Note that for a compact gauge group the invariant measure is periodic in the angle variable $\Delta t A_0^a(x)$ and the time-like cut-off, Δt, remains present explicitly in the renormalized path integral.

There is a fundamental difference between the static A_0 path integral, (2.24), and (2.28) for confining theories. The manifest center symmetry means that the functional integration in the projection operator $\mathcal{P}$ includes the center transformations. The insertion of $\mathcal{P}$ several times in the time evolution does not change this feature and the center transformations are performed during the infinitesimal time interval, Δt, in (2.28). Thus

$$\omega(\mathbf{x},t) = e^{i \int_t^{t+\Delta t} dt' A_0^a(\mathbf{x},t') \lambda^a}, \qquad (2.29)$$

which is the gauge transformation performed between the time slices can be a finite distance away from the identity, $\omega(x) \approx z$, and one finds the singularities (1.3) in the the renormalized theory. In a manifest covariant regularization we find the singularities

$$A_\mu(x) \approx a_\mu(x) \delta(x^\mu - x_0^\mu). \qquad (2.30)$$

10

These singularities lead immediately to further complications. In fact, the gauge invariance is lost at the singularities and a more sophisticated action is needed. The more careful repetition of the steps leading to (2.28) shows that the kinetic energy density is actually

$$\frac{1}{2g^2}\mathbf{E}^2(\mathbf{x},t) = \frac{1}{2}\left(\frac{\mathbf{A}(\mathbf{x},t+\Delta t) - \mathbf{A}^\omega(\mathbf{x},t)(\mathbf{x},t)}{\Delta t}\right)^2,\tag{2.31}$$

which reduces to the usual expression,

$$\frac{1}{2}(\partial_0\mathbf{A}(\mathbf{x},t) - \mathbf{D}A_0(\mathbf{x},t))^2,\tag{2.32}$$

only for $\omega \approx 1$. For ω finite distance away from the identity one must retain the higher order terms in the transformation law (2.9) which generate the link variable ω in the action.

One can write (2.28) in terms of the flat integration measure as

$$\int D[h] < \mathbf{A}_f^h|e^{-itH}|\mathbf{A}_i> = \int D[\Delta t A_0(\mathbf{x})]D[\mathbf{A}(\mathbf{x},t)]$$
$$e^{i\int d^4x\{L_{\mathrm{YM}}[\mathbf{A},A_0]+\frac{1}{\Delta t a^3}\ln\rho(\Delta t A_0)\}}.\tag{2.33}$$

Thus the cut-off lagrangian is

$$L = L_{\mathrm{YM}} + \frac{1}{a^4}\ln\rho(aA_0),\tag{2.34}$$

where the isotropic lattice cut-off, a, is used. It is instructive to consider the case of SU(2) gauge theory where

$$L = L_{\mathrm{YM}} - \frac{1}{a^4}\ln\sin^2 aA_0 + \frac{1}{a^4}\ln(aA_0)^2.\tag{2.35}$$

What is important here is the second, periodic term in the lagrangian. The center symmetry is just the invariance under the shift $aA_0^a \to aA_0^a(1+\pi)$ of the field by the period length of the second term. In the lacking the compact measure term the flat integral measure breaks the center symmetry and the confinement is lost.

To summarize, we found that the center symmetry is necessary for the gluonic system in order to generate the confining static potential. The manifest center symmetry generates Dirac delta singularities on the gluon field and the derivation of the QCD action in the path integral formalism requires more care. The new pieces of the path integral expression, the invariant Haar-measure and the non-polynomial link variables in the action are essential for the confinement mechanism and represent a serious challenge to our understanding of gauge theories. Is the center symmetrical, confining gauge theory renormalizable and asymptotically free ? The answer suggested by the numerical results of lattice gauge theory is affirmative. In fact, the invariant Haar-measure and our link variables are standard pieces of the lattice regulated Yang-Mills theories which appear to have ultraviolet fixed point with asymptotically free beta function. But in this case what happens with perturbative QCD ? How can it satisfy the Slavnov-Taylor identities with the flat integral measure which misses Gauss' law ? How do these new pieces of the path integral influence the dynamics ? We attempt to answer these questions below.

3. Singular Trajectories in Quantum Physics

The appearance of the singularities on the field configurations is the rule rather than the exception for renormalized quantum field theories. To understand this better we start with the case of a free non-relativistic particle when the path integral is of the form

$$\prod_j \int dx_j e^{\frac{i}{2\Delta t}\sum_j (x_{j+1}-x_j)^2}. \tag{3.1}$$

The typical trajectories are where each contribution in the sum is $O(1)$. In fact, for smaller or larger kinetic energy the trajectory has too small 'entropy' or is suppressed by the kinetic energy. Thus the typical trajectories are non-differentiable,

$$\frac{\Delta x}{\Delta t} = O\left(\frac{1}{\sqrt{\Delta t}}\right). \tag{3.2}$$

The more precise statement which can be proven for Euclidean space-time is that the Wiener integral measure which corresponds to the diffusion problem is concentrated on nowhere differentiable functions [6]. The better the resolution we use to follow the evolution of the particle, larger are the fluctuations of the trajectories. One can write (3.2) in the form

$$\Delta p \sqrt{\Delta t} = O(1), \tag{3.3}$$

involving the uncertainty of the momentum, Δp, determined by the trajectories in time Δt. One may associate the fractal dimension 1.5 to the particle moving in 1 dimension. This inherent disorder of the quantum particle is not a mathematical subtlety. One can easily verify that the canonical commutation relations, the loss of the canonical invariance in Quantum Mechanics and the special role played by the Cartesian coordinate system in the quantization process depend crucially on the presence of the singularities (3.2).

In higher dimensions, i.e. in quantum field theories the singularities become more violent. One can show that the configurations in the d dimensional path integral for which the kinetic energy is order one for each time slice are continuous only if one averages them over a d-2 dimensional region [7]. Thus the typical field configurations are nowhere continuous in field theories. The applicability of the perturbation expansion depends on the magnitude of these disordered fluctuations. For weak coupling one expects small fluctuations and the typical configurations of the semiclassical expansion are the sum of a smooth saddle point configuration and the small discontinuous fluctuations. If the observable one is interested in is a smooth functional of the field configuration then its expectation value is dominated by the smooth saddle point configuration and the semiclassical approximation should work. But for strong couplings the fluctuations are as important as the saddle point and the non-smooth behavior of the configurations can not be neglected. In this case the topological notions are useless for the field theory. It is usually a rather complicated energy-entropy balance which determines the density and other characteristic parameters of the singularities in the renormalization process. As an example consider point singularities in a scale invariant theory where the action of a point singularity is logarithmically divergent,

$$S_{\text{sing}} = C \log \frac{L}{a}. \tag{3.4}$$

Here L and a are the infrared and the ultraviolet cut-off, respectively. The contribution of one point singularity to the path integral is

$$\left(\frac{L}{a} \right)^d e^{-C \ln \frac{L}{a}} = L^{d-C} a^{C-d}. \tag{3.5}$$

For strong coupling, $C < d$, such kind of point singularities should play role in the renormalized theory despite their divergent action.

A point singularity is called anomalous if its presence influences the long distance features of the configurations. It is easy to find such singularities in theories with a naively conserved topological current, such as the d-dimensional O(d) sigma model or the sine-Gordon model. Topological defects, i.e. the point singularities with topological flux are of this kind because they usually lead to the non-conservation of the the topological current. The presence or the absence of these singularities in the renormalized theory which is of interest for us. The three dimensional non-linear sigma model and the two dimensional sine-Gordon model have been analyzed from this point of view [8]. It has been found that the anomalous singularities are present or are suppressed for strong or weak coupling, respectively. Similar scenario can be found in compact U(1) gauge theory, too.

One can usually split the field configuration into a 'smooth', non-compact and a singular part in the Villain approximation. In this scheme a periodic function, $\frac{1}{\epsilon} s(x) = \frac{1}{\epsilon} s(x+1)$, is approximated for small ϵ as

$$s_{\text{Vill}}(x) = -\epsilon \ln \sum_{n=-\infty}^{\infty} e^{-\frac{s''}{2\epsilon}(x-n)^2}, \tag{3.6}$$

with $s'' = \frac{d^2 s(0)}{dx^2}$. This method allows us to approximate the path integral with a periodic action, $S[\phi(x)]$, as

$$\int D[\phi(x)]e^{-S[\phi(x)]} \approx \int D[\phi(x)]e^{-S_{\text{vill}}[\phi(x)]}$$
$$= \sum_{\{n_\alpha(x)\}} \int D[\phi(x)]e^{-S[\phi(x),n_\alpha(x)]}. \tag{3.7}$$

Here the variable $\phi(x)$ is unconstrained, there is an integer valued Villain field, $n_\alpha(x)$, for each periodic piece in the action and $S[\phi(x), n_\alpha(x)]$ is quadratic in $\phi(x)$. In theories with dimensionless coupling constant the quadratic part of $S[\phi(x), n_\alpha(x)]$ is the massless kinetic energy for $\phi(x)$. Thus the field $\phi(x)$ is slowly varying in x. The singularities appear at the space-time locations x where $n_\alpha(x) \neq 0$. The integration over $\phi(x)$ can be performed and one finds Coulomb interaction between the singularities. The strong coupling phase of the original compact theory with singularities can not be described in continuum regularization. Thus the continuum limit, if exists, represents a continuum quantum field theory which can only be described in discrete space-time. The usefulness of the Coulomb-gas representation of the singularities is that it rescues the continuum formalism by the price of having new degrees of freedom with point-like Coulomb singularities. One may eventually develop improved approximations for this Coulomb gas guided by our knowledge in classical electrodynamics.

The singularities are absent in the perturbation expansion if their kinetic energy diverges and the perturbative renormalization can be formulated without taking them into account. When a non-perturbative lattice regularization is employed then theory may develop a different continuum limit than with the continuum regulators. This is because by putting the theory on the lattice we automatically allow the presence of the singularities. The systematical renormalization has to preserve their effects which usually requires additional fine tuning in the action. Thus the non-perturbative regulator reveals a new, strong coupling continuum limit by allowing new modes which are absent in the perturbative continuum regularizations to appear in the cut-off theory. These modes may make new operators, such as the density of the point singularities, relevant or marginal at the infrared fixed point of the theory. In this case these operators have to be present in the cut-off theory as well. Since the ultimate, non-perturbative definition of the renormalized quantum field theories are given in terms of the ultraviolet fixed points of the lattice regulated cut-off theories these new continuum limits are as physical as the usual, perturbative ones.

By moving over gauge theories one suspects the relevance of more complicated singularities. In compact U(1) gauge theory with mixed action exhibits critical behavior and it may possesses a Coulombic and a confining continuum limit for weak and strong coupling, respectively. The confinement is achieved by the condensation of Dirac monopoles in the Villain approximation [9]. In the same approximation one can see the generation of a fugacity term for the area of the world-sheets swept by the Dirac strings which connect the monopole-anti monopole vacuum polarization pairs. Thus the unusual strong coupling phase is confining by the help of the proper fine tuning of this fugacity parameter.

The center symmetry discussed above introduces the volume singularity (1.3) for the Yang-Mills systems with unbroken gauge symmetry. It is an anomalous singularity which can be understood in the following manner. Suppose that we have configurations where $a_0^a(x)$ of (1.3) is non-vanishing in a finite space-time region, R_κ. Consider then a Wilson loop, $W[\Gamma]$, corresponding to the contour, Γ, in the space-time whose size is larger than the typical size of the region R_κ. Then the 'gas' of these singularities influences the Wilson loop because $W[\Gamma]$ changes by a finite phase when Γ pierces one additional region R_κ. The correlation of the regions R_κ governs the decay of the Wilson loop by this mechanism. This is actually a specific realization of the stochastic confinement mechanism which can be introduced in a more general manner [10]. The important difference between the center symmetry related singularities (1.3) and the point singularities mentioned briefly above is that the former has smaller action. In fact, the global, i.e. space independent center transformations form a symmetry of the system and they leave the action invariant. The contribution of the center transformations to the action when it is performed in a finite region, R_κ, comes from a surface effect. Thus it is more likely that its suppression is overwritten by the entropy factor and the singularities remain present in the continuum limit. It is the lesson of the numerical simulations of lattice gauge theory that the center symmetry is preserved in the continuum limit. Thus the continuum limit, i.e. renormalization of the QCD vacuum must be rather involved in the continuum description.

4. Continuum vs. lattice QCD

We learned in the previous Chapter that the singularities of the field configurations may survive the renormalization procedure and drive the system toward a different ultraviolet fixed point. Furthermore, the direct numerical evaluation of the lattice regulated path integral without any gauge fixing indicates the presence of this mechanism in the QCD vacuum. Let us consider first perturbative QCD with the covariant gauge fixing,

$$L_{\text{gf}} = -\frac{1}{\xi}\text{tr}(\partial_\mu A_\mu)^2. \tag{4.1}$$

This is the gauge where the theory with flat integration measure was studied most intensely and the renormalization has been established in every finite order of the loop expansion. The use of the invariant Haar-measure amounts to the additional terms (2.34) in the lagrangian. Their coupling constant, $1/a^4$, is the volume of the momentum space and can be written as

$$\frac{1}{a^4} = \int \frac{d^4q}{(2\pi)^4}.$$ (4.2)

The integral of the type $I(z) = \int dq\, q^z$ is vanishing in the dimensional regularization. Thus the invariant measure which is needed for the formal gauge invariance gives no contributions in this regularization. This property of the dimensional regularization, namely that the Feynman graphs containing the insertions of the measure term can be neglected might appear as an advantage. In fact, the gauge invariant theory can be developed with the flat measure in perturbation expansion.

The lattice regularization seems clumsier with its measure terms in the action. But these complicated looking terms are needed to fix up the gauge invariance of the theory which is broken by the specific choice of the action. To see this consider for example the Ward identity in QED,

$$\frac{\partial}{\partial p_\mu} \Sigma(p) = \Gamma^\mu(p, p, 0),$$ (4.3)

where $\Sigma(p)$ and $\Gamma^\mu(p, q, r)$ are the electron self energy and the electron-photon vertex function, respectively. The perturbative proof of (4.3) is based on the observation that the insertion of a zero momentum photon exchange into an electron line of a Feynman graph,

$$\frac{1}{ip_\nu \gamma^\nu} \rightarrow \frac{1}{ip_\nu \gamma^\nu} i\gamma^\mu \frac{1}{ip_\kappa \gamma^\kappa},$$ (4.4)

is equivalent by taking the derivative of the original electron line with respect of the momentum,

$$\frac{1}{ip_\nu \gamma^\nu} \rightarrow \frac{\partial}{\partial p_\mu} \frac{1}{ip_\nu \gamma^\nu}.$$ (4.5)

In the lattice regularization the degrees of freedom are located finite distance apart in the space-time, the electron propagator is periodic in the momentum space,

$$\frac{1}{i \sum_\mu \frac{1}{a} \sin(p_\mu a) \gamma^\mu},$$ (4.6)

and the argument fails. The Ward identities are now rescued by adding further electron-photon vertices. These can be generated by replacing the naive non-compact electron lagrangian,

$$L_{\mathrm{nc}} = \bar{\psi}(x) \big[\frac{1}{2a} \sum_\mu \gamma^\mu ((1 + ieaA_\mu(x))\psi(x + \hat{\mu}) - (1 - ieaA_\mu(x))\psi(x - \hat{\mu})) + m\psi(x) \big],$$ (4.7)

by the compact one,

$$L_{\mathrm{c}} = \bar{\psi}(x) \big[\frac{1}{2a} \sum_\mu \gamma^\mu (e^{ieaA_\mu(x)}\psi(x + \hat{\mu}) - e^{-ieaA_\mu(x - \hat{\mu})}\psi(x - \hat{\mu})) + m\psi(x) \big].$$ (4.8)

The necessity of the change $L_{\mathrm{nc}} \to L_{\mathrm{c}}$ can be seen on the classical level as well since L_{nc} is not gauge invariant.

This example shows that the local regularization in space-time which yields periodic Green functions in the momentum space fails gauge invariance unless higher order vertices are inserted in the theory. With non-Abelian, curved gauge group the naive non-compact formalism violates gauge invariance, too. The vertices generated by the invariant measure terms and the link variables of the action cure this problem in every order of the loop expansion. Thus the genuine lattice vertices which arise due to the non-polynomial interactions lead to more Feynman graphs than in dimensional regularization and the perturbation expansion is more involved. But the bonus of this complication is that one may go beyond the perturbation expansion in the regulated theory. In contrast, the dimensional regularization has fewer graphs but it is tied to the perturbation expansion.

The center symmetry which is realized by the help of the periodicity of the measure term is definitely lost in the perturbative treatment. In fact, if we truncate the polynomial approximation of a periodic function at a finite order we fail to reproduce its periodicity. Thus the center symmetry and the confining features of the gauge theory are missed in the dimensional regularization despite the manifest perturbative gauge invariance. Our goal which is to obtain a description of the center symmetrical renormalized vacuum in terms of the concepts of the continuum field theory must be achieved by starting with the lattice regularization and by going beyond the standard perturbative QCD framework. The resulting renormalized theory, we believe, will belong to a different renormalized trajectory then the conventional perturbative QCD. On the one hand, the renormalized trajectories of the center symmetrical and the perturbative theory are similar in the ultraviolet since both describe asymptotically free systems. But on the other hand, the renormalized trajectories should separate as we move towards the infrared. The direct argument we propose to support this conjecture will be presented below where we show that the center symmetrical renormalized theory develops linear potential between static color charges. Similar behavior has been observed numerically in lattice QCD [11] and analytically [12] in the sine-Gordon model [8].

5. Renormalization of QCD

The renormalization of quantum field theories can be discussed conveniently in terms of the Kadanoff-Wilson blocking procedure [13]. The quantity one studies in this scheme is the transfer matrix,

$$e^{-ia_\tau \mathcal{H}[\Phi_f(a_\sigma), \Phi_i(a_\sigma); a_\sigma, a_\tau]} = <\Phi_f(a_\sigma)|e^{-ia_\tau H(a_\sigma)}|\Phi_i(a_\sigma)>, \tag{5.1}$$

which is the regulated matrix element of the time evolution operator between the states with the space-like cut-off, a_σ. One increases the time-like and the space-like cut-offs, $a_\tau \to a'_\tau$ and $a_\sigma \to a'_\sigma$, respectively in (5.1) and finds the new 'hamiltonian', $\mathcal{H}[\Phi_f(a'_\sigma), \Phi_i(a'_\sigma); a'_\sigma, a'_\tau]$, which reproduces the same physics as the original one. The coupling constants of this family of 'hamiltonians' give the renormalization group flow of the theory. The coupling constants which decrease as we move in the infrared direction are called irrelevant. In the vicinity of the ultraviolet fixed point where the blocking transformation can be linearized these coupling constants increase exponentially as one goes along the ultraviolet direction and drive the system away from the ultraviolet fixed point. Thus the irrelevant coupling constants are just the non-renormalizable ones.

The theories with mass gap possess infrared fixed point because the interactions beyond the Compton wavelength of the elementary mass are screened. One may find different set of relevant operators in the vicinity of the infrared fixed point than at the ultraviolet one. In this case one has to keep both set of relevant operators in the cut-off action in order to understand the infrared behavior of the theory. We shall illustrate this phenomenon in QCD where a local periodic potential must be added to the lagrangian.

In order to demonstrate the shortcomings of the perturbative renormalization in QCD by a hand-waving manner we consider a much more simple, but related quantity,

$$Z(a_t au) = \int D[\Phi] < \Phi(a_\sigma)|e^{-a_\tau H(a_\sigma)}|\phi(a_\sigma) >, \qquad (5.2)$$

for fixed $a_\sigma(<< a_\tau)$. This is nothing else than the partition function at temperature $T = 1/a_\tau$. The coupling constants of $\mathcal{H}[A_f(a'_\sigma), A_i(a'_\sigma); a'_\sigma, a'_\tau]$ for the gluonic system follows the perturbative scaling for $a_\tau, a_\sigma << \Lambda_{QCD}^{-1}$ due to the asymptotic freedom. The long range modes appear in the trace (5.2) and the perturbation expansion fails to reproduce $Z(a_\tau)$ for any value of a_τ. Despite this obvious non-perturbative effect we can learn an important property of the scale dependence from $Z(a_\tau)$: The center symmetry is broken dynamically and the confining behavior is lost for $a_\tau < 1/T_{\mathrm{dec}} = O(\Lambda_{QCD}^{-1})$. A lesson to be learned from this phenomenon is the following: The perturbative treatment which is based on weakly interacting gluons might be correct up to the length scale Λ_{QCD}^{-1}. This is the region where linearization around the perturbative ultraviolet fixed point applies. The parton model is useful in this regime after taking into account the radiative QCD corrections. As the observational length scale exceeds Λ_{QCD}^{-1} the individual partons are not observable any more and the system confines. The effective 'hamiltonian', (5.1), of this length scale is in the vicinity of the infrared fixed point and it is made mainly of hadronic operators. The cut-off dependence of (5.1) can only be understood in a center symmetrical treatment.

The usual perturbative QCD with dimensional regularization works for the deconfined regime, in the vicinity of the ultraviolet fixed point because the center symmetry is broken dynamically. As we move towards the infrared and arrive to the confined regime the center symmetry is restored in the matrix element (5.1). The dimensionally regulated theory is not applicable in this regime any more. The complicated structure of the lattice regulated theory with the irrelevant coupling constants which are generated by the invariant integral measure and the link variables have important role in this regime. In fact, though they are decreasing as we move towards the infrared in the vicinity of the ultraviolet fixed point, their presence is necessary to restore the center symmetry. In the vicinity of the ultraviolet fixed point the lattice regulated theory happens to be renormalizable despite of the apparently non-renormalizable lattice vertices [14]. This is because these vertices are governed coherently during the renormalization by the gauge invariance. When we arrive into the vicinity of the infrared fixed point a new relevant operator set is developed and these vertices become relevant.

6. Small instantons and renormalization

We saw in the previous Chapter that one has to go beyond the perturbation expansion and to restore center symmetry in the renormalization procedure as the subtraction scale approaches the infrared fixed point. We outline a possible procedure to achieve this result. Consider the gauge transformation,

$$\Omega(\mathbf{x}) = P e^{\int_0^t dt' A_0(\mathbf{x},t')}, \tag{6.1}$$

which is performed between the initial and the final time in (2.33). The center symmetry is realized if, together with $\Omega(\mathbf{x})$,

$$e^{ij\frac{2\pi}{N}}\Omega(\mathbf{x}), \tag{6.2}$$

with $j = 1, \cdots, N$ occur as well in the functional integration in SU(N) gauge theory. One encounters similar situation in the ising model. The global, spin up-spin down symmetry is realized there by forming spin up and spin down domains with equal weight. Similarly, a possible mechanism to restore the center symmetry would be if one could argue that domains with different j occur in the path integration with equal weight. We shall show that (i) instantons may realize such domains and (ii) the density of the instantons with the size of the ultraviolet cut-off requires an additional adjusting in the renormalization procedure. The condition to find this additional fine tuning should be the realization of the center symmetry. This is in line with the fine tuning of the fugacity for the anomalous singularities mentioned before.

(i) The gauge transformation (6.1) gives a map of the three-space into the gauge group. Suppose that the three-space is compact and the gauge group is SU(2). The maps (6.1) then form the homotopy classes characterized by the winding number. It is

$$\nu[\Omega] = -\frac{1}{16\pi^2} \int d^3x \epsilon^{ijk} \mathrm{tr}\Omega(\mathbf{x})\partial_i\Omega^\dagger(\mathbf{x})\mathrm{tr}\Omega(\mathbf{x})\partial_j\Omega^\dagger(\mathbf{x})\mathrm{tr}\Omega(\mathbf{x})\partial_k\Omega^\dagger(\mathbf{x}), \tag{6.3}$$

and takes the values 1 on an instanton configuration. Consider an SU(2) instanton with size $\rho \ll t$ in the path integral (2.33). Since $\Omega(\mathbf{x})$ has unit winding it covers the whole gauge group once and the center elements, +1 and -1, must be taken at some locations. The instanton field configuration is pure gauge for large distances and $\Omega(\mathbf{x})$ corresponding to these space locations is close to the identity. The value -1 is actually taken at the center of the instanton in three-space. Thus this instanton represents a domain of -1 with size ρ embedded into the 'vacuum' +1.

For larger SU(N) groups the situation is more involved. The instantons carry unit winding for an SU(2) subgroup of SU(N). $\Omega(\mathbf{x})$ is -1 only in that subgroup at the center of the instanton and +1 in the rest of the N-dimensional internal space. Thus the single instanton does not represent a non-trivial center domain. But it is easy to imagine that $\binom{N}{2} = \frac{(N-1)N}{2}$ properly placed instantons can create such a domain.

(ii) An instanton or a 'molecule' of the instantons describes a domain with non-trivial orientation in regards of the center symmetry. Thus we need a precise fine tuning of the instanton density to create the same volume for the domains of the N different center orientations in SU(N) gauge theory in order to realize the center symmetrical matrix elements. For the restoration of the center symmetry in a manner outlined here we need the precise knowledge of the instanton density up to the size $O(\Lambda_{\mathrm{QCD}}^{-1})$. The usual semiclassical expansion, [15], yields an infrared divergence in the instanton density. It will be shown in the next Chapter that the center symmetrical theory has non-vanishing mass gap. Thus it is reasonable to assume that the blocking procedure which includes the instanton effects properly and leads to the center symmetrical matrix elements in the infrared will in the same time screen the infrared singularity of the naive expansion.

The density of the instantons with size ρ contains the exponential suppression factor

$$e^{-\frac{2\pi^2}{g^2(\rho)}}. \tag{6.4}$$

This makes the contributions of the instantons of the perturbative domain too small to create a center symmetrical matrix element. But the usual semiclassical expansion is not entirely correct in finding the exponent in (6.4) for classically scale invariant theories. The path integral we intend to approximate for small g^2 is

$$\int D[\Phi]e^{-\frac{1}{g^2}(S_{\mathrm{ren}}[\Phi]+S_{\mathrm{ct}}[\Phi])}, \tag{6.5}$$

where S_{ren} and S_{ct} denotes the renormalized action and the counterterms. In the selection of the saddle point of (6.5) one ignores S_{ct} and takes only into account S_{ren} according to the usual logic of the perturbation expansion. This approximation might be justified for asymptotically free theories by noting that the counterterms are small compared to the renormalized action. The renormalized action has no scale parameter and the rescaling of the saddle point yields another saddle point with the same value of S_{ren}. The dilatation becomes a zero mode as a result. This picture naturally changes at the one loop order where the ultraviolet divergences of the quantum fluctuations requires the introduction of the regulator which in turn violates the classical scale invariance. This effect survives the renormalization and is known as the 'anomalous' breakdown of scale invariance. The fluctuation determinant remains dependent on the subtraction scale i.e. the instanton size and the scale invariance is lost in the renormalized theory. The large instantons are preferred after taking these corrections into account.

The complete integrand includes the counterterms and their effect is not negligible for classically scale invariant theories. The counterterms break scale invariance and even an infinitesimally weak splitting of a highly degenerated situation may have large effect. The quantum hall effect is a spectacular demonstration of this phenomenon. Thus the dilatation is actually not a zero mode for the complete bare action, $S_{\text{ren}} + S_{\text{ct}}$. But the manner the degeneracy is lifted on the tree level is different than in the renormalized perturbation expansion.

To illustrate this point consider the lattice regularized theory. For instantons $\rho \gg a$ the lattice action is close to the continuum one, $S_{\text{latt}} = S_{\text{ren}} + O(\frac{a}{\rho})$. But as the instanton size is decreased and the region $\rho \ll a$ is reached the lattice action approaches zero. In fact, the the gauge invariant part of the instanton field is $O(\frac{1}{r^2})$ and its contribution to large Wilson loops and to the lattice action is negligible. This tree level scale dependence is just the opposite of the one observed in the renormalized perturbation expansion. In fact, it prefers small instantons, the density of instantons with size comparable of the cutoff, $a \sim \rho$, is exponentially larger. Since the systematical semiclassical expansion is based on the bare rather than the renormalized perturbation expansion we have to follow the former in case of any discrepancy.

One arrives at an ill-defined problem here: The instantons which are much smaller than the lattice spacing, a, should not be counted as instantons in the lattice regulated theory since the link variables computed with this configuration are almost pure gauge. This is because there is no topological structure on the lattice. In other words, the small instantons fall through the lattice without contributing to anything which might resemble the winding number density. It is worthwhile noting at this point that there is no Θ-parameter in lattice gauge theory either. In fact, the space of the static gauge transformations, $\mathcal{G}$, forms a connected set in the lattice regularization and Gauss' law enforces the invariance of the wave functionals over the whole $\mathcal{G}$. The definition of the winding number density is rather arbitrary in lattice regulated theories. What is important is that the all different local expressions for the lattice winding number density yield divergent expectation value in the numerical simulations. The divergences are always due to the small, i.e. $\rho \sim a$ instantons.

The ill-defined nature of the instanton density can easily be demonstrated in the Pauli-Villars regularization. The bare lagrangian is of the form

$$L_{\mathrm{PV}} = -\frac{1}{4g^2} F^a_{\mu\nu} \left[1 + \mathcal{P}\left(\frac{D^2}{\lambda^2}\right) \right] F^a_{\mu\nu}, \tag{6.6}$$

where $\mathcal{P}(x)$ is a polynom starting with the quadratic term. $\mathcal{P}$ represents operators which are irrelevant according to the classification of the ultraviolet fixed point. But the instanton contributions are influenced in a major manner by $\mathcal{P}$. In fact, the theory with $\mathcal{P}(x) = c_2 x^2$ has no instanton solution at all. The choice $P(x) = -c_2 x^2 + c_4 x^4$ with $c_j = O(1) > 0$ breaks the scale invariance of the bare theory in such a manner that only the instantons with size $\rho = O(\Lambda^{-1})$ remain saddle points. Thus the irrelevant terms of the regulator determines the saddle point structure of the theory. If the instantons are to play any role in forming the long range structure of the theory then the operators represented by $\mathcal{P}$ must become relevant in the vicinity of the infrared fixed point.

To summarize the problem we have just found in the instanton approximation: The counterterms can not be treated perturbatively in the semiclassical expansion of a scale invariant theory. They must be taken into account in the selection of the saddle point together with the renormalized action. The cut-off dependence in the counterterms lifts the degeneracy of the rescaled saddle points. Depending on the actual choice of the regulator the saddle points may completely be destabilized against the dilatation mode or can be stabilized only at some length scale which is close to the ultraviolet cut-off.

The numerical studies of different expressions for the topological charge in the lattice regulated theory showed an ultraviolet divergence in the instanton density. Since the instantons have domain structure from the point of view of the center symmetry their density must strictly be controlled in the center symmetrical matrix elements. Thus one expects the need of the fine tuning of additional coupling constant(s) in QCD. Unfortunately the present accuracy of the verification of the continuum limit in lattice QCD is not enough to support or disprove this conjecture.

We argued in this Chapter that the small instantons, i.e. instantons with size close to the cut-off form a dense gas and should represent the modes which restore the center symmetry. The saddle point expansion around the gas of small instantons simultaneously solves another puzzle of the perturbation expansion. The careful reader might have noticed already that the gauge fixing (4.1) breaks the center symmetry. The simple perturbation expansion is clearly inadequate to study the center symmetry aspects of the theory. The situation is different if one expands around the gas of instantons. Though each instanton gas configuration breaks the center symmetry the summation over the configurations should restore it. The conventional covariant perturbation expansion can be used in this manner in the investigation of the renormalization properties.

7. The effective theory

In the previous Chapters we overviewed the ingredients what appear to be important in the understanding the confining properties of the QCD vacuum in the light of the numerical experiences accumulated in lattice gauge theory. Now we make an attempt to bring these considerations together in an effective theory which is defined in the continuum where our analytical and intuitive power is more developed. We argued that the gas of small instantons should be sufficient to keep the center symmetry for the matrix elements in the infrared. These small instantons are anomalous point singularities of the gauge theories in the sense that the large Wilson loops are influenced by them. Following the analogy with simpler theories with compact variables one might develop the Villain approximation for the center symmetry, based on an integer valued field, $n_0(x)$. The 'smooth' link variable, $u_0(x)$, is chosen from the group $SU(N)/Z_N$ and the variable $U_0(x)$ which appears in the action is $U_0(x) = u_0(x)\exp i m_0(x)\frac{2\pi}{N}$. The path integral in this approximation is written as

$$\sum_{\{n_0\}} \int D[u_0]D[U_i]\exp\left(-S_{\text{latt}}[U_i, U_0 = e^{in_0\frac{2\pi}{N}}u_0]\right). \tag{7.1}$$

The locations x with $n_0(x) \neq 0$ contain small instantons in this parametrization. One can integrate out the field u_0 in the leading order of the weak coupling expansion. The small instantons develop a four dimensional Coulomb interaction in this manner.

This description is not well suited to analytical studies because it is given in lattice regularization. It is remarkable that one can derive a similar Coulomb gas picture by the help of formal manipulations in the continuum, too. We start with the bare lagrangian, (2.34), and write it as the sum of the renormalized and the counterterm part,

$$L = L_{\text{ren}} + L_{\text{ct}}. \tag{7.2}$$

The parameters in L_{ren} possess physical i.e. infrared meaning but the perturbation expansion for L_{ren} has ultraviolet divergences. These are removed by the counterterms, L_{ct}. The bare lagrangian is the sum of the usual quartic term, L_{YM}, and a local potential,

$$V(A_0) = \frac{1}{a^4} \ln \rho(aA_0).$$ (7.3)

The center symmetry is reflected in the periodic structure of $V(A_0)$. Our main assumption is that the renormalized lagrangian which is used to describe the center symmetrical matrix elements has similar form in the gauge where A_0 is diagonal,

$$L_{\text{ren}} = -\frac{1}{4} F_{\mu\nu}^{a2} - V_r(A_0).$$ (7.4)

The invariance under the diagonal gauge group is removed by choosing the Feynman gauge in the maximal Abelian subgroup. $V_r(A_0)$ has similar periodic structure as in the bare lagrangian but with finite values and period length. We shall consider SU(2) theory for simplicity where we have $V_r(A_0) = V_r(\mathcal{A}_0)$, with $\mathcal{A}_0 = \sqrt{(A_0^{a2})}$ and

$$V_r(x) = V_r(x + \frac{2\pi}{\kappa}).$$ (7.5)

We shall show that the potential V_r in (7.4) reproduces the expected long distance features of the QCD vacuum. To this end we set $A_i = 0$ and keep only the component of the gluon field which appear in the potential. In SU(2) gauge model we are left with a single component, $A_0^3(x) = u(x)$. In the Feynman gauge the kinetic energy of the field $u(x)$ consists of the four dimensional Laplacian. Thus our effective Euclidean lagrangian is choosen to be

$$L = \frac{1}{2}(\partial u)^2 + V_r(u),$$ (7.6)

where the potential, $V_r(u)$, is kept as a phenomenological quantity in the lacking of the control of the renormalization properties of the bare lagrangian (2.34). A complete potential, with its infinitely many parameters would be too much to introduce phenomenologically. This model has predictive power only if few parameters of the potential determine the long distance structure of the vacuum. It turns out that the leading infrared contributions in the effective theory depend on the behavior of the potential in the vicinity of the origin. In this manner it is enough to fix two constants, $V_r(0)$ and $V_r''(0)$, by the experimental input.

The interesting feature of the sine-Gordon type theory, (7.6), is its Coulomb gas representation. To find it we first write the periodic potential as

$$V_r(x) = \sum_{m=-\infty}^{\infty} v_m e^{i\kappa m x},$$ (7.7)

and after that expand the generator functional,

$$Z[J] = \int D[u] e^{-\int d^4 x (L - iJu)},$$ (7.8)

24

in the coefficients v_m,

$$Z[J] = \sum_{n=0}^{\infty} \frac{1}{n!} \prod_{j=0}^{n} \int d^4 z_j \sum_{m(j)=-\infty}^{\infty} v_{m(j)}$$
$$\int D[u] e^{-\frac{1}{2}\int dx (\partial u)^2} e^{i\kappa \sum_j m(j) u(z_j) + i \int dy J(y) u(y)}. \tag{7.9}$$

The functional integral is Gaussian and can easily be carried out,

$$Z[J] = \sum_{n=0}^{\infty} \frac{1}{n!} \prod_{j=0}^{n} \int d^4 z_j \sum_{m(j)=-\infty}^{\infty} v_{m(j)} I_{mm} I_{mJ} I_{JJ}, \tag{7.10}$$

where

$$\begin{cases} I_{mm} = e^{-\frac{\kappa^2}{2} \sum_{jj'} m(j) G(z_j - z_{j'}) m(j')} \\ I_{JJ} = e^{i \int dy\, dy' J(y) G(y - y') J(y')} \\ I_{mJ} = e^{-\frac{1}{2} \int dy \sum_j m(j) G(z_j - y) J(y)}, \end{cases} \tag{7.11}$$

and

$$G(x) = \int \frac{d^4 p}{(2\pi)^4} \frac{e^{-ipx}}{p^2} = \frac{1}{4\pi^2 x^2}. \tag{7.12}$$

I_{mm} in (7.10) is as if there were point particles with charge $m(j)$, fugacity $v_{m(j)}$, at the location of the sine-Gordon vertices z_j interacting via the Coulomb potential, $G(x)$. The external source is coupled directly to their Coulomb potential according to I_{mJ}. The last factor in (7.10) describes the Coulombic self interaction the source with itself and is uninteresting from the point of view of the confining interaction.

We call these Coulombic pseudoparticles haarons since they represent the effects of the periodicity of the invariant Haar-measure in the path integral. One can verify that a haaron with the charge m at location z contributes to the generator functional in a manner similar to the Villain field, $n_0(x)$, with $n_0(z) = m$. The specific choice of the Villain action corresponds to a certain choice of the fugacity parameters, v_m, and vice versa. In other words, there is a correspondence between the choice of the Villain action and the local potential, $V_r(u)$.

We shall use the free haaron gas approximation to compute the generation functional. The polarization effects are neglected in this scheme and $I_{mm} = 1$ is used. The summation over n can then be carried out in (7.10) yielding

$$Z[J] = I_{JJ} e^{-\int dz V_r(i \int dy G(z-y) J(y))}. \tag{7.13}$$

The infrared behavior of the Green functions can be studied by evaluating the generator functional for the source which is non-vanishing only at well separated locations,

$$J(x) = \sum_a J_a \delta(x - x_a), \tag{7.14}$$

with $|x_k - x_l| \gg \Lambda_{\text{QCD}}^{-1}$. For source with zero net cahrge, $\sum_a J_a = 0$, we have

$$Z[J] = I_{JJ} e^{-\int dz V_r(i \sum_a G(z-x_a)J_a)} \approx I_{JJ} e^{-\frac{1}{2} V_r''(0) \sum_{ab} J_a G^2(x_a - x_b) J_b}, \qquad (7.15)$$

with

$$G^2(x - y) = \int dz G(x - z) G(z - y) = \int \frac{d^4p}{(2\pi)^4} \frac{e^{-ip(x-y)}}{p^4}. \qquad (7.16)$$

The leading infrared part of the gluon propagator is $\frac{V_r''(0)}{p^4}$ in the free haaron approximation.

One can evaluate (7.13) for the static source

$$J(x) = g\delta(\mathbf{x}) - g\delta(\mathbf{x} - \mathbf{r}). \qquad (7.17)$$

One finds after some algebra that the static quark-anti quark potential determined by the logarithm of the generator functional at (7.17) is

$$V_{\bar{q}q}(r) = -\frac{g^2}{8\pi^2} V_r''(0) r + \sum_{n=1}^{\infty} \frac{c_n}{r^n}. \qquad (7.18)$$

The coefficients c_n are exponentially ultraviolet divergent due to the partial resummation which was achieved at the re-exponentialization, (7.13). But the ultraviolet divergent pieces of the potential are sub-leading for large r compared to the first term and will be ignored.

The renormalized center symmetry, (7.5), excludes the mass generation for the field $u(x)$. In fact, any mass term would suppress the large scale, long range fluctuations of $u(x)$ and would violate the invariance of the vacuum under (7.5). When the effective theory is not protected against the mass generation then the propagator which is used in the partial resummation of the perturbation expansion should be the massive one. It is an interesting lesson of the computation leading to (7.18) that the static quark-anti quark potential starts with the Coulomb term, $\frac{c_1}{r}$, and the string tension is vanishing in this case.

One would object the use of the massless propagator in the haaron gas picture by pointing out that the semiclassical Thomas-Fermi screening of the electron gas generates an infrared mass gap under any circumstances. Analogously, the non-vanishing curvature of the potential, $V_r''(0) \neq 0$, indicates the presence of the mass gap which in turn eliminates the linear piece in (7.18). These objections can be avoided by noting that the fugacity, v_m, may be negative. The negative fugacity particle has negative 'probability' and its contribution to the polarization has the opposite sign. The appropriate choice of the fugacities, i.e. the potential $V_r(u)$, may lead to complete cancellation between the positive and the negative fugacity haarons in the polarization. One can verify that the self energy of the field $u(x)$ can be canceled at any finite order of the perturbation expansion with the suitable choice of the fugacities. The condition $V_r''(0) = 0$ is not needed for such kind of cancellation.

One can include dynamical quarks, too. The effective lagrangian

$$L = \frac{1}{2}(\partial u)^2 - V_r(u) - \bar\psi i\gamma^\mu[\partial_\mu - ig\sigma^3 u]\psi, \tag{7.19}$$

yields

$$L = -\bar\psi(x)i\gamma^\mu\partial_\mu\psi(x) + V_r(ig\int dy\bar\psi(y)\gamma^0\sigma^3\psi(y)G(x-y)), \tag{7.20}$$

in the free haaron gas approximation. The derivation of (7.20) proceeds similar to the argument leading to (7.13) except that the source is $J = \bar\psi\gamma^0\sigma^3\psi$. The leading infrared part of this effective theory can be obtained by retaining the quadratic part of the potential V_r,

$$\begin{aligned}
S_{\text{ir}} = \int dx L_{\text{ir}} = &-\int dx\bar\psi(x)i\gamma^\mu\partial_\mu\psi(x) \\
&-\frac{g^2}{2}V_r''(0)\int dxdy\bar\psi(x)\gamma^0\sigma^3\psi(x)G^2(x-y)\bar\psi(y)\gamma^0\sigma^3\psi(y),
\end{aligned} \tag{7.21}$$

as in (7.15). This non-local Jona-Lasinio model can be simplified by introducing the Hubbard-Stratanovich auxiliary field, $v(x)$,

$$L_{\text{ir}} = -\bar\psi[i\gamma^\mu\partial_\mu + \gamma^0\sigma^3]\psi - \frac{1}{8\pi^2\sigma}v(\partial^2)^2 v, \tag{7.22}$$

where we used the string tension, $\sigma = -\frac{g^2}{8\pi^2}V_r''(0)$, to eliminate V_r in the lagrangian.

The model (7.22) is not Euclidean covariant. The covariance which would be preserved in an exact computation in the temporal gauge was violated by the numerous approximations we have committed in 'deriving' (7.22). But is is easy to find the simplest covariant extension of this effective theory. It contains a vector field, A_μ, and the lagrangian is

$$L_{\text{cov}} = -\bar\psi i\gamma^\mu(\partial_\mu + iA_\mu\sigma^3]\psi - \frac{1}{16\pi^2\sigma}\mathcal{F}_{\mu\nu}\partial^2\mathcal{F}^{\mu\nu}, \tag{7.23}$$

with

$$\mathcal{F}_{\mu\nu} = \partial_\mu A_\nu - \partial_\nu A_\mu. \tag{7.24}$$

The Abelian gauge field, A_μ, represents a collective mode of the gluon field with propagator $\frac{1}{p^4}$ and minimal coupling to the quarks. The extension of (7.23) which includes the perturbative gluon exchange as well is

$$L_{\text{cov}} = -\bar\psi i\gamma^\mu(\partial_\mu + iA_\mu\sigma^3]\psi - \frac{1}{4}\mathcal{F}_{\mu\nu}\frac{\partial^2}{g^2\partial^2 - 4\pi\sigma}\mathcal{F}^{\mu\nu}. \tag{7.25}$$

Note that the dielectric tensor of this model is

$$\epsilon(k) = \frac{k^2}{g^2k^2 + 4\pi\sigma}, \tag{7.26}$$

which is $O(k^2)$ in the infrared as postulated in the dual superconductor model to the QCD vacuum [16].

The chiral symmetry is broken dynamically in the free haaron approximation. This can be seen by pointing out the existence of a localized zero mode of the massless Dirac equation on the background of a haaron field, similarly to the argument followed in the dilute instanton gas approximation. The chiral condensate turns out to be proportional to $V_r(0)$.

8. Conclusion

A description of the confining vacuum of QCD which is suggested here is supported by the numerical experiences of lattice gauge theory. The lattice gauge theory computations shows that the presence or the breakdown of the center symmetry goes together with the presence or the absence of the string tension. The center symmetrical path integral involves gluon field configurations with Dirac delta singularities. At the location of these singularities the gauge invariance is lost unless some higher order terms are inserted into the QCD lagrangian. These higher order terms are present in lattice QCD and apparently do not destroy the asymptotic freedom. We made the conjecture that the renormalization properties of this theory could be understood by taking into account the effects of instantons with size of the ultraviolet cut-off.

Our main assumption in describing an effective theory of the QCD vacuum is that the renormalized part of the lagrangian reflects the center symmetry in the same manner as the bare theory. The renormalized lagrangian is the sum of the usual quartic term and a local periodic potential to the gauge field according to this assumption. We investigated the physical content of this renormalized theory by keeping only the diagonal components of A_0. It is known that the field theory with a cosine potential is equivalent with a classical Coulomb gas problem. Similarly, our effective theory with the periodic local potential is equivalent with the problem of point charges, haarons, interacting via the four dimensional Coulomb potential. The generator functional has been obtained in the free haaron approximation. This allows us to compute the string tension and to derive different effective theories for dynamical quarks. One obtains a non-local, confining version of the Jona-Lasinio model and a dual superconductor type model in this manner.

The physics of the confining interaction is similar to the Anderson localisation [17]. This can be seen clearly in the haaron gas representation. The charges can be localized in solid states by random impurities because the fluctuation of the random phase shift generated by the collision off the impurities cancel the scattering amplitude in the average. The pole of the resolvent is at imaginary location for such system and the imaginary part of the pole is proportional to the momentum of the charged state. The role of the impurities is played by the haarons in QCD. The pole in the charged particle propagator is shifted to imaginary location by the interactions with the Coulomb field of the haarons and the colored states are not propagating. The haaron potential is instantaneous thus the localisation takes place in the space-time rather than in the three space and excludes static charges, too.

The binding of the quark-anti quark pair in a meson can be understood in a similar qualitative manner. The values of the potential of the haaron gas at different space-time locations are not independent and there is a correlation length, ξ, extracted from the fluctuations of the potential. If the quark and the anti quark run closer that the correlation length then the phase shifts suffered by the quark and the anti quark tend to cancel. The charges propagate as free particles as long they are close to each other. When they separate more than ξ then the cancellation mechanism due to the stochastically independent potentials turns on and suppresses the amplitude. Thus the quark-anti quark pair is bound and the confinement radius is approximatively the correlation length, ξ.

Our approach to the confinement problem opens more questions than answers. The most pressing general questions waiting for resolution in our opinion are

(i) What are the effects of the singular configurations in general in the renormalized quantum field theories ? If these configurations survive the removal of the cut-off then the continuum regulators are not sufficient to describe a continuum field theory. How to find the Wick rotation for such theories ?

(ii) What are the ultraviolet effects of the operators which are relevant at the infrared fixed point ? More specifically, can it happen that an operator which is irrelevant at the ultraviolet fixed point becomes relevant at the infrared fixed point ? In this case the renormalization of the theory should include this operator.

The advantage of the picture outlined here is its novelty. By placing the difficulties from the infrared to the ultraviolet sector we have a new chance to try to understand the problem of confinement. We believe that there are more powerful methods, such as the renormalization group, to deal with issues in the ultraviolet rather than in the infrared because the former represent localised problems.

9. Acknowledgements

This work is supported in part by funds provided by the U. S. Department of Energy (D.O.E.) under contract #DE-AC02-76ER03069 and the NSF-PYI grant

PHY-8958079.

10.References

[1] V. Gribov, unpublished.

[2] K. Johnson, L. Lellouch and J. Polonyi, *Nucl. Phys.* **B367** (1991) 675.

[3] L. McLerran and B. Svetitsky, *Phys. Lett.* **98B** (1981) 195; J. Kuti, J. Polonyi and K. Szlachanyi, *Phys. Lett.* **98B** (1981) 199.

[4] J. Polonyi, *Phys. Lett.* **213B** (1988) 340; J. Polonyi, in the *Quark-Gluon Plasma*, edt. R. Hwa (World Scientific, 1990).

[5] G. C. Rossi and M. Testa, *Nucl. Phys.* **B163** (1980) 109; **B176** (1980) 477; **B237** (1984) 442; J. P. Leroy, J. Micheli and C. G. Rossi, *Nucl. Phys.* **B323** (1984) 511; *Il Nuovo Cim.* **84A** (1984) 270; H. S. Chan, *Phys. Rev.* **D34** (1986) 2433; G. Rossi and Y. Yoshida, *Il Nuovo Cim.* **11D** (1989) 101; J. P. Leroy, J. Micheli, G. C. Rossi and K. Yoshida, *Z. Phys.* **C48** (1990) 653;

[6] I. M. Gel'fand and A. M. Yaglom, *J. Math. Phys.* **1** (1960) 48.

[7] L. S. Schulman, *Techniques and Applications of Path Integration*, (John Wiley, 1981).

15. K. Huang and J. Polonyi, *Int. J. Mod. Phys.* **A6** (1991), 409; K. Huang, Y. Koike and J. Polonyi, *Int. J. Mod. Phys.* **A6** (1991) 1267.

[9] T. Banks, J. Kogut and R. Meyerson, *Nucl. Phys.* **B129** (1977) 493; E. Fradkin and L. Susskind, *Phys. Rev.* **D17** (1978) 2637; T. DeGrand and D. Toussant, *Phys. Rev.* **D22** (1980) 2478.

[10] M. B. Voloshin, *Nucl. Phys.* **B154** (1979) 365; H. Leutwyler, *Phys. Lett.* **B98** (1981) 447; V. N. Baier and Yu. F. Pinelis, *Phys. Let.* **B116** (1982) 179; D. Gromes, *Phys. Lett.* **B115** (1982) 482; G. Cerci, A. Di Giacomo and G. Paffuti, *Z. Phys.* **C18** (1983) 135; I.I Balitsky, *Nucl. Phys.* **B254** (1985) 166; H. Campostrini, A. Di Giacomo and S. Olejnik, *Z. Phys.* **C31** (1986) 577; L. S. Celenza and C. H. Shakin, *Phys. Rev.* **D34** (1986) 1591; H. D. Dosch, *Phys. Lett.* **B190** (1987) 177; V. Marquard and H. G. Dosch, *Phys. Rev.* **D35** (1987) 2238; H. G. Dosch and Yu. Simonov, *Phys. Lett.* **B205** (1988) 339; P. V. Landshoff and O. Nachtmann, *Z. Phys.* **C35** (1987) 405.

[11] A. Patriasioiu, E. Seiler and I. Stamatescu, *Phys. Lett.* **107B** (1981) 364; K. Cahill, *Nucl. Phys. B (Proc. Suppl.)* **9** (1989) 529.

[12] J. M. Kosterlitz and D. J. Thouless, *J. Phys.* **C6** (1973) 118; J. M. Kosterlitz, *J. Phys.* **C7** (1974) 1046; J. V. Jose, L. P. Kadanoff, S. Kirkpatrik and D. R. Nelson, *Phys. Rev.* **B16** (1977) 1217.

30

[13] K. Wilson and J. Kogut, *Phys. Rep.* **C12** (1974) 75.

[14] T. Reisz, *Nucl. Phys.* **B318** 417 (1989).

[15] A. A. Belavin, A. M. Polyakov, A. S. Schwartz and Yu. S. Tyupkin, *Phys. Lett.* **59B** (1975) 85; G. 'tHooft, *Phys. Rev. Lett.* **37** (1976) 8, *Phys. Rev.* **D14** (1976) 3432, Erratum *Phys. Rev.* **D18** (1978) 2199; R. Jackiw and C. Rebbi, *Phys. Rev. Lett.* **37** (1976) 172; C. Callen, R. Dashem and D. Gross, *Phys. Lett.* **63B**, *Phys. Rev.* **D17** (1977) 2717; S. Coleman, "The Uses of Instantons", Erice Lectures, 1977.

[16] H. B. Nielsen and P. Olesen, *Nucl. Phys.* **B61** (1973); Y. Nambu, *Phys. Rev.* **D10** (1974) 4262; S. Mandelstam, *Phys. Lett.* **53B** (1974) 476; G. Parisi, *Phys. Rev.* **D11** (1975) 970; Z. F. Ezawa and H. C. Tze, *Nucl. Phys.* **B100** (1975) 1; R. Brout, F. Englert and W. Fischler, *Phys. Rev. Lett.* **36** 1976) 649; F. Englert and P. Windey, *Nucl. Phys.* **B135** (1978) 529; G. 'tHooft, *Nucl. Phys.* **B190** (1981) 455, *Phys. Scr.* **25** (1981) 133; H. B. Nielsen and A. Patkos, *Nucl. Phys.* **B195** (1982) 137; V. P. Nair and C. Rosenzweig, *Phys. Rev.* **D31** (1985) 401; C. Rosenzweig, *Phys. Rev.* **D38** (1988) 1934; H. J. Pirner, J. Wroldsen and M. Ilgenfritz, *Nucl. Phys.* **B294** (1987) 905.

[17] P. W. Anderson, *Phys. Rev.* **109** (1958) 1492.

THE METHOD OF VACUUM CORRELATORS

H. G. Dosch

Institut für Theoretische Physik
Universität Heidelberg
Philosophenweg 16, D-6900 Heidelberg, BRD

Abstract

The method of vacuum correlators, which allows to calculate non-perturbative quantities is shortly presented. Applications to hadron spectroscopy and soft high-energy scattering are discussed.

1   The Method

In this lecture I want to present a phenomenological treatment of the QCD vacuum which has been developed and advocated by Yuri Simonov and myself for several years [1],[2],[3],[4]. Our method is best described in terms of functional integration.

Any Green function depending on fields of quarks (ψ) and gluons (A_μ) can be expressed as an integral over all field configurations with the weight given by the QCD action S_{QCD}.

$$\langle G(\bar{\psi}\psi, A)\rangle = \int G(\bar{\psi}\psi, A)e^{-S_{QCD}}\mathcal{D}\psi\mathcal{D}\bar{\psi}\mathcal{D}A \qquad (1.1)$$

A Wick rotation to a Euclidean space-time continuum: $ct \to ix_4$ has been performed, i.e. $x^2 = \sum_{i=1}^{4} x_\mu^2$. Unless otherwise stated, I shall always evaluate Green functions in the Euclidean space-time continuum, physical quantities can be obtained at the end by analytic continuation $x_4 \to -ict$.

The fermion integration can (formally) be performed analytically, leading to

$$\langle G[\bar{\psi}\psi, A]\rangle = \int F[A]e^{-S_{PG}}\mathcal{D}A \qquad (1.2)$$

where now the weight at the integration over the gluon fields is given by the pure gauge Lagrangian

$$S_{PG} = \int F_{\mu\nu}^i(x)F_{\mu\nu}^i(x)d^4x; \quad F_{\mu\nu}^i = \partial_\mu A_\nu^i - \partial_\nu A_\mu^i + ig_s f_{ike}A_\mu^k(x)A_\nu^e(x). \qquad (1.3)$$

$F[A]$ contains the fermion determinant which is put equal to 1 in the quenched approximation. In processes where only high energies and large momenta transfers are involved we know how to treat the problem: Since due to asymptotic freedom in that case the coupling g_s becomes small, we can apply the machinery of perturbation theory.

Unfortunately these conditions of high energies and momentum transfer are not so typical for many interesting questions of strong interaction physics:

The problems of bound states of quarks or high energy scattering at low momentum transfer are important examples for that case in which not only (virtual) gluons with high momenta, i.e. with large wave number k and hence rapidly varying in space time are involved, but also soft gluons which correspond to fields slowly varying in space time. It is just the averaging over these slowly varying gluon fields, corresponding to soft gluon exchange, for which we have no analytical handle. We have heard in Janos Polonyi's [5] talk a very interesting approach to handle this problem starting from the QCD Lagrangian. I shall be more modest here and make the following two assumptions:

1) that the formal expressions (1.1) and (1.2) make sense, and that the method of cumulant (or linked cluster) can be applied here;

2) that the cumulants occurring in this expansion have a finite correlation time.

Let me shortly recapitulate the cumulant expansion [6]:

Be $\xi(t)$ a stochastic variable, depending on one (or several) parameter(s) t. It is distributed according to some distribution which determines the expectation values:

$$\langle \xi(t) \rangle, \quad \langle \xi(t_1)\xi(t_2) \rangle \ldots \quad \langle \xi(t_1)\ldots\xi(t_n) \rangle \tag{1.4}$$

A particularly simple distribution is that of a <u>Gaussian process</u>, where <u>all</u> expectation values with $n > 2$ factorize:

$$\langle \xi(t_1)\xi(t_2\xi(t_3) \rangle = \langle \xi(t_1)\xi(t_2) \rangle \langle \xi(t_3) \rangle + \langle \xi(t_1)\xi(t_2) \rangle \langle \xi(t_3) \rangle + \langle \xi(t_2)\xi(t_3) \rangle \langle \xi(t_1) \rangle$$
$$\langle \xi(t_1)\ldots\xi(t_4) \rangle = \langle \xi(t_1)\xi(t_2) \rangle \langle \xi(t_3)\xi(t_n) \rangle + \ldots \tag{1.5}$$

The cumulants (or linked clusters) are defined by

$$\langle \exp\{ \int \xi(t)dt \} \rangle$$
$$= \exp\left\{ \int \ll \xi(t) \gg dt + \frac{1}{2!} \int \int \ll \xi(t_1)\xi(t'_2) \gg dt_1 dt'_2 \right.$$
$$\left. + \frac{1}{3!} \int \int \int \ll \xi(t_1)\xi(t_2)\xi(t_3) \gg dt_1 dt_2 dt_3 + \ldots \right\} \tag{1.6}$$

Expansion of both sides shows the following relations:

$$\ll \xi(t) \gg = < \xi(t) > \tag{1.7}$$

$$\ll \xi(t_1)\xi(t_2) \gg = < \xi(t_1)\xi(t_2) > - \ll \xi(t_1) \gg \ll \xi(t_2) \gg$$
$$\ll \xi(t_1)\xi(t_2)\xi(t_3) \gg = < \xi(t_1)\xi(t_2)\xi(t_3) > - \ll \xi(t_1)\xi(t_2) \gg \ll \xi(t_3) \gg$$
$$- \ll \xi(t_1)\xi(t_3) \gg \ll \xi(t_2) \gg - \ll \xi(t_2)\xi(t_3) \gg \ll \xi(t_1) \gg \tag{1.8}$$

i.e. for a Gaussian process all cumulants with $n \geq 3$ vanish. If the process is centered, i.e. $\ll \xi(t) \gg = 0$, than one obtains for a Gaussian process:

$$< e^{\int \xi(t)dt} > = \int e^{\frac{1}{2}\int <\xi(t_1)\xi(t_2)>dt_1 dt_2} \tag{1.9}$$

A first meaningful approximation for any process for which the cumulant expansion converges is thus a Gaussian stochastic process, determined by the correlator $< \xi(t_1)\xi(t_2) >$. We shall make this Gaussian approximation for the process defined through (1.3).

We shall first discuss an extremely important quantity, namely the Wegner-Wilson loop introduced by Wegner in 1971 [7] in a Z_2 gauge theory on the lattice and recognized by Wilson [8] as the relevant order parameter for confinement in QCD.

The W-loop is the exponential of a line integral along a closed path C:

$$W[C] = \int \exp\{ig \oint e^P A_\mu(x)dx_\mu\}e^{-S_{PG}}\mathcal{D}A. \tag{1.10}$$

If C is a rectangle with R and length T, one obtains the relation:

$$W(c) = e^{-V(R)\cdot T} \tag{1.11}$$

where $V(R)$ is the potential between two static quarks at distance R. It is an easy exercise to show that applying perturbation theory one can obtain from the W-loop the Coulomb potential $\frac{1}{R}$. If especially the area low holds,

$$W[c] = e^{-\alpha R \cdot T} \tag{1.12}$$

one has $V(R) = \alpha R$, i.e. the potential increases linearly with distance and the quarks can never be separated, which means confinement.

We now want to apply our assumptions stated earlier to the W-loop, namely that the averaging over the slowly varying (i.e. soft gluon) fields makes sense, and secondly that the cluster expansion coverges. A very desired result of our method is that even in the simplest approximation there is a crucial difference between Abelian gauge theories like QED and non-Abelian ones like QCD, but nevertheless I shall start for pedagogical reasons with an Abelian gauge theory.

If we want to describe the expectation values entering the cluster expansion through measurable quantities, we have to express them in a gauge-invariant way.

For that we apply Stoke's theorem to the W-loop:

$$W = \langle \exp\{ig \oint_{\partial \mathcal{F}} A_\mu dx_\mu\}\rangle = \langle \exp\{ig \int_{\mathcal{F}} F_{\mu\nu}(x)d\sigma_{\mu\nu}(x)\}\rangle \tag{1.13}$$

where $F_{\mu\nu}(x)$ is the (Abelian) field tensor

$$F_{\mu\nu}(x) = \partial_\mu A_\nu(x) - \partial_\nu A_\mu(x) \tag{1.14}$$

34

and $d\sigma_{\mu\nu}$ is the surface element with its normal perpendicular to the $\mu\nu$ direction: We now apply the cluster expansion to the right-hand side of (1.15):

$$W = \langle e^{-ig\int_{\mathcal{F}} F_{\mu\nu}(x)d\sigma_{\mu\nu}(x)}\rangle = e^{-\frac{g^2}{2}\int_{\mathcal{F}}\langle F_{\mu\nu}(x)F_{\kappa\lambda}(x')\rangle\sigma_{\mu\nu}(x)d\sigma_{\kappa\lambda}(x')+\cdots} \tag{1.15}$$

Sharpening for the moment the requirement and demanding a Gaussian process, we ignore the dots.

Lorentz invariance imposes $\langle F_{\mu\nu}(x)\rangle = 0$ and translational invariance allows as most general form for the correlator:

$$\langle F_{\mu\nu}(x)F_{\kappa\lambda}(x')\rangle =$$
$$\frac{1}{12}\langle F_{\mu\nu}(0)F_{\mu\nu}(0)\rangle \cdot \Big\{(\delta_{\mu\kappa}\delta_{\nu\lambda} - \delta_{\mu\lambda}\delta_{\nu\kappa})D(z^2)\cdot\kappa$$
$$+\Big[\Big(\frac{1}{2}\frac{\partial}{\partial z_\mu}(z_\kappa\delta_{\nu\lambda} - z_\lambda\delta_{\nu\kappa}) + \frac{1}{2}\frac{\partial}{\partial z_\nu}(z_\lambda\delta_{\mu\kappa} - z_\kappa\delta_{\mu\lambda})\Big)D_1(z^2)(1-\kappa)\Big] \tag{1.16}$$

with $z = x - x'$, $D(0) = D_1(0) = 1$.

We can perform the integration over $d\sigma_{\mu\nu}(x)$ and $d\sigma_{\kappa\lambda}(x')$ if we make further assumptions on the funcitons $D(z^2)$ and $D_1(z^2)$, but we can also draw some general conclusions even without specific knowledge of these functions. Let us assume that the loop is large compared to the finite correlation length Λ, then using translational invariance, one of the surface integrations $d\sigma$ can be performed, yielding only contributions if $|z| = |x - x'| \leq \Lambda$, hence yielding a contribution $\sim \Lambda^2$. The next surface integration is then unconstrained, and we obtain from it a factor $|\mathcal{F}|$, the area of the loop, hence we get

$$\langle W\rangle = \exp\{-K\frac{g^2}{12}\langle F_{\mu\nu}(0)F_\mu\nu(0)\rangle\cdot\Lambda^2\cdot|\mathcal{F}|(1+O(\frac{1}{|\mathcal{F}|}))\} \tag{1.17}$$

where K is a number depending on the shape of D. One can also show that

$$K = 0 \quad \text{for} \quad D = 0, \tag{1.18}$$

i.e. the correlator function D_1 of eq. (1.16) does not contribute to the term proportional to the area of the loop, and hence does not contribute to confinement.

The assumptions about the cumulant expansion thus lead to the area law (i.e. linear confinement) if the tensor structure with D in (1.18) does not vanish. A similar consideration shows that the higher cumulants do not alter this result. Thus we seemingly obtain linear confinement quite naturally also for Abelian gauge theories.

But in Abelian gauge theory we also have the homogeneous Maxwell equations

$$\partial_\rho\epsilon_{\rho\mu\nu\sigma}F_{\mu\nu} = 0 \tag{1.19}$$

and a simple calculation shows that these imply

$$D = 0 \tag{1.20}$$

i.e. we have <u>no</u> confinement in an Abelian gauge theory since the correlator responsible for the area law vanishes.

Now let us turn to non-Abelian gauge theory.

The first, and most <u>cherished</u> difference is

$$\partial_\rho \epsilon_{\rho\mu\nu\sigma} F^i_{\mu\nu} = -ig f_{ikl} \partial_\rho (A^k_\mu A^l_\nu) \epsilon_{\rho\mu\nu\sigma} \neq 0 \tag{1.21}$$

due to gluon self-coupling. Hence there is no reason for the function $D(z^2)$ to vanish, and hence linear confinement occurs naturally. But the situation is more complicated. In a non-Abelian gauge theory the field tenssors $F^i_{\mu\nu}$ are no longer gauge-invariant. In order to form gauge-invariant correlators, we have to transport all fields to a reference point y

$$\mathbf{F}_{\mu\nu}(x,y) \equiv \phi^{-1}(x,y) \mathbf{F}_{\mu\nu}(x) \phi(x,y) \tag{1.22}$$

where $\phi(x,y)$ is the parallel-transporter

$$= \mathbf{P} \int_x^y e^{ig \int \mathbf{A}_\mu(z) dz_m u} \tag{1.23}$$

with the su(3)-Lie algebra-valued fields

$$\mathbf{A}_\mu = \Sigma \lambda^i A^i_\mu; \quad \mathbf{F}_{\mu\nu} = \Sigma \lambda^i F^i_{\mu\nu} \tag{1.24}$$

$\lambda^i : i = 1...8$ are the Gell-Mann matrices.

Now the correlator $\langle \mathbf{F}_{\mu\nu}(x',x), \quad \mathbf{F}_{\rho\sigma}x) \rangle$ is a gauge-invariant expression and can be treated in the same way as an Abelian correlator. It has been investigated on the lattice by DiGiacomo [9], and he finds indeed that $D(z^2)$ is indeed $\neq 0$ and is even the dominant contribution. Hence our simple model assumptions lead to confinement in the non-Abelian gauge theory.

In the next part I want to explore some quantitative consequences.

2 Applications

2.1 Central potential between heavy quarks

A straightforward calculation yields for the central potential between heavy quarks [2]:

$$\begin{aligned}
\epsilon(r) \;=\; & \frac{\pi^2 \cdot G_2}{12 N_c} \\
& \left\{ 2r \int_0^r d\rho \int_0^\infty d\tau D(\tau^2 + \rho^2) \right. \\
& \left. \int_0^r \rho d\rho \int_0^\infty d\tau [-2D(\tau^2 + \rho^2) + D_1(\tau^2 + \rho^2)] \right\}
\end{aligned} \tag{2.25}$$

where the Coulomb (i.e. perturbative) contribution can be incorporated in D_1. G_2 is the gluon condensate [10]:

$$\langle \frac{\alpha_s}{\pi} F^i_{\mu\nu}(0) F^i_{\mu\nu}(0) \rangle \tag{2.26}$$

If we make the ansatz

$$D(z^2) = e^{-z^2/\Lambda^2} \tag{2.27}$$

we can relate the gluon condensate G_2, the correlation length Λ, and the phenomeno-logically known slope of the confining potential, obtaining for the correlation length

$$\Lambda \sim 0.3./.0.5 \; fm,$$

depending on the values of the slope and G_2. These values are very reasonable values and well compatible with the lattice results of DiGiacomo [10].

2.2 Spin-dependent potentials

From phenomenology [11] we know that the spin-dependent contributions of the confining part of the $q\bar{q}$-potential must be specifically different from the ones of a Coulomb potential. Very good results are obtained assuming that the confining potential leads to the same spin-dependent forces as a scalar exchange (as opposed to the vector exchange). One can show [12],[13], by making e.g. an $1/m$ expansion, that the structure $D(z^2)$, which is responsible for the linear confinement, leads indeed to that result [12],[13]. Again, the spin-dependent potentials can be expressed through the correlators D and D_1, leading specifically to [13]:

$$
\begin{aligned}
V_{SD} &= \frac{\vec{L} \cdot \vec{S}}{2m^2} \cdot V_{LS} + \frac{3(\vec{S}_i \hat{r})(\vec{S}_2(\hat{r}) - \vec{S}_1 \vec{S}_2}{3m^2} V_T \\
&\quad \frac{\vec{S} \cdot \vec{S}}{3m^2} \cdot V_{HF} \\
V_{LS} &= \frac{1}{r}(\epsilon'(r) + \frac{2\pi^2 G_2}{12 N_c} \int_0^r d\rho \int_0^\infty d\tau \times \\
&\quad (D(\rho^2 + \tau^2) + D_1(\rho^2 + \tau^2))(\frac{2\rho}{r} - 1) \\
&\quad -\tau^2 \frac{\partial}{\partial \tau^2} D_1(\rho^2 + \tau^2)) \\
V_T &= -\frac{2\pi^2 G_2}{12 N_c} \int_0^\infty d\tau \frac{\partial}{\partial r^2} D_1(r^2 + \tau^2)
\end{aligned}
\tag{2.28}
$$

2.3 Soft high energy scattering

The last application I want to make is that of high energy hadron-hadron scattering at small momentum transfer ($|t| \leq 1\text{GeV}^2$). Here we consider the hadrons as composed of quarks on lightlike paths [14], the transversal extension governed by a transversal wave function [15].

The scattering amplitude can then be expressed as the expectation value of two Wilson loops

$$T_{hh} = \langle W_+ W_- \rangle - \langle W_+ X W_- \rangle \tag{2.29}$$

smeared over with transversal wave functions $\psi(x_\perp)$. In that case we cannot work in the Euclidean space time, since two sides of the loops (representing the quark paths), are lightlike, i.e.

$$x_0^2 = x_3^2 \tag{2.30}$$

and thus would shrink to the point

$$-x_4^2 = x_3^2 = 0 \tag{2.31}$$

upon continuation to Euclidean space time. We therefore have to work in Minkowski space time and to choose a form of the correlator, which can be continued from Minkowski to Euclide and vice versa. A class of these correlation functions is [16]

$$D(z^2) \int \frac{e^{ikz}}{(k^2 + a^2)^n} d^4k \tag{2.32}$$

which falls off exponentially like $e^{-z^2 a^2}$. The evaluation is straightforward, but tedious, and we obtain e.g. for the total hadron-hadron cross section [15]

$$\sigma_{tot} = G_2^2 \cdot \Lambda^{10} F(S_T/\Lambda) \tag{2.33}$$

where the number $F(S_T/\Lambda)$ depends on the choice of the transversal wave function, specifically on the size of the hadrons parametrized by S_T. The dependence on that size is very interesting. If the size of the hadrons is small as compared to the correlation length Λ, then $F(S_T/\Lambda) \propto S_T^4$. If, however, $S_T \geq \Lambda$, we obtain specifically different contributions from the (confining) correlator $D(z^2)$ and the (nonconfining) $D_1(z^2)$. The contribution of D_1 yields that in the limit of hadron sizes large as compared to the correlation length Λ the cross section becomes independent of the hadron-hadron size.

For the contribution of the confining correlator, however, we obtain

$$F_{D_1}(S_T/\lambda) \longrightarrow S_T^3 \quad \text{if} \quad S_T \gg \Lambda, \tag{2.34}$$

i.e. the cross section increases. Thus the correlator D_1 (present in Abelian theories and in lowest order perturbation theory) induces a quark-quark interaction, whereas the typically non-Abelian correlator D responsible for confinement yields a geometrical picture, which can be interpreted as due to string-string interaction.

Due to the dramatic dependence of the cross seciton on Λ, namely Λ^{10}, the correlation length is very stringently fixed to $\Lambda \sim 0.4 \pm 0.04 \; fm$, where the error includes very conservative variations in the other parameters. It is in good agreement with the value determined from the slope of the confining potential. The size dependence of the cross section yields a natural explanation of the flavour dependence of total cross sections at high energy [17].

Acknowledgement

It is a pleasure to thank the organizers of the conference for creating such a pleasant atmosphere.

1. H. G. Dosch, *Phys. Lett.* **B156** (1987) 365.
2. H. G. Dosch and Yu. A. Simonov, *Phys. Lett.* **B205** (1988) 339.
3. Yu. A. Simonov, *Nucl. Phys.* **B307** (1988) 512.
4. H. G. Dosch, *Nucl. Phys.* **23B** (Proc. Suppl.) (1991) 186.
5. J. Polonyi, Contribution to this conference.
6. see e.g. N. G. van Kampen, *Phys. Rep.* **C24** (1976) 172.
7. F. Wegner, *J. Math. Phys.* **12** (1971) 2259.
8. K. Wilson, *Phys. Rev.* **D10** (1976) 2445.
9. A. Di Giacomo and H. Panagopoulos, *Phys. Lett.* **B285** (1992) 133.
10. M. A. Shifman, A. J. Vainshtein, and V. J. Zakharov, *Nucl. Phys.* **B147** (1979) 385, 448.
11. H. J. Schnitzer, *Phys. Rev.* **D18** (1978) 3482;
    D. Gromes, Z. Phys. **C26** (1984) 401.
12. M. Schiestl and H. G. Dosch, *Phys. Lett.* **B209** (1988) 85.
13. Yu. A. Simonov, *Nucl. Phys.* **B324** (1989) 67.
14. O. Nachtmann, *Ann. Phys.* (N.Y.) **209** (1991) 436.
15. H. Krämer and H. G. Dosch, *Phys. Lett.* **B272** (1991) 114.
16. H. Krämer and H. G. Dosch, *Phys. Lett.* **B252** (1991) 669.
17. H. G. Dosch, E. Ferreira, H. G. Dosch, *Phys. Lett. B*, to appear.

THE QCD VACUUM IN THE FIELD STRENGTH FORMALISM

PER A. AMUNDSEN
Institutt for Matematikk og Naturvitenskap, Høgskolesenteret i Rogaland
P.O.Box 2557 Ullandhaug, N-4004 Stavanger, Norway

and

MARTIN SCHADEN*
Physik Department, Technische Universität München
James-Franck-Straße, D-8046 Garching, Germany

ABSTRACT

We present results for classical (stationary-phase) solutions for pure SU(3) Yang-Mills theory in the field strength formalism. Altogether 25 inequivalent constant solutions have been found. We also discuss the corresponding non-constant solutions, which are degenerate in action with the constant ones. These solutions can be used to rewrite the classical theory entirely in terms of transitions (flips) between regions of different classical solutions, which obey fermi statistics. The importance of quantum corrections is briefly discussed.

1. Introduction

The subject of the present conference stems from Saviddy's observation[1] that a constant colour-magnetic field lowers the QCD vacuum energy density below that of the perturbative vacuum, corresponding to a non-vanishing vacuum gluon condensate. Subsequent studies by the Copenhagen group[2,3] showed that this uniform state is still unstable, and will break down into domains of different colour field strengths. The description of a such a vacuum in terms of gauge potentials is quite non-trivial, and so it is natural to try to reformulate the theory in terms of field strengths as the more appropriate variables for describing low energy phenomena.

Such an field strength formulation of QCD has been introduced by Halpern[4]. It is obtained by rewriting the path integral as a local field theory in terms of variables which are really the fourier conjugates of the conventional field strengths[5]. In this formulation the breaking of scale invariance becomes non-anomalous, giving rise to a manifest scale μ in the action. The corresponding classical (stationary phase) equations of motion have purely imaginary saddle-point solutions corresponding to non-vanishing field strengths and hence a gluon condensate already at the tree level.

These constant (in a suitable gauge) classical field strength solutions are rather unsatisfactory as candidates for an (approximate) description of the physical vacuum, as they explicitly break global O(4) symmetry, even after averaging over gauge configurations. But from any constant vacuum solution one can construct

---

* Supported by: Gesellschaft für Schwerionenforschung, Darmstadt

an infinity of non-constant solutions degenerate in action with it[6]. These solutions, which totally dominate the path integral in the classical limit, lead to a rather cubistic picture of the QCD vacuum, with temporally constant but spatially alternating chromoelectric flux-tubes and temporally unstable (flipping) chromomagnetic flux sheets ('Braque vacuum'). These flipping fields can be described by solutions of the (abelian) dual Maxwell equations being sustained by chromomagnetic currents.

In this talk, we review the known exact classical solutions to QCD in the field strength formalism. While the SU(2) solutions can be found analytically[5], for SU(3) numerical methods have proven necessary. To fix the notation, we sketch in section 2 the derivation of the field strength action. In section 3 we discuss the constant solutions to the corresponding equations of motion. In section 4 we discuss the flipping solutions, and show how the classical contributions to the path integral can be rewritten completely in terms of the flips, which obey Fermi statistics.

2. The field strength action

The generating functional for an SU(N) Euclidean Yang-Mills theory in D dimensions for a compact gauge group with structure constants f^{abc} is:

$$Z[j_\mu^a, J_{\mu\nu}^a] = \int \mathrm{D}[A_\mu^a] \, \exp\left(-\int \mathrm{d}^D x \, \mathcal{L}\right), \quad \mathcal{L} = \tfrac{1}{4}\mathcal{F}_{\mu\nu}(A)\mathcal{F}_{\mu\nu}(A) - \tfrac{1}{2}j_\mu^a A_\mu^a - \tfrac{1}{2}J_{\mu\nu}^a G_{\mu\nu}^a, \quad (1)$$

where $\mathcal{F}_{\mu\nu}^a(A) = \partial_\mu A_\nu^a - \partial_\nu A_\mu^a + g f^{abc} A_\mu^b A_\nu^c$. No gauge fixing will be necessary for our discussion, and the quark contributions have been left out, as we shall only present results for a pure Yang-Mills theory. In addition to the vector sources j_μ^a, it is convenient to introduce antisymmetric tensor sources $J_{\mu\nu}^a$ (background fields), which allow for easy extraction of correlation functions of the field strengths $G_{\mu\nu}^a = \mathcal{F}_{\mu\nu}^a(A)$. One then introduces a new set of fields, $T_{\mu\nu}^a$, fourier conjugates of the physical field strengths $G_{\mu\nu}^a$, by inserting into the integral in the standard manner:

$$1 = \int \mathrm{D}[G_{\mu\nu}^a] \, \delta(G_{\mu\nu}^a - \mathcal{F}_{\mu\nu}^a) = \int \mathrm{D}[G_{\mu\nu}^a] \int \mathrm{D}[T_{\mu\nu}^a] \, \exp\left[\tfrac{i}{2}\int \mathrm{d}^D x \, T_{\mu\nu}^a (G_{\mu\nu}^a - \mathcal{F}_{\mu\nu}^a(A))\right]. \quad (2)$$

Carrying out the integrals over A_μ^a and $G_{\mu\nu}^a$, one obtains [4,5]:

$$Z = \int \mathrm{D}[T_{\mu\nu}^a] \, \exp\left\{-S_T + \tfrac{i}{2}\int \mathrm{d}^D x \, J_{\mu\nu}^a T_{\mu\nu}^a + \tfrac{1}{4}\int \mathrm{d}^D x \, J_{\mu\nu}^a J_{\mu\nu}^a\right\},$$

$$S_T = \int \mathrm{d}^D x \, \tfrac{1}{4}T_{\mu\nu}^a T_{\mu\nu}^a + \tfrac{\mu^4}{2}\mathrm{Tr}\,\ln(\tfrac{i\widehat{T}}{g\mu^2}) - \tfrac{i}{2g}(\partial_\lambda T_{\mu\lambda}^a + ig\, j_\mu^a)(\widehat{T}^{-1})_{\mu\nu}^{ab}(\partial_\sigma T_{\nu\sigma}^b + ig\, j_\nu^b), \quad (3)$$

where $\widehat{X}_{\mu\nu}^{bc} := f^{abc} X_{\mu\nu}^a = \widehat{X}_{\nu\mu}^{cb}$ for any field strength-like quantity $X_{\mu\nu}^a$. For $D > 2$ $\widehat{X}$ is generically a non-singular symmetric matrix in the index pairs $(b,\mu)(c,\nu)$[7]. The logarithmic term in the action density arises from the change from A_μ^a to $T_{\mu\nu}^a$ as integration variables[5]. This term promotes the anomalous breaking of scale invariance in QCD into an explicit breaking. The physical value of the associated dimensionful parameter μ must be fixed by a renormalization condition. The presence of the

logarithmic means that some loop corrections to the original action are included in
in S_T, but this is not occuring in terms of a systematic $\hbar$-expansion.

The stationary points of the action S_T are purely imaginary saddle points at
$T^a_{\mu\nu} = iG^a_{\mu\nu}$ satisfying the equations of motion:

$$G^a_{\mu\nu} - \mu^4 f^{abc}(\widehat{G}^{-1})^{bc}_{\mu\nu} - \mathcal{F}^a_{\mu\nu}(\chi) = 0, \tag{4}$$

where $\chi^a_\mu = (\widehat{G}^{-1})^{ab}_{\mu\nu}(\partial_\sigma G^b_{\sigma\nu} + ig\, j^b_\nu)$. For field-strength configurations gauge equivalent
to constant fields, Eq. (4) reduces to

$$G^a_{\mu\nu} - \mu^4 f^{abc}(\widehat{G}^{-1})^{bc}_{\mu\nu} = 0. \tag{5}$$

That S_T has only imaginary saddle points makes sense, since one expects
the QCD action to have minima for non-vanishing physical field strengths, where
thus the vacuum wave-functional is concentrated. The corresponding path integral
Fourier transform will then have the same minima as purely imaginary saddle points,
like in ordinary quantum mechanics.

The value of μ at the classical (saddle point) level can be related to the gluon
condensate, because any solution of Eq. (5) satisfies

$$G^a_{\mu\nu}G^a_{\mu\nu} = DN_A\mu^4 \equiv\, < G^2 >_{\text{classical}}, \tag{6}$$

where $N_A = N^2 - 1$ is the dimension of the adjoint representation of the group. The
scale anomaly of (pure) QCD can then be used to relate the parameter μ for small
coupling, g, to a renormalization group invariant scale[5].

Near the saddle points one can also find the effective classical action for $G^a_{\mu\nu}$
a Legendre transform, $\Gamma[G] = -\ln Z[J] + \int G^a_{\mu\nu}J^a_{\mu\nu}$ where $J^a_{\mu\nu} = \delta \ln Z/\delta G^a_{\mu\nu}$. One finds:

$$\Gamma[G] = \int d^D x\, \tfrac{1}{4}G^a_{\mu\nu}G^a_{\mu\nu} - \tfrac{\mu^4}{2}\text{Tr}\,\ln\left(\frac{\widehat{G}}{g\mu^2}\right) + \frac{1}{2g}\chi^a_\mu \widehat{G}^{ab}_{\mu\nu c}\chi^b_\nu. \tag{7}$$

Note that this is not what one obtains by naively substituting $T^a_{\mu\nu} = iG^a_{\mu\nu}$ in S_T.

3. Constant solutions

Since $G^a_{\mu\nu}G^a_{\mu\nu} = \tfrac{1}{N}\text{Tr}\widehat{G}\widehat{G}$, we can express the action density at the saddle points
for constant fields (obtained from Eq. (7) with with $\chi^a_{\mu\nu} = 0$) in terms of the eigen-
values of $\widehat{G}$ alone, say $\{\epsilon_i, i = 1,\ldots, M = DN_A\}$:

$$\gamma_G = \frac{\Gamma[G = const]}{V_D} = \frac{1}{4}\sum_{i=1}^{M}\left(\frac{\epsilon_i^2}{N} - \ln \epsilon_i^2\right) \tag{8}$$

where V_D is the volume of space-time (here, and in the following, we have absorbed
a factor μ^{-2} in $G^a_{\mu\nu}$). Since Γ diverges if an eigenvalue is zero, the subspaces of
vanishing eigenvalues divides the classical field configurations into disjoint regions,
where the difference $2m = M_+ - M_-$ between the number of positive and negative

Table 1.

Distribution of 10^7 random SU(3) field strengths according to m-sector for D=4.

$|m|$	probability
4	0.0002 %
3	0.0122 %
2	0.8964 %
1	32.1821 %
0	66.9091 %

eigenvalues of $\widehat{G}$ remains fixed. Note, however, that these potential barriers are integrable, so quantum tunnelling is allowed between these m-sectors.

For SU(2) for $D = 3$ and $D = 4$ all solutions of Eq. (5) are known analytically[5]. For $D = 3$, $\epsilon_i = \pm 3/\sqrt{2}$ for all i, and $m = \pm\frac{1}{2}$. For $D = 4$, we have two set of solutions, with $m = 0$ and with $m = 2$, which are degenerate in action ($\gamma_G = 1.6137$). The solutions are completely characterized by their duality properties. Thus, in terms of the chromoelectric and -magnetic components, $E_i^a = G_{0i}^a$ and $B_i^a = \frac{1}{2}\epsilon_{ijk}G_{jk}^a$, one can find a coordinate system for any solution of Eq. (5) such that the field strengths are orthonormal and independently (anti-) selfdual in each colour separately, $\mathbf{B}^a = \epsilon^a \mathbf{E}^a$, $\epsilon^a = \pm 1$. The $m = 2$ sector corresponds to selfdual solutions, the $m = -2$ solutions to the anti selfdual ones, while the solutions of mixed duality (with both selfdual and anti-selfdual components) have $m = 0$. The corresponding eigenvalues are $\epsilon_i = \pm 2/\sqrt{3}$, and the chiral condensate $G_{\mu\nu}^a \widetilde{G}_{\mu\nu}^a = 4\mathbf{E}^a \cdot \mathbf{B}^a = \pm 4(|m| + 1)$

For $N > 2$, Eq. (5) is less amendable to analytical investigations, and we have resorted to numerical methods. Originally solutions with $|m| = 0, 1$ and 2 were reported[5], of which the parity doublet of lowest action has $m = 0$, $\gamma_G = 2.1836$ and $\mathbf{E}^a \cdot \mathbf{B}^a = \pm 3.603$. Subsequently a completely (anti-) selfdual doublet (hence $\mathbf{E}^a \cdot \mathbf{B}^a = \pm 8$) with $m = \pm 4$ and a much lower action density of $\gamma_G = 1.3041$ was reported by Alkofer[8]. This solution is also interesting because it has the colour structure of a conventional SU(2) instanton, but in the natural embedding of the spin 1 representation in SU(3)[9,10].

As a first step in a a systematic search for solutions of Eq. (5) we generated 10^7 random field strengths uniformly on S_{47} (cf. Eq. (6)), diagonalized the corresponding $\widehat{G}_{\mu\nu}^{ab}$, and determined m. Since the action must have at least one local minimum in each m-sector, this also yields a lower bound on the total number of solutions. The ensuing statistics is summarized in Table 1, from which one can make several observations. First, no configuration with $|m| > 4$ has been found. Although we have heuristic arguments for the impossibility of such configurations, we have found no rigorous bound better than the trivial one, $|m| < 16$, which follows from $\widehat{G}_{\mu\mu}^{ab} = 0$. Second, there also must exist an $m = 3$ solution of Eq. (5). Third, although the $m = 4$ sector has the lowest minimum of the action, the contribution to the path integral from this sector is severely constrained by the small phase-space (low entropy).

To get more detailed information, we have searched for solutions of Eq. (5), by searching for fixpoints of the iterated equation

$$\left[G_{\mu\nu}^a \right]_{\text{new}} = \left[\alpha G_{\mu\nu}^a - (1 - \alpha) f^{abc} (\widehat{G}^{-1})_{\mu\nu}^{bc} \right]_{\text{old}} , \tag{9}$$

Table 2.

Constant classical vacuum solutions of SU(3) in the field strength formalism in D=4, according to m-sector and action density (γ_G). Listed is the chiral condensate ($|\mathbf{E}^a \cdot \mathbf{B}^a|$), the symmetry of the eigenvalue spectrum, the number of zero-modes of the mass matrix (n_0) (? means not investigated), and the probability for the solution (%). Based upon 144 699 matrices.

$|m|$	γ_G	$|\mathbf{E}^a \cdot \mathbf{B}^a|$	symmetric	n_0	%	comments
4	1.3041	8.000	No	8	0.006	selfdual solution
0	2.1836	3.603	Yes	12	17.959	
2	2.2714	0.000	No	12	2.255	
1	2.4752	4.227	No	14	6.561	
2	2.5421	6.284	No	13	0.393	
0	2.6383	1.860	No	12	17.677	
0	2.6736	1.786	Yes	13	11.604	
0	2.7401	0.000	No	14	9.174	
1	2.7422	0.743	No	14	4.783	
0	2.7662	0.232	Yes	13	6.950	
0	2.7946	4.000	Yes	13	3.761	
0	2.7946	0.000	Yes	?	0.016	
1	2.8039	0.000	No	14	10.560	
1	2.8076	0.363	No	14	2.509	
1	2.8167	0.555	No	?	0.004	
0	2.8277	0.583	Yes	?	0.004	
0	2.9978	0.129	Yes	14	1.702	
2	3.0118	4.834	No	10	0.076	
0	3.0285	2.786	No	13	0.142	
0	3.0302	4.000	Yes	?	0.422	
0	3.0302	2.35–6.76	Yes	?	0.159	$\mathbf{E}^a \cdot \mathbf{B}^a$ not fixed
(0	3.1222	1.305	Yes		0.073	iteration incomplete?)
1	3.1540	3.714	No	14	1.425	
0	3.1834	0.000	No	14	1.763	
2	3.2335	2.218	Yes	14	0.044	
3	3.575	6.24	No	?	< 0.001	

where $0 < \alpha \leq 1$ (for numerical stability it is advantageous to rescale $G^a_{\mu\nu}$, so that all eigenvalues are substantially less than 1). The starting value is again a random (normalized) configuration. Ideally the iteration should conserves m, but it is numerically expensive to diagonalize $\hat{G}$ for each iteration, rejecting m-changing steps. Thus we have instead constrained the crossing of sector boundaries by introducing α in Eq. (9) ($\alpha = 0.2 - 0.5$). A summary of the results based upon 144 699 matrices is given in Table 2 (The $m = 3$ solution did not show up in this search, but was found in a dedicated search, rejecting all m-changing iteration steps). The numerical accuracy of the calculations is at least 10^{-5}, and the iteration was continued until convergence in all of the gauge invariants $G^a_{\mu\nu} G^a_{\mu\nu}$, γ_G, $\mathbf{E}^a \cdot \mathbf{B}^a$, and $\sum_{i=1}^{M} |\epsilon_i^{\text{new}} - \epsilon_i^{\text{old}}|$.

In Table 2 we have not listed doublets under parity ($\mathcal{P}$), time reversal ($\mathcal{T}$) (both with $\mathbf{E} \cdot \mathbf{B}^a \to -\mathbf{E}^a \cdot \mathbf{B}^a$, same m) and charge conjugation ($\mathcal{C}$) ($G^a_{\mu\nu} \to -G^a_{\mu\nu}$ and $m \to -m$) separately. For $m = 0$ there are two types of solutions, the self-conjugate ones, with $G^a_{\mu\nu}$ and $-G^a_{\mu\nu}$ related by an SO(32) transformation, and hence with a symmetric eigenvalue spectrum, and those with an asymmetric eigenvalue spectrum.

In the symmetric case, we do not in general know whether the transformation relating $\widehat{G}$ and $-\widehat{G}$ lies in the physical $SO(4)\times SU(3)$ subgroup, although for the $m = 0$ solution of lowest action ($\gamma_G = 2.1836$), which has a symmetric eigenvalue spectrum, they are related by an $O(4)$ transformation[5].

It is seen from Table 2 that in most cases a unique value of $\mathbf{E}^a \cdot \mathbf{B}^a$ is associated with a given γ_G, with only two exceptions (both for $m = 0$). For the solution with $\gamma_G = 2.7946$ there is a small probability for having $\mathbf{E}^a \cdot \mathbf{B}^a = 0$ instead of ± 4, and for $\gamma_G = 3.0302$ we find a continuum of allowed values for $|\mathbf{E}^a \cdot \mathbf{B}^a|$, with embedded discrete peaks at ± 4. Also observe that Alkofer's $m = 4$ solution is the only selfdual one found, and that there are several solutions with $\mathbf{E}^a \cdot \mathbf{B}^a = 0$.

We have also investigated the stability of most of the solutions with respect to small fluctuation about the saddle point. Since the path integral is defined in terms of $T^a_{\mu\nu}$, we insert $T^a_{\mu\nu} = -iG^a_{\mu\nu} + \theta^a_{\mu\nu}$ in Eq. (3), and expand to quadratic order in $\theta^a_{\mu\nu}$, obtaining for $j^a_\mu = 0$:

$$S_T \approx -\gamma_G V_D + \tfrac{1}{4}\int \mathrm{d}^D x\, \theta^a_{\mu\nu}\theta^a_{\mu\nu} - \mu^4 \mathrm{Tr}\left(\widehat{\theta}\widehat{T}^{-1}\widehat{\theta}\widehat{T}^{-1}\right) + \tfrac{1}{2g}(\partial_\lambda \theta^a_{\mu\lambda})(\widehat{T}^{-1})^{ab}_{\mu\nu}(\partial_\sigma \theta^b_{\nu\sigma})\,. \tag{10}$$

By diagonalizing, the corresponding mass matrix (with respect to $\theta^a_{\mu\nu}$) was found to be positive semidefinite in all cases investigated, except for the $m = 0$ solution with $\gamma_G = 3.1222$. The latter has a single marginally, but numerically significant, unstable mode. We suspect that this solution will disappear if the accuracy of the iteration procedure is improved. In Table 2 we list the number of zero eigenvalues of the mass matrix, n_0, which thus counts both massless modes and true zero modes of S_T. Since a gauge is not fixed, the minimum number is $n_0 = 8$, obtained for Alkofer's solution only. Thus, in this case the space-time orientation is completely locked to the gauge configuration, like in an instanton crystal (cf. Ref. 11). For all other solutions investigated $n_0 = 10-14$. If this corresponds to true zero modes, which may well be the case, there is partial or even complete freedom of orientation between the space-time and gauge degrees of freedom, more reminiscent of a liquid or gas than a crystal.

In order to get a rough idea of the entropy contribution for each solution, we also give in Table 2 the relative frequency of the solutions. As noted above, the iteration scheme is not strictly m-conserving, so these frequencies depend somewhat on the choice of α in Eq. (12). Indeed, by comparing with the results shown in Table 1, one sees there has indeed been some, but not a substantial, flux from the odd to the even m-sectors.

Most notable about these results is that *all* solutions found are almost degenerate in action, as compared with the purely classical density $\gamma_G^0 = \tfrac{1}{4}G^a_{\mu\nu}G^a_{\mu\nu} = 8$. Since the difference between γ_G and γ_G^0 is a quantum effect, further quantum corrections are likely to modify these small differences significantly. Preliminary calculations including the field strength one-loop effective potential[11] (but without the measure term), indicate that this is indeed the case. Furthermore, comparing Eqs. (7) and (10), it is seen that the solutions of lowest γ_G are *suppressed* in the contribution to the path-integral over $T^a_{\mu\nu}$. Indeed, since different solutions behave very differently under $O(4)$ transformations, the relative importance of the various solutions may depend strongly on the operator whose expectation value is being calculated.

Inclusion of the kinetic term will not destroy the stability of the solutions in the static limit. However, for some finite momentum the stability matrix will develop negative eigenvalues, as shown by Reinhardt[12] for SU(2) and $D = 3$. This naively implies acausal propagation, but, as Fried has argued at this meeting[13], this is not necessarily a catastrophe for a confining theory. Anyhow, in order to solve this problem, one must also understand the dynamical aspects of the non-constant 'flip-flop' solutions.

4. The flip-flop ("Braque") vacuum

To see how one can make non-constant solutions of Eq. (4) out of constant ones[6], observe that derivatives only enter S_T as $\partial_\nu T^a_{\mu\nu}$, and hence that any configuration with a non-vanishing $\partial_\nu \tilde{G}^a_{\mu\nu}$ only has no kinetic energy. In order not to generate additional potential energy either such a solution must still fulfill Eq. (5) at each space-time point. Now by combing an improper $O(4)$ transformation which flips just one euclidean coordinate axis (the λ-th, say) with charge conjugation, one obtains a (non-orthogonal) transformation $T^\lambda = iP^\lambda = i(\delta_{\mu\nu} - 2\delta_{\lambda\mu}\delta_{\lambda\nu})$ with the property

$$T^\lambda_{\mu\sigma} G^a_{\sigma\rho} T^\lambda_{\rho\nu} = \begin{cases} -G^a_{\mu\nu} & \lambda \neq \mu \text{ and } \lambda \neq \nu, \\ G^a_{\mu\nu} & \lambda = \mu \text{ or } \lambda = \nu, \end{cases} \quad (\text{ no sum over } \lambda), \qquad (11)$$

¿From an arbitrary constant solution of Eq (5), $G^a_{\mu\nu}(0)$ say, one can then generate a new solution by transforming $G^a_{\mu\nu}$ by T^λ for all points *on one side only* of the hyperplane $x^\lambda = constant$. Actually, these transformations are not all independent, since $T^1 T^2 T^3 T^4 = 1$, and so there are altogether eight distinct field configurations connected by flips in different directions. Furthermore, the transformed solution is *not* gauge equivalent to the original one (as noted in the previous section, it can be $O(4)$ equivalent in some cases).

In general, such flips are allowed at an arbitrary set of hyperplanes. Using the origin as a reference position, let the n-th flip along the μ-axis be situated at the hyperplane $x^\mu = y_n^{(\mu)}$, and correspondingly denote by $n_\mu(x)$ the number of flips in this direction between 0 and x. The resulting field-strength is:

$$G^a_{\mu\nu}(x) = (-1)^{\sum_{\lambda \neq \mu,\nu} n_\lambda(x)} G^a_{\mu\nu}(0). \qquad (12)$$

These field-strengths satisfy

$$\partial_\nu G^a_{\mu\nu}(x) = 0, \qquad (13)$$

and so solve the equations motion, Eq. (3), and with the same action as $G^a_{\mu\nu}(0)$. Actually, this construction does not depend on any particularities of $G^a_{\mu\nu}(0)$, but only on the fact that it is not gauge-equivalent to $-G^a_{\mu\nu}(0)$. Thus, such flip-flop vacua can equally well be constructed from constant vacuum solutions of improved effective potentials.

For SU(2), flipping a single component $G^a_{\mu\nu}$ brings us from one vacuum domain to another degenerate in action with it (but with a different value of m). Hence in this case G and $-G$ are connected by series of flips. In three dimensions there are eight non-equivalent configurations connected by flips, whereas for SU(2) in four dimensions there are 2^6 independent components.

For SU(3) the field-strengths in each domain can be picked among a collection of eight different configurations. In order to calculate the contribution to the path integral from these vacua in the classical approximation, we have to sum over all such solutions. A lattice regularization will make this sum well defined. One can then introduce variables c^μ_j which take the values 1 or 0 depending on whether a flip occurs or not at the j-th hyperplane $x^\mu = ja$ (with the lattice constant a). Hence $n_\mu(x) = \sum_{j<x^\mu/a} c^\mu_j$, and the classical contribution to the lattice path integral reduces to a sum over random variables $\{c^\mu_i\}$, which take the values 0 and 1 with equal probability, and a sum over the different vacuum solutions. In addition one must of course sum over all space-time directions. This makes it quite simple to evaluate the classical contribution to correlation functions. Thus e.g. $\mathbf{E}^a \cdot \mathbf{B}^a$ is effectively a two-valued random variable (unless it vanishes) and its classical correlation function between different points vanishes, as do the expectaion values of other operators which breaks the Euclidean O(4) symmetry.

In addition to Eq (13), the vacuum field-strengths satisfy the abelian dual Maxwell equations:

$$\partial_\nu \widetilde{G}^a_{\mu\nu} = 4\pi k^a_\mu \tag{14}$$

with the conserved chromomagnetic currents k^a_μ given explicitly by:

$$4\pi k^a_0(x) = 2(-1)^{n_0(x)} \sum_i B^a_i(0) \sum_m (-1)^{n_i(x)} \delta(x^i - y^{(i)}_m),$$

$$4\pi k^a_i(x) = 2(-1)^{n_i(x)} \Big[\sum_{j,k} \epsilon_{ijk} E_k(0) \sum_m (-1)^{n_j(x)} \delta(x^j - y^{(j)}_m) \tag{15}$$
$$- B_i(0) \sum_m (-1)^{n_0(x)} \delta(x^0 - y^{(0)}_m) \Big].$$

These currents and charges are concentrated on the 'flips' $x^\lambda = y^{(\lambda)}_n$ such that $\mathbf{k}^a$ circulates around stable strings of chromoelectric flux with fields in neighboring domains oppositely directed. The magnetic fields are organized in sheets with the field lines ending on sheets of magnetic charges, just like in a stack of (chromomagnetic) plate condensers. At random (Euclidean) times all magnetic charges flips their signs, and so the constant chromomagnetic fields are unstable. This picture is also in at least qualitative agreement with several other approaches which describe the structure of the QCD vacuum in terms of a dual superconductor[14,13].

A notable feature of the flips is that they obey *Fermi* statistics: making two flips at a given hyperplane is the same as making no flip, and so at least the classical contribution to the field strength path integral can be rewritten in terms of Grassmann fields. This also means that a flip can be characterized as a fermionic zero mode of the theory, being a solution which spontaneously break translational invariance. Of course, averaging over all gauges, the invariance will be restored. It is in this connection intrigueing that similar gauge-dependent fermionic zero modes, breaking BRS symmetry, has recently also been found in the ghost-sector in the operator product expansion of pure QCD[15]

In order to proceed, one must learn how to handle quantum fluctuations in the flip-flop vacua. In this connection it is important to observe that $\hat{T}^{-1}$, and in particular its diagonal matrix elements, is discontinous at the flip positions. On the other hand the action density must be continous, so that partial integrations can be performed without introducing surface terms, which will break gauge invariance. In order for the kinetic energy term to be gauge invariant, one must thus require that also the fluctuations satisfy $\partial_\nu \theta^a_{\mu\nu} = 0$ at the flip positions. These boundary conditions means that the transverse fluctuations in the flip-sectors (chromoelectric flux-tubes and chromomagnetic flux sheets) will be quantised, like in a wave guide, and that there will be a Casimir contribution to the action density. The interplay between this effect and the unstable modes of the constant solutions within each flip-sector is presently being investigated.

5. Acknowledgements

We would like to thank the organizers of this workshop for their kind hospitality and the very good atmosphere, and the participants for numerous enlightening discussions. We also thank The Norwegian Research Council for Science and the Humanities (NAVF) for travel grants.

6. References

1. G. K. Savvidy *Phys. Lett.* **B71** (1977) 133.
2. N. K. Nielsen and P. Olesen, *Nucl. Phys.* **B144** (1978) 376.
3. P. Olesen, *Physica Scripta* **23** (1981) 1000, and references therein.
4. M. B. Halpern, *Phys. Rev.* **D16** (1977) 1798; 3515.
5. M. Schaden, H. Reinhardt, P. A. Amundsen and M. J. Lavelle, *Nucl. Phys.* **B339** (1990) 595.
6. P. A. Amundsen and M. Schaden, *Phys. Lett.* **B252** (1990) 265.
7. S. Deser and C. Teitelboim, *Phys. Rev.* **D13** (1976) 1592.
8. R. Alkofer, *J. Phys.* **G17** (1991) 310.
9. A. Actor, *Rev. Mod. Phys.* **51** (1979) 461, (appendix B).
10. H. Reinhardt *these proceedings*.
11. C. Pietmaaier, *PhD Thesis, Universty of Regensburg* (1991).
12. H. Reinhardt, *Phys. Lett.* **B257** (1991) 375.
13. H. M. Fried, *Institute Non Lineaire de Nice preprint* INLN 92,21 (1992), and these proceedings.
14. M. Baker, J. Ball and F. Zachariasen, *Phys. Rep.* **209** (1990) 73; M. Baker, *these proceedings*.
15. J. Abbach, A. Streibl and M Schaden, *these proceedings*, and to be published.

Instantons in the Field Strength
Approach to QCD [*]

H. Reinhardt

Institut für Theoretische Physik

Universität Tübingen

D-7400 Tübingen, FRG

Abstract:

Field strength formulated Yang-Mills theory is confronted to the traditional formulation in terms of gauge fields. It is shown that both formulations yield the same semiclassics, in particular the same instanton physics. The field strength formulation is, however, superior at the tree level where it includes already a good deal of quantum fluctuations of the standard formulation. These quantum fluctuations break the scale invariance of classical QCD and give rise to an instanton interaction. The latter causes the instanton to condense and to form a homogeneous instanton solid. These instanton solids show up in the field strength approach as homogeneous (constant up to gauge transformations) vacuum solutions. A new class of $SU(N)$ instantons is presented which are not embeddings of $SU(N-1)$ instantons but have non-trivial $SU(N)$ color structure and carry winding number $n = \frac{N}{6}(N^2 - 1)$. These novel instantons generate (after condensation) the lowest action homogeneous solutions of the field strength approach.

1. Introduction

The QCD vacuum is known to have a complicated structure. It was first observed by Savvidy [1] that the perturbative vacuum with zero gauge field is unstable and that a constant color magnetic field lowers the energy. Later on it was shown by the Copenhagen group [2] that a constant color magnetic field configuration is also unstable and breaks into domains with randomly orientated magnetic fields remmiscent to a ferro magnet. But even such a field configuration turns out to be unstable and decays into a spaghetti-like vacuum of color magnetic flux tubes.

---

[*] Supported in part by DFG under contract Re 856/1-1 and by COSY under contract 41170853

Soon after the discovery of the instantons by Belavin et al. [3] it was realised that they must play an important role in the structure of the QCD vacuum. Callen, Dashen and Gross considered the QCD vacuum as a gas of weakly interacting instantons [4]. It was later pointed out by Shuryak that the instantons must strongly interact and perhaps form something like an instanton liquid [5].

The instanton picture of the QCD vacuum has proven very successful in the sense that it can explain both the gluon condensation (leading to the unnormalous breaking of scale invariance) and the quark condensation (which is a manifestation of the spontaneous break-down of chiral symmetry.)

The instantons exist due to the self-interaction of the gluons which is strong at low energies and is presumably responsible for confinement. It is clear that any approach to the QCD vacuum has to treat the gluon self-interaction in a non-perturbative manner. The only rigorous approach to the QCD ground state is perhaps provided by the lattice Monte-Carlo calculations. These calculations are, however, rather time-consuming and still the physical insights one gains are pretty meager. Therefore any alternative approach which does not rely heavily on numerical simulations but which makes the physical structure of the QCD vacuum transparent would be very welcome.

I would like to present here an approach which treats the gluon self-interaction in a non-perturbative way from the very beginning, but which at the same time is sufficiently simple so that most of the calculations can be done analytically. This is the so-called field strength approach [6,7]. In this approach the gauge potential is completely eliminated in favor of the field strength, i.e. in favor of the color electric and magnetic fields. The upshot is an effective tensor theory in the field strength which in many respects is similar to the Ginsburg-Landau theory of superconductors.

The first attempt to formulate Yang-Mills theory in terms of field strength was made by Halpern [8] but unfortunately he stayed essentially at the perturbative level and did not realise that this approach offers a simple non-perturbative description of the Yang-Mills vacuum. The non-perturbative field strength approach was formulated first in refs. [6] where a classical analysis of the corresponding ground state was carried out. Quantum fluctuations in the field strength approach were considered in ref.[7]. The fermion sector was studied within the field strength approach in refs. [9-12]. I will concentrate in my talk on a recent work which was done together with K. Langfeld where we establish the connection of the field strength approach to the more traditional instanton physics [13, 14].

50

The plan of my talk is as follows: In the next section I will briefly sketch the reformulation of Yang-Mills theories in terms of field strength. I will then show that the field strength formulation yields the same semi-classics as the standard approach formulated in terms of gauge fields. We will see that the field strength formulation provides the same instantons as stationary points of the action as the standard formulation. Moreover, even the quantum corrections are the same in both approaches [13]. However, the field strength formulation is superior at the tree-level since at this level it includes already quantum fluctuations of the standard formulation.

I will present a new class of instantons which play a very important role in the ground state of Yang-Mills theories. I will show that in the field strength approach these instantons condense and form a color and Lorentz aligned instanton solid. This instanton solid shows up as homogeneous solutions in the effective tensor theory of the field strength approach. Due to the condensation of instantons an effective quark interaction is formed which at low energy reduces to a Nambu-Jona Lasinio type of interaction i.e. to a simple contact force but which at high energies has the correct asymptotic behavior of the one-gluon exchange. This effective quark interaction has chiral symmetry and can be shown to lead a spontaneous break-down of chiral symmetry in the QCD ground state [9-12]. I will conclude my talk with a short summary and a brief outline of open questions and future projects to be studied in the field strength approach to Yang-Mills theories.

2. Field strength formulated Yang-Mills theory

QCD is defined by the following classical Lagrange density

$$\mathcal{L} = \bar{q}(i\not{\partial} - m_0)q - \frac{1}{4g^2}F^a_{\mu\nu}F^a_{\mu\nu} + A_\mu j^a_\mu$$
$$\equiv \mathcal{L}_0 + \mathcal{L}_{yM} + \mathcal{L}_{int} \tag{1}$$

where q denotes the quark field and A_μ is the gauge potential whose field strength is given by

$$F^a_{\mu\nu} = \partial_\mu A^a_\nu - \partial_\nu A^a_\mu + f^{abc}A^b_\mu A^c_\nu. \tag{2}$$

Furthermore m_0 denotes the current mass of the quarks and g is the coupling constant. Finally j denotes the color current the quarks defined by

$$j^a_\mu(x) = \bar{q}(x)\gamma_\mu t^a q(x) \tag{3}$$

where t_a denotes the generators of the gauge group, which satisfy the lie algebra

$$[t^a, t^b] = i f^{abc} t^c \tag{4}$$

with f^{abc} being the structure constant of the gauge group. The quantum theory is defined by the following functional integral

$$Z = \int DA\mathcal{D}q\mathcal{D}\bar{q}e^{i\int d^4x \mathcal{L}}.$$
(5)

If I collect all pieces which depend on the gauge field I can rewrite the generating functional of QCD in the following form:

$$Z = \int \mathcal{D}q\mathcal{D}\bar{q}e^{i\int d^4x \mathcal{L}_0}Z_{YM}[j]$$
(6)

where

$$Z_{YM}[j] = \int DA\, exp[-\frac{1}{4g^2}\int d^4x F_{\mu\nu}^a F_{\mu\nu}^a - \int d^4x j_\mu^a A_\mu^a]$$
(7)

is the generating functional of the gluon sector. In the gluon generating functional the color current of the quarks figures as an external current so that the quarks are passive. To solve the quantum mechanics of the gluon sector we have to carry out the functional integral over the gauge field. This cannot be done exactly due to the presence of the self-interaction of the gauge field since we can do only Gaussian functional integrals. On the other hand since the self-interaction is strong, at least at low energies, a perturbative treatment of the gluon self-interaction is ruled out. The trick which helps here to exactly integrate out the gauge field while keeping fully the gluon self-interaction is to linearize the exponent by means of an auxiliary field χ via the following identity

$$exp[-\frac{1}{4}\int d^4x F_{\mu\nu}^a(A)F_{\mu\nu}^a(A)]$$

$$= \int D\chi_{\mu\nu}^a exp[-\frac{1}{4}\int d^4x \chi_{\mu\nu}^a \chi_{\mu\nu}^a - \frac{i}{2g}\int d^4x \chi_{\mu\nu}^a F_{\mu\nu}^a(A)]$$
(8)

The auxiliary field $\chi_{\mu\nu}^a$ has the same color and Lorentz tensor structure as the field strength of the gauge field. It is antisymmetric in the Lorentz indices and lives in the adjoint representation of the gauge group. The linearization of the square in the exponent by means of an integration over an auxiliary field is sometimes referred to (wrongly!) as Hubbard-Stratonowich transformation. This identity has been very fruitful in other branches of physics as e.g. in solid state physics or nuclear many-body theory [15].

Inserting the identity (8) into the Yang-Mills amplitude (7) the integral over the gauge field becomes Gaussian and can be exactly carried out. One finds then for the Yang-Mills amplitude (7)

$$Z_{YM}[j] = \int D\chi \, Det^{-1/2}(i\hat{\chi}) exp[-S_{FS}(\chi) + \int d^4x \{ j_\mu^a(x) J_\mu^a(x) + \frac{g^2}{Zi} j_\mu^a(x)(\hat{\chi}^{-1}) j_\mu^a(x) \}]$$

(9)

Here the preexponential factor, the functional determinant $det^{-1/2}(i\hat{\chi})$, arises from the integral over the gauge field. The quantity $\hat{\chi}$ denotes the tensor field in the adjoint representation of the gauge group

$$\hat{\chi}_{\mu\nu}^{ab} = f^{abc} \chi_{\mu\nu}^c.$$

(10)

The effective action of the tensor field is given by

$$S_{FS}(\chi) = \frac{1}{4} \int d^4x \chi_{\mu\nu}^a \chi_{\mu\nu}^a + \frac{i}{2g} \int \chi_{\mu\nu}^a F_{\mu\nu}^a(J(\chi))$$

(11)

The explicit form of this action is obvious from the fact that for Gaussian functional integrals the stationary phase approximation is exact and this action is precisely the negative exponent of the right-hand side of eq. (8) taken at the stationary phase point. Now the stationary phase value of the gauge potential A_μ is given by the following induced vector potential

$$J_\mu^a(\chi) = (\hat{\chi}^{-1})_{\mu\nu}^{ab} \partial_\lambda \chi_{\lambda\nu}^b \sim A_\mu^a(x)$$

(12)

which under gauge transformation behaves precisely like the original gauge field $A_\mu^a(x)$ and represents thus the counter part of the gauge potential in the field strength formulation.

The exponent of the effective tensor theory (eq. (9)) contains in addition a coupling of the quark current j to the induced gauge potential $J(\chi)$. This term is necessary in order to describe phenomena like the Bohm-Aharonov effect which do not just measure the field strength but are also sensitive to the gauge potentials. Finally the last term in the exponent on the right-hand side of eq. (9) represents a current-current interaction of the quarks which is mediated by the tensor field χ.

Eq. (9) gives an exact mapping of the Yang-Mills sector of QCD into an effective tensor field theory. In the following let me discuss this effective tensor theory for vanishing external quark currents $j = 0$

$$Z_{YM}[0] = \int D\chi (Det \, i\hat{\chi})^{-1/2} e^{-S_{FS}(\chi)}$$

(13)

Let me now show that the effective tensor theory has the same semiclassical expansion as the original theory defined in terms of the gauge field. In particular the same instanton physics is obtained.

3. Instantons in field strength formulated Yang-Mills theories

The classical equation of motion following by varying the effect action (11) is given by

$$g\chi_{\mu\nu}^a = -iF_{\mu\nu}^a(J(\chi)) \tag{14}$$

It is not difficult to show that this equation is solved by the field strength of an instanton. More precisely: if A_μ^{inst} denotes the gauge potential of an instanton, i.e. a stationary point of the standard Yang-Mills action, then the classical solution is given by

$$g\chi_{\mu\nu} = -iF_{\mu\nu}(A^{inst}) \tag{15}$$

Furthermore for this classical field strength solution the corresponding induced gauge potential (eq. (12)) becomes just the gauge potential of the instanton

$$J_\mu(\chi = -\frac{i}{g}F(A^{inst})) = A_\mu^{inst}. \tag{16}$$

Finally the effective action of the field strength for the instanton solution reduces precisely to the action of the instanton in the standard Yang-Mills approach

$$S_{FS}(\chi = -iF(A^{inst})) = S_{YM}(A^{inst})) = \frac{1}{4g^2}\int d^4x(F_{\mu\nu}^a(A^{inst}))^2 \tag{17}$$

So we observe a one-to-one correspondence at the classical level between the field strength formulation of Yang-Mills theory and the standard formulation in terms of the gauge potential. This equivalence holds, however, also beyond the classical level. If we include the leading order quantum fluctuations of the tensor field χ around the classical instanton solution

$$\chi = -\frac{i}{g}F(A^{inst}) + \chi' \tag{18}$$

the Yang-Mills amplitude (eq. (13)) becomes with the identity (eq. (17))

$$Z_{YM}[0] = \left(Det(i\hat{\chi})Det'(\frac{\delta^2 S_{FS}(\chi)}{\delta\chi\delta\chi})_{\chi=-iF_{inst}}\right)^{-1/2}e^{-S_{YM}(A^{inst})} \tag{19}$$

54

On the other hand in the standard Yang-Mills approach we would obtain in the semiclassical approximation, expanding the gluon field around the instanton, the following expression for the Yang-Mills amplitude

$$Z_{YM}[0] = \left(Det'(\frac{\delta^2 S_{FS}}{\delta A \delta A})_{A_\mu = A_\mu^{inst}} \right)^{-1/2} e^{-S_{YM}(A^{inst})} \tag{20}$$

The here arising functional determinants are defined after a proper elimination of the zero eigenvalues, which is indicated by the prime. Recently it has been shown that the two functional determinants are identical [13]

$$Det(i\hat{\chi})Det'(\frac{\delta^2 S_{FS}(\chi)}{\delta \chi \delta \chi})_{\chi = -iF_{inst}} = Det'(\frac{\delta^2 S_{YM}(A)}{\delta A \delta A})_{A = A^{inst}} \tag{21}$$

so that the field strength approach reproduces exactly the semiclassics of the standard formulation of Yang-Mills theory in terms of the gauge potential. The field strength formulation is, however, superior if one sticks to the tree-level. This is because of the determinant of $\hat{\chi}$ (cf. eq. (37)) it contains already at the tree level quantum fluctuations of the standard Yang-Mills formulation. Remember that this term arose precisely from the integration over the gauge field after the χ-field had been introduced. A little bit later we will see that this functional determinant contains in fact the dominating part of the quantum fluctuations around an instanton.

Before I discuss these quantum effects let me present some new instantons which have been recently discovered [14] and which are very relevant in the context of the field strength formulation of Yang-Mills theory.

4. Novel SU(N) Instantons

It is well-known that in four-dimensional non-Abelian Yang-Mills theories any gauge field configuration which has finite action and either self-dual or anti-self-dual field strength represents an instanton solution [3]. There exists also a general method for constructing instantons which goes back to work by Atiyah and Ward and Atiyah et al. [16]. This general method is, however, difficult to apply in order to find explicit instanton solutions. Therefore I would like to follow a different route which as a by-product will provide useful algebraic relations which are relevant for the field strength approach discussed in section 5.

All instanton physics of QCD has been based so far on the SU(2) instanton discovered by Belavin et al. [3], more precisely on the embedding of this SU(2) instanton into the SU(3) group. It is also well-known that multi-instanton solutions found by 'tHooft [17] exist but to my knowledge so far no explicit instantons with non-trivial SU(N) color structure i.e. which are not embeddings of the standard SU(2) instanton into the larger gauge group have been reported. Recently a new class of SU(N) instantons have been found which are not embeddings of SU(N-1) instantons but carry a non-trivial SU(N) color structure [14]. These instantons can be easily found when one uses the Schwinger-Fock gauge

$$x_\mu A_\mu = 0. \tag{22}$$

In this gauge the gauge potential can be expressed in terms of the field strength by the relation

$$A_\mu(x) = -\int_0^1 d\alpha \alpha F_{\mu\nu}(\alpha x) x_\nu. \tag{23}$$

For the field strength of an instanton I make the following ansatz

$$F_{\mu\nu}^a(x) = G_{\mu\nu}^a \phi(x^2), \tag{24}$$

where $G_{\mu\nu}^a$ is a constant field strength matrix and $\phi(x^2)$ is a space-time dependent function of the four-dimensional radius $x^2 = x_\mu x_\mu$.

When we start with an arbitrary field strength $F_{\mu\nu}$ and insert it into the equation (23) for the gauge potential the resulting gauge potential A_μ will have in general a field strength $F[A]$ different from the starting field configuration $F_{\mu\nu}$ which we have plugged into equation (23). To ensure that the gauge potential obtained from a given field strength configuration $F_{\mu\nu}$ via eq. (23) has in fact the starting field strength we have to impose the constraint

$$F_{\mu\nu}(A(F)) = F_{\mu\nu}. \tag{25}$$

Solving this constraint with the ansatz (24) for F leads to a differential equation for the function $\phi(x^2)$ with the solution

$$\phi(x^2) = \frac{\rho^2}{(x^2 + \rho^2)^2}. \tag{26}$$

where ρ represents the size of the instanton. This is precisely the space-time dependence of the field strength of the well-known Polyakov-t'Hooft instanton [3]. The constraint (25) gives furthermore rise to an algebraic equation for the constant field strength matrix $G^a_{\mu\nu}$, which can be easily solved by expanding this matrix in terms of the 'tHooft symbols $\{\eta^i, \bar{\eta}^i\}$

$$G^a_{\mu\nu} = G^a_i \eta^i_{\mu\nu} + \bar{G}^a_i \bar{\eta}^i_{\mu\nu} \tag{27}$$

The 'tHooft symbols η^i and $\bar{\eta}^i$ are self-dual and anti-self-dual, respectively, and fulfil a SU(2) algebra.

$$\begin{aligned} [\eta^i, \eta^j] &= -2\epsilon_{ijk}\eta^k \\ [\bar{\eta}^i, \bar{\eta}^j] &= -2\epsilon_{ijk}\bar{\eta}^k \end{aligned} \tag{28}$$

The expansion coefficients in eq. (27) are related to the color electric and color magnetic fields by

$$\begin{aligned} G^a_i &= \frac{1}{2}(E^a_i + B^a_i) \\ \bar{G}^a_i &= \frac{1}{2}(E^a_i - B^a_i) \end{aligned} \tag{29}$$

For simplicity in the following let me consider only self-dual field configurations for which $\bar{G} = 0$. Furthermore for the self-dual field configurations let me define the three SU(N) valued matrices

$$G_i = G^a_i t^a \tag{30}$$

In ref. 14 it was shown that the constraint (25) requires these matrices to fulfil a SU(2) algebra

$$[G_i, G_j] = i\epsilon_{ijk}G_k \tag{31}$$

This result is an agreement with a general theorem by Bott [18] which states that any continuous mapping of S^3 into a semi-simple Lie group e.g. SU(N), can be continuously deformed into a mapping into a SU(2) subgroup.

Eq. (31) has far-reaching consequences. It implies that any spin S representation of the SU(N) valued matrices (30) yields an instanton solution (24) with the space-time dependence given by (26). Furthermore it can be shown that the winding number of these instantons is given by

$$n = \frac{2}{3}tr(G_iG_i) = \frac{2}{3}S(S+1)d_{(S)} \tag{32}$$

where we have used the fact that G_iG_i represents the quadratic Casimir operator of the SU(2) spin group whose eigenvalue is $S(S+1)$, and $d_{(S)}$ denotes the dimension of the spin representation.

Let me now construct explicit instanton solutions (24) from the SU(2) algebra (31). For the gauge group SU(N) the G_i (30) are hermitian $N \times N$ matrices. The maximum spin which can be realised by N-dimensional hermitian matrices is spin $S = \frac{N-1}{2}$. For this spin we find from (32) the following winding number

$$n = \frac{2}{3}\frac{N-1}{2}N\frac{N+1}{2} = \frac{N}{6}(N^2 - 1) \tag{33}$$

Let me now consider a few illustrative examples which are relevant for the field strength approach discussed below.

For the gauge group SU(N=2) the G_i can represent only spin $S = 1/2$. The corresponding spin representation is found by identifying the G_i by the Pauli matrices

$$S = \frac{1}{2} : \quad G_i = \frac{1}{2}\tau_i \quad , \quad G_i^a = \delta_i^a \tag{34}$$

The color vectors $\mathbf{G^a}$ represent then orthonormal 3-beins in color space as is illustrated in fig. 1 (a). From the general expression for the winding number (33) we find that this instanton has winding number $n = 1$. This instanton is in fact the standard instanton found by Belavin et al. [3].

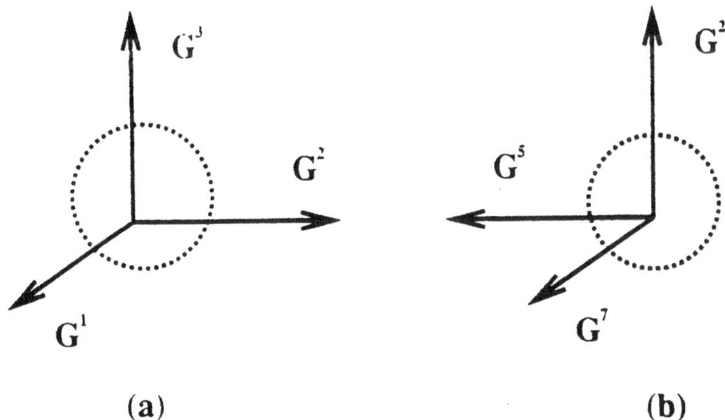

Fig. 1
Graphical illustration of (a) the SU(2) instanton and (b) the SU(3) instanton.

For the gauge group SU(N=3) the maximum spin which can be realised by the matrices G_k is $S = 1$. In this case the G_k matrices are given up to a factor of $\pm i$

58

by the totally antisymmetric tensor ϵ_{klm} which are antisymmetric generators of the SU(3) group and form the $\bar{3}$ representation. The only non-zero color components are then given by

$$S = 1: \quad G_1^7 = -G_2^5 = G_3^2 = 2 \tag{35}$$

From (32) we find for this instanton winding number $n = 4$. The color structure of this instanton is illustrated in fig. 1(b) although it has again the form of a color dreibein it has a non-trivial SU(3) color structure which cannot be interpreted as an embedding of the SU(2) instanton.

We can go on and consider higher gauge groups e.g. in the case of SU(4) the maximum spin instanton carries spin 3/2 and winding number $N = 10$. Furthermore since SU(4) has a SU(2) x SU(2) subgroup there exist also non-maximum spin instantons which again are not embeddings of a standard SU(2) instanton. These instantons are obtained by identifying the matrices G_k (30) with the t'Hooft symbols $(\frac{-i}{2}\eta^k), (\frac{-i}{2}\bar{\eta}^k)$ which are four-dimensional representations of spin 1/2. From (33) follows that these instantons carry winding number $n = 2$.

5. Instanton interaction

Let us now consider the contribution of an instanton to the partition function in the standard Yang-Mills approach and in the field strength formulation. We have seen above that their contribution is the same in both approaches on the semiclassical level. If we stick, however, to the tree-level

$$Z_{YM}[j = 0] = \int \mathcal{D}A e^{-S_{YM}(A)} \approx e^{-S_{YM}(A_{inst})} \tag{36}$$

we get an extra contribution in the field strength formulation from the pre-exponential functional determinant

$$Z_{YM}[j = 0] = \int \mathcal{D}\chi Det^{-1/2}(i\hat{\chi})e^{-S_{FS}(\chi)} \approx (Det(F_{inst})^{-1/2}e^{-S_{YM}(F_{inst})} \equiv e^{-\tilde{S}(F_{inst})} \tag{37}$$

which we have included into the action by using

$$Det^{-1/2}(i\hat{\chi}) = exp[-\frac{1}{2}Trlog(i\hat{\chi})] \tag{38}$$

This determinant is a diverging object and needs regularization, which introduces a scale μ

$$Trlog i\hat{\chi} \rightarrow \mu^4 trlog(i\hat{\chi}/\mu^2)$$

As I have already discussed this determinant arises from integration over the gauge field and represents hence quantum fluctuations which appear in the standard Yang-Mills approach only after including explicit fluctuations around the instanton. So we expect the field strength formulation to describe quantum effects already the tree-level. To illustrate this let us consider the interaction between two instantons. For the field strength of the two instantons we make the ansatz

$$F_{\mu\nu}^{(12)}(R) = F_{\mu\nu}^{(1)}(x + \frac{R}{2}) + F_{\mu\nu}^{(2)}(x - \frac{R}{2}) \qquad (39)$$

and define the instanton interaction by

$$V(R) = \tilde{S}(F^{(12)}) - \tilde{S}(F^{(1)}) - \tilde{S}(F^{(2)}) \qquad (40)$$

where $\tilde{S}(F)$ is defined by eq. (37). In the standard Yang-Mills approach we do not obtain any interaction between two instantons at the classical level if we allow for instanton deformations (though an interaction between instanton and antiinstanton exists). This is because there exists an exact two-instanton solution whose action is precisely twice the action of a single instanton [17]. We find, however, an instanton-instanton interaction at the tree level in the field strength formulation from the pre-exponential determinant factor (see eqs. (37), (39)). If we consider for simplicity the interaction induced by this term between two instantons for fixed shape and the same color and Lorentz orientation we find the interaction shown in fig. 2. This interaction looks similar to the nucleon-nucleon interaction. It has a short range repulsion and a medium range attraction. This interaction also depends somewhat on the instanton radius. Fig. 2 shows the instanton-instanton interaction between two aligned instanton for instanton radii $\rho\mu = 1.3, 2 and 2.5$. From the shape of the instanton interaction it is clear that this interaction favours a condensation of instantons analogously to the condensation of nucleons in nuclear matter.

6. The non-perturbative field strength approach

The above example shows that the pre-exponential determinant arising in the field strength approach describes already a good deal of quantum fluctuations which in the standard approach show up only after expanding the function integral around the classical field configuration. If this term has such important effects on the instanton physics one should not treat this term perturbatively i.e. on top of the classical approximation but rather one should include it into the action by using the relation (38) One obtains then an effective action which is given by

$$S_{FSA}(\chi) = \frac{1}{4} \int \chi^2 d^4 x + \frac{\mu^4}{2} Tr log i \frac{\hat{\chi}}{\mu^2} + \frac{i}{2g} \int \chi F(J(\chi)) d^4 x \qquad (41)$$

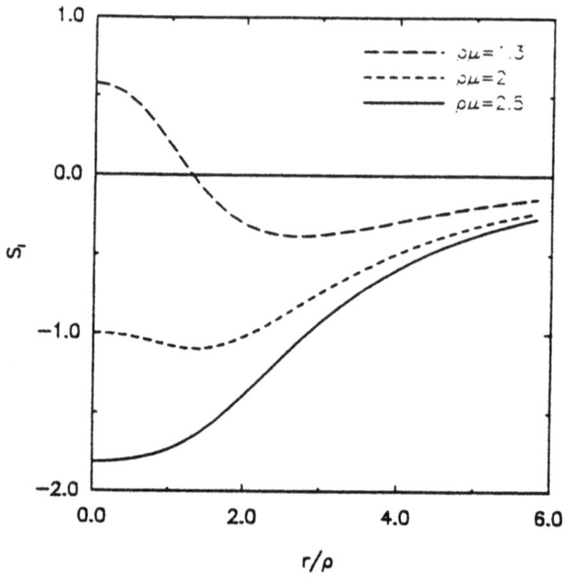

Fig. 2

The instanton-instanton interaction calculated from eq. (40).

The generating functional of the gluon sector is then given by the path integral

$$Z[j = 0] = \int D\chi e^{-S_{FSA}(\chi)} \tag{42}$$

Please note that the new term in the effective action $trlog(i\hat{\chi})$ is of order $\hbar$ compared to the remaining two terms. Similar action functionals occur in the path integral description of many-body systems. For example in the description of Fermi systems one has to include the so-called fermion determinant into an effective action in order to obtain already at the tree-level a decent description of the many-body system like the Hartree-Fock approximation [15].

Due to the appearance of the energy scale μ the effective action breaks explicitly the scale invariance which was originally present in the classical Yang-Mills action. So while in the standard Yang-Mills approach the scale invariance is broken in an anomalous fashion by quantum fluctuations in the effective theory this invariance is already broken at the tree-level. This is not surprising since the determinant term arose precisely from quantum fluctuations in the standard approach, namely from integrating out the gauge field.

Variation of the effective action (41) leads to the following equation of motion

$$\chi_{\mu\nu}^a = -\mu^4 f^{abc}(\hat{\chi}^{-1})_{\mu\nu}^{bc} - \frac{i}{g}F_{\mu\nu}^a(J(\chi)) \tag{43}$$

Comparing this equation of motion with the one we obtained previously when the determinant was not included in the action (see eq. (14)) we find a new term with an explicit energy scale μ. From the presence of the energy scale it is already clear that this equation can no longer tolerate instantons with arbitrary scale as ρ solutions (see eq. (26)). Instead of localized instantons equation (43) has now constant (up to gauge transformations) solutions, for which the induced gauge potential $J(\chi)$ (12) vanishes. In view of the fact that the determinantal term gave rise to an instanton interaction which favours the condensation of instantons and furthermore that the same term when included into the effective action spoils the existence of instantons and at the same time triggers the appearance of the constant field strength solutions, we are led to the conclusion that the latter represent condensates of instantons. In fact, one can show that each maximum spin instanton of the form $\chi_{\mu\nu}^a = -iG_\mu^a \phi(x^2)$ discussed above leads precisely to a constant field strength solution of the form $\chi_{\mu\nu}^a = -iG_{\mu\nu}^a const$ with the same color and Lorentz structure definded by $G_{\mu\nu}^a$. This suggests to interpret indeed the constant field strength solutions as solids of aligned instantons. Furthermore one can show that the constant field strength solutions generated by the maximum spin instantons are precisely those constant solutions which have lowest action.

One can easily imagine that one cannot just glue together aligned instantons but also instanton and anti-instantons. In fact, the equation of motion (43) allows for constant solutions which are not generated by instantons and which can be interpreted as a condensate of aligned instantons and anti-instantons. These field configurations are neither dual nor anti-self dual in contrast to the solutions which are generated by the instantons, which are either self-dual or anti-self-dual. Furthermore these solutions have usually higher classical action but at the same time a larger entropy, so they might give important contributions to the functional integral. A more detailed discussion of the homogeneous vacuum solutions is presented in [20].

A few comments are in order. An individual homogeneous field strength solution breaks CP-invariance. It was first pointed out by H.B. Nielsen that such field configurations cannot form the QCD ground state since the inclusion of the electroweak interaction would lift the degeneracy between CP-conjugated vacua. One would face then a strong CP-problem. He conjectured that there should exist domain-wall-like solutions (flips) which interpolate between CP-conjugated vacua. Field configurations which flip from one homogeneous state to a CP-conjugated state in fact exist and were reported in ref.19. There these field configurations were interpreted as classical solutions to eq. (43) i.e. as stationary points of the effective action (41). It was claimed [19] that the presence of such a flip does not increase the action i.e. this flip has zero action. This is, in fact, true for the kinetic energy. However, as one can easily see, the potential energy receives from the flip an infinite imaginary phase.

Furthermore if the flips (kinks) were true classical solutions of zero action i.e. large amplitude zero modes interpolating between CP-conjugate vacua one should see in the spectrum of the small amplitude fluctuations around these corresponding vacua already zero modes in the direction of the kinks. However, these zero modes are not seen. This further supports the observation that the kinks are not true classical solutions.

Concluding remarks

In my talk I have shown that there exists an alternative formulation of Yang-Mills theories in terms of field strength. We have seen that the field strength formulation provides the same instanton physics as the standard formulation in terms of the gauge field and furthermore yields also the same semiclassical approximation. However, the field strength formulation includes already a good deal of quantum fluctuations of the standard Yang-Mills approach at the tree-level. In the field strength approach these quantum effects are included already into an effective action (41). Then individual instantons cease to exist. They condense and form a homogeneous vacuum which can be interpreted as a color and Lorentz aligned instanton solid. I expect that this solid melts when quantum fluctuations are included. This conjecture is supported by the fact that in the field strength approach at high energies one recovers the usual perturbative QCD regime [13].

Due to the condensation of the instantons forming a homogeneous field strength vacuum an effective quark interaction is induced, which at low energies reduces to a Nambu-Jona Lasinio type of model. It has chiral symmetry but a more complicated color and Lorentz structure. At high energies where the instanton solid is presumably destroyed the effective quark interaction reduces to the one gluon exchange. It was shown in refs. 9-12 in the strong coupling limit that the induced effective quark interaction leads to spontaneous break-down of chiral symmetry.

In ref. [21] the effective potential for the field strength was evaluated in three space-time dimensions for the SU(2) gauge group. It was found that this potential has a minimum corresponding to the homogeneous vacua discussed before and remains finite as the field strength tends to zero.

Note added

During the Paris meeting an interesting alternative field strength formulation of Yang-Mills theory in Minkowski space was presented by H.M. Fried, which treats only the non-Abelian part of $(F_{\mu\nu}^a)^2$ in terms of field strength while keeping the Abelian part in an effective gluon propagator.

References

[1] G.K. Savvidy, Phys. Lett. **B 71** (1977)133

[2] N.K. Nielsen and P. Olesen, Nucl. Phys. **B 144** (1978)376

[3] A.A. Belavin et al, Phys. Lett. **59 B** (1975)15

[4] Callan, Dashen and D. Gross, Phys. Rev. D

[5] E.V. Shuryak, Phys. Rep. **115** (1984)151

[6] M. Schaden, H. Reinhardt, P. Amundsen and M. Lavelle, Nucl. Phys. **B 339** (1990)595

[7] H. Reinhardt, Phys. Lett. **B 248**(1990)365

[8] M. B. Halpern, Phys. Rev. **D 16** (1977)1798

[9] H. Reinhardt, Phys. Lett. **B 257**(1991)375

[10] K. Langfeld and M. Schaden, Phys. Lett.**B 272**(1991)100

[11] R. Alkofer and H. Reinhardt, Z. Phys. A (1992), in press

[12] K. Langfeld, R. Alkofer and H. Reinhardt, Phys. Lett. B

[13] H. Reinhardt and K. Langfeld, Instantons in field strength formulated Yang-Mills theory, Uni-Tübingen preprint, June 1992, subm. to Phys. Lett. B

[14] K. Langfeld and H. Reinhardt, Instanton condensation in field strength formulated Yang-Mills theories, Uni-Tübingen preprint, July 1992, subm. to Nucl. Phys. B

[15] H. Reinhardt, J. Phys. **G 5**(1979) L 91, H. Reinhardt, Nucl. Phys. **A 298** (1978)77, H. Reinhardt, Nucl. Phys. **A 346**(1980) 1 A

[16] M.F. Atijah, R.S. Ward, Comm. Math. Phys. **55** (1977)177, M.F. Atiyah, N.J. Hitchin, V.G. Drinfeld, Yu I. Manin, Phys. Lett. **A 65** (1978)185

[17] G. 'tHooft, Phys. Rev. Lett. **37**(1976)8

[18] R. Bott, Bull. Soc. Math. France **84**(1956)251

[19] P. Amundsen and M. Schaden, Phys. Lett. **B 252**(1990)265

[20] H. Reinhardt in proceedings of X. International Seminar on High Energy Physics Problems, Relativistic Nuclear Physics and Quantum Chromodynamics, Dubna, 24-29 September 1990 (eds. A.M. Baldin, V.V. Barov and L.P. Kaptari)

[21] H. Reinhardt, The effective potential for Yang-Mills field strength, Uni-Tübingen preprint (1991)

[22] H.M. Fried, Transversally confined acausal gluons, preprint INLN 92.21 and this meeting.

CONFINEMENT WITH GAUSSIAN GLUON FIELDS[1]

TAMÁS S. BIRÓ

Institut für Theoretische Physik der JLU Gießen

Heinrich-Buff-Ring 16, DW-6300 Gießen, Germany

Abstract

A phenomenological model of the QCD vacuum is presented, which describes a linearly rising variational energy of a quark antiquark pair. It is also speculated about quasi hadrons above the color deconfinement temperature due to non-Debye forces.

A model of the QCD vacuum structure and quark confinement - if it can not be obtained by a direct solution of the quantummechanical equations of motion - is usually described by its assumption about the dominant gluon field configuration in the ground state and by its momentum-scale dependence. In the present model the coupling constant is treated perturbatively, while rather specific assumptions are made about the gluon field configuration.

After presenting these underlying assumptions the variational action will be minimized in order to determine the mass-scale of the vacuum state self-consistently. In the second step the extra energy of heavy quarks and antiquarks is calculated on this background, which defines an effective $q\bar{q}$ potential in terms of two mass scales. Finally this potential is investigated in a hot QCD plasma and a possibility of a formation of hadron-like clusters will be pointed out.

---

[1]Work supported by BMFT and GSI Darmstadt

1. Assumptions about the A-field

The following assumptions are made about the vector potential in the vacuum state:

- $A_0 = 0$ fixes the Lorentz-frame and partially the gauge.

- The classical component of the vector potential, $< A_i^a(x) >= \overline{A}_i^a(x)$ describes a non-vanishing vacuum expectation value in the presence of external color charges ρ^a. It vanishes for $\rho^a = 0$, it is longitudinal and its value can be obtained as

$$\overline{A}_i^a(x) = \int \frac{d^3k}{(2\pi)^3} \frac{\rho^a(k)}{k^2 \omega_L(k)} k_i e^{i\vec{k}\vec{x}}, \tag{1}$$

which is motivated by an abelianized Gauss law. The spectrum $\omega_L(k)$ of these fields, however, is not necessarily determined by the classical movement of the color charge, it will be treated as a parameter in this model.

- Furthermore we assume that correlations of the gluon field have a classical, extremly stochastic (incoherent) component and a coherent, homogenous one, which can be treated perturbatively. Since the 'classical' fields are longitudinal and the perturbative ones are purely transverse no double counting is involved here.

$$< A_i^a(x)A_j^b(y) >= \overline{A}_i^a(x) \cdot \overline{A}_j^b(y) + \delta^{ab} \int \frac{d^4k}{(2\pi)^4} P_{ij}^T(k)S(k)e^{ik(x-y)} \tag{2}$$

with the transverse projection $P_{ij}^T = \delta_{ij} - k_i k_j / k^2$.

2. Variational action at $\rho^a = 0$

The QCD action in our particular gauge can be written as

$$W = \int d^4x \frac{1}{2} (E_i^a E_i^a - B_i^a B_i^a) \tag{3}$$

using the following field strength definitions

$$E_i^a = -\partial_0 A_i^a \tag{4}$$

$$B_i^a = \epsilon_i^{jk} \left(\partial_j A_k^a + \frac{g}{2} f_{bc}^a A_j^b A_k^c \right). \tag{5}$$

In case of no external charge the longitudinal part of the vector potential and henceforth the expectation value of the chromoelectric and chromomagnetic field in the vacuum state vanish. The electric and magnetic contributions to the QCD action, however, do not vanish. We get

$$< E_i^a E_i^a > = 16 \int \frac{d^4k}{(2\pi)^4} k_0^2 S(k) \tag{6}$$

and

$$< B_i^a B_i^a > = 16 \int \frac{d^4k}{(2\pi)^4} k_i^2 S(k) + 32g^2 \int \frac{d^4p}{(2\pi)^4} S(p) \int \frac{d^4k}{(2\pi)^4} S(k) \tag{7}$$

after substituting our gaussian assumption (2) and averaging over the angle between the three-momenta $\vec{p}$ and $\vec{k}$ ($< \cos^2\theta > = 1/3$).

The variational action we finally get is proportional to an infinite space-time volume due to our homogenous assumption and has an apparently Lorentz-invariant form

$$\mathcal{L}_{eff} = 8 \left[\int \frac{d^4k}{(2\pi)^4} \left(k_0^2 - k_i^2 \right) S(k) - 2g^2 \int \frac{d^4p}{(2\pi)^4} S(p) \int \frac{d^4k}{(2\pi)^4} S(k) \right]. \tag{8}$$

with $W_{eff} = \int d^4x \mathcal{L}_{eff}$. Using

$$M^2 = g^2 \int \frac{d^4p}{(2\pi)^4} S(p) \tag{9}$$

as variational parameter the variation of the effective action,

$$\frac{\delta W_{eff}}{\delta S(k)} = \int d^4x\, 8 \left(k_0^2 - k_i^2 - M^2 \right) = 0, \tag{10}$$

leads to an implicit mass-gap equation

$$M^2 = g^2 \int \frac{d^3k}{(2\pi)^3} \frac{1}{\sqrt{k^2 + M^2}} \left(n(k) + \frac{1}{2} \right) \tag{11}$$

due to the general on-shell solution for the spectral function

$$S(k) = (2n(k) + 1)\, \delta \left(k_0^2 - k_i^2 - M^2 \right). \tag{12}$$

The contribution of the gluon phase space densuty, $n(k)$, is zero at zero temperature and taken as the Bose-Einstein distribution with mass M at finite temperature. The vacuum contribution $\frac{1}{2}$ will be dimensionally regularized and renormalized by substituting the effective action at $M = 0$. We get an effective energy density of the vacuum

$$w_{eff} = \frac{W_{eff}}{8 \int d^4x} = \frac{3}{2\pi^2} \frac{M^4}{4} \left(\frac{1}{g^2} + \ln \frac{M^2}{M_0^2} - \frac{1}{2} \right), \tag{13}$$

where it is convenient to choose the physical reference point so that $M_0 = M(1/g^2 = 0)$. This variational potential has a non-vanishing minimum at $M = M_0$, which can be fitted to the MIT bag constant. Above a given temperature this nontrivial solution disappears leaving us with the usual perturbative QCD at $M = 0$. The values obtained are $M_0 = 500 - 600 MeV$ and $T_c = 230 MeV$ without light quarks [1] and $T_c = 160 MeV$ or $\mu_c = 1230 MeV$ with 2 light quark flavors which have an effective chemical potential $\mu_{eff} = \mu - g\sqrt{<A^2>}$.

3. Extra energy due to $\rho^a \neq 0$

Putting color charges on this vacuum background it suffices to calculate the change of the energy in the rest frame of these charges. In fact new contributions arise due to our, by the Gauss law motivated, assumption. The electric energy will be modified independently of the spectrum $\omega_L(k)$ and can be interpreted as the Coulomb-energy

$$\int d^3x < E_i^a E_i^a > = \int \frac{d^3k}{(2\pi)^3} \frac{|\rho^a(k)|^2}{k^2}. \tag{14}$$

The magnetic energy has no purely classical contribution, but a mixed longitudinal-transverse one due to the non-abelian magnetic field

$$\int d^3x < B_i^a B_i^a > = M^2 \int \frac{d^3k}{(2\pi)^3} \frac{|\rho^a(k)|^2}{k^2 \omega_L(k)^2} \tag{15}$$

We study this extra energy $< H >_\rho - < H >_{\rho=0}$ in the following two extreme cases

- $\omega_l(k) = k$, and

- $\omega_L(k) = \sqrt{k^2 + M^2}$.

While the energy of a single charge (or any uncompensated color charge) diverges in the first case it is regularized by a finite longitudinal mass. The energy of a $Q\overline{Q}$ pair on the other hand rises linearly with the distance in the $\omega_L(k) = k$ limit

$$< H >_{Q\overline{Q}} = \alpha \left(-\frac{1}{R} + \frac{1}{2}M^2 R \right), \tag{16}$$

while using the same longitudinal as transverse mass M gives an effective potential for high temperatures

$$< H >_{Q\overline{Q}} = \alpha \left(M - \frac{2}{R} + \frac{e^{-MR}}{R} \right), \tag{17}$$

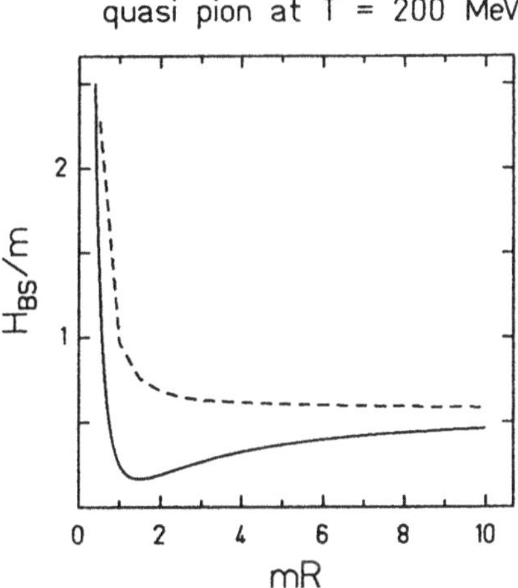

Figure 1. The Bohr-Sommerfeld energy of a $Q\overline{Q}$ pair at $T = 200$ MeV for a pure Debye (dashed line) and for the gaussian effective potential (full line).

which differs qualitatively from the Debye-potential.

4. Quasi pions?

Although a numerical comparison of the above result with lattice QCD calculations does not show any dramatic difference, a long range attractive force, like $-\frac{1}{R}$, in the hot QCD plasma has serious consequences: it allows namely for bound states of $Q\overline{Q}$ pairs.

In order to estimate the ground state energy of such a quasi meson we use the Bohr-Sommerfeld approximation, which gives for the realtive motion

$$H_{BS} = \frac{1}{4m_Q R^2} + V(R, m_Q, M). \tag{18}$$

Its minimum as a function of R describes the ground state. Fig.1 shows a comparison between this effective potential and a Debye-screened Coulomb one.

Using the mass scale obtained as $M = gT$ and an effective coupling constant $\alpha = 0.57$ at $T = 200$ MeV temperature we get a 'pion' mass of 200

MeV, a binding energy of -216 MeV and a size of 0.55 fm.

We note finally, that if long range attractive forces occur in hot QCD plasma and therefore quasi-hadrons - in accordance to some recent lattice QCD calculations too - exist, then not only the pion, tze rho and fi meson, but even the J/ψ can be bound. A more elaborate picture of the QCD vacuum structure, confinement and quark-gluon plasma may in general modify our expectations regarding specific signals of quark matter in relativistic heavy-ion collisions as well.

References

1. T.S.Biró *Phys.Lett.* **B278**, 15 (1992).

Renormalization in the QCD Hamiltonian Method

B. DIEKMANN, M. LANGER AND D. SCHÜTTE
Institut für Theoretische Kernphysik
Nussallee 14-16, D-5300 Bonn, Germany

Abstract: It is shown that the coefficients of Lüscher's effective model space Hamiltonian can be obtained within a finite volume lattice regularization and in the continuum limit where the lattice spacing goes to zero (applying standard coupling constant renormalization). This method avoids a formal treatment of the (unregularized) Yang-Mills Hamiltonian as it is neccessary within the framework of Lüscher using dimensional regularization.

1 Introduction and Overview

The Hamiltonian method as alternative to lattice Monte Carlo QCD suffers from a number of complications: After introducing a *suitable gauge fixing* a consistent *regularization* has to be defined guaranteeing the correct *beta function* when going to the continuum limit. (Correct scaling or - within perturbative renormalization - finiteness of observables after reordering in terms of the universally defined expansion of the bare coupling g_0 in terms of g.) A reliable *many-body technique* has to be developed allowing the incorporation of regularization and renormalization.

Despite of many attempts[1], presently the only formulation which appears to give controllable results is that of Lüscher, van Baal et al.[2, 3]. In the framework of the Coulomb gauge SU(n) Yang Mills theory they are able to calculate quantitatively and essentially analytically glueball spectra if the theory is restricted to a finite volume with dimension smaller than about .8 fm[4]. This success makes this formulation an attractive starting point for (hopefully) more realistic extensions for larger volumes[5].

Is is the purpose of this talk to elucidate a formal structure which corroborates the controlled nature of this framework: We are able to show that the coefficients of Luescher's "fundamental" effective model space Hamiltonian can be computed from a finite volume Kogut-Susskind lattice Hamiltonian when performing the continuum limit. This method involves only the application of many-body techniques to a *well-defined*, regularized Hamiltonian and avoids the *formal* treatment of the (unregularized) Hamiltonian as it is neccessary in Lüscher's framework using dimensional regularization.

That both methods yield the same results especially for expressions which are formally quadratically divergent sums of positive terms (tadpole terms) and which become finite, negative numbers after renormalization is another example of the rather non-trivial equivalence between dimensional and lattice regularization[6].

At the end of the talk, some structures of the vacuum within Lüscher's formulation will be discussed.

2 Lüscher's Yang–Mills theory

Lüscher's basic assumption is the restriction of the SU(n) Yang-Mills theory to a torus[2] (finite volume of size L^3 with periodic boundary conditions). This corresponds the the introduction of a gauge-invariant infrared cutoff.

Consequently, all fields are to be expressed within a Fourier series with the discrete momenta ($n_j =$ integer)

$$\mathbf{k} = 2\pi/L(n_1, n_2, n_3) \tag{1}$$

Lüscher's many-body technique is the folded diagram expansion of a model space Hamiltonian[7] whose spectrum formally reproduces the true (low energy) sprectrum of the full theory when all orders are taken into account. Choosing the Fock space of *zero momentum modes* as model space, one can show that the folded diagram expansion converges quickly if the volume is small enough. Using Bloch's[8] algebraical method, Lüscher computes the effective Hamiltonian on the basis of the formal Coulomb gauge Yang Mills Hamiltonian[9] up to one loop order yielding

$$
\begin{aligned}
H_{eff} &= \lambda_0 H_0 + \lambda_0^2 H_1 + + \lambda_0^4 (H_2 + H_3 + O(\lambda^5) \ , \\
H_0 &= \frac{1}{2} e_j^a e_j^a + \frac{1}{4} f^{abc} f^{ade} c_j^b c_k^c c_j^d c_k^e \ , \\
H_1 &= \frac{n}{3} a_1 c_j^a c_j^a \ , \\
H_2 &= a_2 H_0 \ , \quad H_3 \propto (c_j^a)^4 \ .
\end{aligned}
\tag{2}
$$

Here, c_j^a, ($j = 1, ., 3, a = 1, .., n^2 - 1$) stand for the expansion coefficients of the constant (zero momentum) SU(n) gauge fields, $e_j^a = -i\partial/\partial c_j^a$ for the canonically conjugates and - due to the structure of the truncated zero momentum Hamiltonian H_0 - the expansion has to be written in terms of an unconventional power of the standard (unrenormalized) coupling g_0 by setting $\lambda_0 = g_0^{2/3}$.

Whereas H_0 and H_3 are finite, the coefficients a_1 and a_2 are divergent:

$$
a_1 = \frac{1}{L} \sum_{k \neq 0} \frac{1}{|\mathbf{k}|} \ , \quad a_2 = \frac{11n}{36L^3} \sum_{k \neq 0} \frac{1}{|\mathbf{k}|^3} \ .
\tag{3}
$$

Lüscher determines these coefficients by extending them for general dimension $d \neq 3$ allowing an analytic continuation for arbitrary d. As techniques he uses hereby the structures of the heat kernel on S^1, the zeta function and the Poisson sum formula. The result for the coefficient a_2 is

$$a_2 \;=\; \frac{\beta}{3\epsilon} + O(1) \;, \quad \epsilon = (3-d)/2 \;, \tag{4}$$

where $\beta = 11n/(3(4\pi)^2)$ is the standard one-loop term of the SU(n) beta function. Consequently, this divergence is removed by the standard coupling constant renormalization

$$g_0^2 \;=\; \mu^{2\epsilon}g^2(1 - \beta g^2/\epsilon + O(g^4)) \;, \tag{5}$$

i.e. a_2 is finite as a power series in g up to order g^4. Since this renormalization is *uniquely defined* and since it does not change the "tadpole" coefficient a_1, this whole procedure is only consistent if a_1 becomes *finite* within this process. This is, in fact, the case for the dimensional method: The analytic function $a_1(d)$ has only a pole for $d = 1$, the expansion (3) is valid for $d < 1$ and the analytic continuation to $d = 3$ yields

$$a_1 \;=\; -\frac{n}{4\pi} 1.8915... \;. \tag{6}$$

Thus a formally quadratically divergent sum of positive numbers, as given in (3), becomes finite and negative!

3 The effective Hamiltonian within the lattice regularization

The question arises whether the step of the *formal* treatment of the Yang-Mills Hamiltonian yielding divergent expressions for the coefficients a_1 and a_2 cannot be avoided, and whether the finite number (6) for a_1 is unique.

An obvious alternative method is to *start* with a suitably *regularized Hamiltonian* allowing the computation of the effective Hamiltonian within a well-defined many-body technique. The continuum limit behaviour of the coefficients should then allow a renormalization of the type (5) yielding finite expressions up to order g^4.

The simplest regularization of the Hamiltonian is to truncate the expansion of the gauge fields

$$\mathbf{A}(\mathbf{x}) \;=\; \sum_{\nu=1}^{N} \mathbf{f}_\nu(\mathbf{x})q_\nu \tag{7}$$

starting from suitable (originally complete) basis of functions $\mathbf{f}_\nu$, $(\nu = 1, .., \infty)$. Just introducing such a truncation into the Hamiltonian, however, can never be consistent since a_1 will be always a sum of positive terms which will never

74

become finite and negative in the continuum limit. This difficulty of removing tadpole terms with simple cutoffs is wellknown from standard Feynman perturbation theory and is related to a loss of gauge invariance in the sense of the Slavnov Taylor identities.

The only known cure is to introduce *counterterms* (CT) into the Hamiltonian and the only systematic method to define such CT is the lattice regularization. It is a *gauge invariant regularization before gauge fixing* and has been demonstrated to be consistent with dimensional regularization of standard Feynman perturbation theory[6].

For our purpose, we start with the *Kogut Susskind Hamiltonion* $H_{KS}(\epsilon)$ defined in terms of the link variables $U_{\mathbf{x},j}$ related to a three dimensional lattice of N^3 points $\mathbf{x} = \epsilon(n_1, n_2, n_3)$ $(n_j = 0, .., N - 1)$. (We have then $3N^3$ links $(\mathbf{x}, j) = \mathbf{x} + t\mathbf{e}_j, 0 \leq t \leq \epsilon$.)

Attemps to solve the corresponding quantum mechanical many top problem where not really succesfull up to now[10]. The determination the one loop order effective Hamiltonian, however, is possible, although in detail technically more involved than in Lüscher's framework. Here we only describe the ideas, details may be found in Ref[11]. In order to define the perturbation expansion of the effective Hamiltonion, the link variables have to be replaced by (continuum limit) field variables $\mathbf{A}$ according to

$$U_{\mathbf{x},j} = \exp(\epsilon g_0 A_j(\mathbf{x})) \tag{8}$$

and the Kogut Susskind Hamiltonian has to be systematically expanded in terms of ϵ and g_0. After introducing a lattice Coulomb gauge, the Hamiltonian corresponds to the continuum limit expression in third order of ϵ. As (uniquely determined) new features, however, there are (infinitely many) *higher order CT*, any differencial operator is replaced by a *difference operator* $(\partial_j \psi(\mathbf{x}) \rightarrow (\psi(\mathbf{x} + \epsilon \mathbf{e}_j) - \psi(\mathbf{x}))/\epsilon)$ and - after introducing lattice plane waves $A_{\mathbf{k},j} = \sum_{\mathbf{k}} A_{\mathbf{k},j} \exp(i\mathbf{k}\mathbf{x})$,($\mathbf{x}$ = lattice point) there is a *momentum cutoff* : $\mathbf{k} = 2\pi L^{-1}(n_1, n_2, n_3), -N/2 \leq n_j \leq N/2 - 1$.

In terms of the lattice plane waves, the perturbation expansion of the effective Hamiltonian is well-defined and - up to one loop order - only a finite number of terms of the Hamiltonian contributes. The result for the most interesting coefficient a_1 is

$$a_1^{lattice}(\epsilon) = \frac{3}{2L} \sum_{\mathbf{k} \neq 0} \left(\frac{1}{\omega_{\mathbf{k}}} - \frac{sin^2(k_1 \epsilon)}{\epsilon^2 \omega_{\mathbf{k}}^3} + \frac{cos(k_1 \epsilon) - 1}{\omega_{\mathbf{k}}} \right) ,$$

$$\omega_{\mathbf{k}}^2 = \frac{4}{\epsilon^2} \sum_j sin^2(k_j \epsilon/2) \tag{9}$$

(9) is finite for any finite ϵ due to the lattice momentum cutoff. However, also the finite volume continuum limit $\epsilon \rightarrow 0$, L fixed , $N = L/\epsilon \rightarrow \infty$ becomes

wellbehaved *due to the CT* (yielding the third term in (9)). Writing (9) as approximate Riemann integral one has the structure

$$a_1^{lattice}(\epsilon) \;=\; \frac{\epsilon}{L}\sum f(\mathbf{k}\epsilon) \;=\; \frac{L^2}{\epsilon^2}(\int_{|\mathbf{p}|\leq\pi} f(\mathbf{p})d^3p + \epsilon^2\Delta + ..) \;. \qquad (10)$$

Inserting the function f from (9), it is easily checked that the *integral* in (10) *vanishes*. The *finite part* Δ of $a_1^{lattice}(\epsilon \to 0)$ can be computed numerically and *agrees with* (6). This demonstrates the equivalence of lattice and dimensional regularisation in a case very different from the examples of[6] and corroborates the consistency of the framework[2, 3].

4 Structure of the vacuum

Starting from standard many theory, one expects that the ground state of a many particle system has the structure $\Psi = \exp S\Phi$ where Φ is a free vacuum, i.e. for bosons a suitable *Gaussian* wave functional.

For the vacuum of a torus non-abelian Gauge field theory this assumption turns out to be false. This is because of the existence topologically non-trivial gauge-symmetries (so-called central conjugations[2]) defined for SU(2) by $g_\mathbf{n} = \exp(i\sigma_3 \mathbf{x} n\pi/L)$. Gauss's law does not fix such gauge transformations. Since the Hamiltonian is also invariant with respect to $g_\mathbf{n}$, one has to construct the spectrum with eigenfunction of this (abelian) symmetry characterized by quantum numbers $(-1)^{n\nu}$, $(\nu_j = 0, 1)$. These eigenfunctions cannot be Gaussian, but have to display a *Bloch wave* structure, at least with respect to certain functional direction. In fact the success of van Baal et al[3] in computing "intermediate volume" glue-ball spectra could only be achieved because this "topological" symmetry could be incorporated within the zero momentum model space by suitable boundary conditions[3].

The present challenge within the discussed framework is the incorporation of the Θ-angle symmetry (related to instantons) which is expected to make possible an enlargement of the volume (this necessitates a generalization of the model space)[5] and a systematic inclusion of Fermions.

References

[1] R.E. Cutkosky et al, *Phys.Rev.D* 42, 1270(1990), *Phys. Rev.D* 42, 1260 (1990); S. Nojiri, *Z. Phys.* **C22** (1984) 245; B. Rosenstein et al,*Phys. Lett.* **B177** (1986) 71; D. Schütte et al, *Phys. Rev.* D 34, 1157 (1986), *Phys. Rev.* D 40, 2692 (1989), *Phys. Rev.* **D43** (1991) 1991, *Annals of Phys.* **211** (1991) 112

[2] M. Lüscher, *Nucl. Phys. B* 219, 233(1983)

[3] J. Koller and P. van Baal, *Ann. of Phys.* **174** (1987) 299; *Nucl. Phys.* **B303** (1988) 1

[4] P. van Baal and A. Kronfeld, *Nucl. Phys* **B9** (Suppl.) (1989) 227

[5] P. van Baal, *Nucl. Phys.* **B351** (1991) 183

[6] A. Hasenfratz and P. Hasenfratz, *Phys. Lett.* **93B** (1980) 165

[7] T.T.S. Kuo et al, *Nucl. Phys* **A176** (1971) 65

[8] C. Bloch, *Nucl. Phys.* **6** (1958) 329

[9] N. H. Christ and T. D. Lee, *Phys. Rev.* **D22** (1980) 939

[10] D. Horn et al., *Phys. Rev.* **D31** (1985) 2589; S. A. Chin et al.*Phys. Rev.* *D* 37, 3001 (1988);

[11] M. Langer, Diplomarbeit Bonn 1992, unpublished

GAUGE CONSISTENCY CONSTRAINTS ON GLUON MASS

J. ELY SHRAUNER

Arthur Holly Compton Laboratory of Physics, Washington University

St. Louis, MO 63130

ABSTRACT

The exact Dyson-Schwinger equation for the effective gluon mass in the temporal axial gauge is shown to reduce to the evaluation of a single two-loop skeleton graph as a consequence of the Slavnov-Taylor-Ward-Takahashi condition for gauge consistency. Since the exact 1-loop skeleton graphs contribute identically zero and almost all previous calculations are included at that level, the exact results presented here bear a self-evident importance. The longitudinal part of the exact effective gluon mass vanishes identically in this gauge.

My purpose is to discuss and demonstrate the importance of maintaining exact gauge consistency throughout calculations of the the effective gluon mass and some new results for the exact gluon mass. The effective gluon mass and its calculation are essential to the understanding of hadron physics in terms of QCD. Self-consistent nonperturbative solutions for the effective gluon mass are involved essentially with rearrangement of the symmetry structure in the QCD ground state. It serves as the principal order parameter reflecting the rearrangement of the ground-state symmetry structure and associated phase transitions. Evidence for gluon condensation implies self-consistent nonperturbative solutions for the effective gluon mass are important at both zero and finite temperatures.

Solutions indicating effective gluon masses of a few hundred MeV have been found in lattice simulations as well as in analytic treatments[1]. Uncertainties in the results are large, of the order 100 percent, and the gauge dependences are severe, as expected where ever gauge fields transmute into effective massive fields.

The Slavnov-Taylor-Ward-Takahashi, STWT, identities are subsidiary relations among vertex functions, exact or bare, expressing the conditions that the action of a quantum field theory remain invariant under infinitesimal gauge transformations. Self-consistent nonperturbative field theories and the STWT relations are conveniently formulated in the temporal axial gauge, TAG, largely because in this gauge ghosts can be ignored. I will limit my considerations within this gauge.

The expansion of the Dyson-Schwinger equation, DSE, for the gluon self-energy/polarization tensor Π when the expectations of the field operators vanish has been shown by Cornwall, Jackiw and Tomboulis[2] to be the sum of all one-particle-irreducible two-point vertex graphs in which all lines are exact propagators and vertices are those of the bare gauge theory. Baker, Ball and Zachariasen[3] have shown a resummation of these graphs to close at the 2-loop level with only four skeleton graphs, as shown in Figure 1.

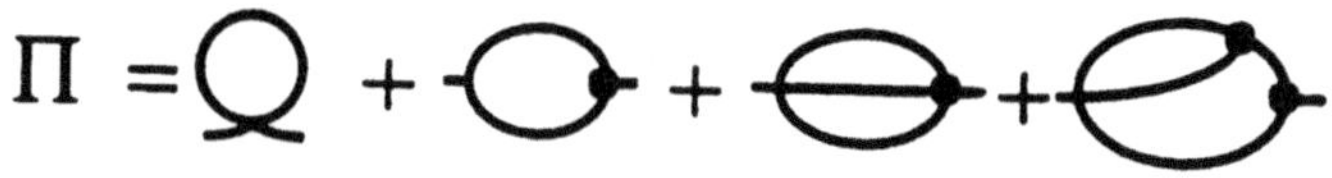

$$\Pi = \bigcirc\!\!\!\bigcirc + \multimap\!\bigcirc\!\!\bigcirc\!\multimap + \multimap\!\ominus\!\!\ominus\!\multimap + \multimap\!\oslash\!\!\oslash\!\multimap$$

Figure 1. Expansion of the Dyson-Schwinger equation for the gluon self-energy/polarization tensor $\Pi_{\mu\nu}^{ab}$ in nonperturbative primitive skeleton graphs. All lines represent the exact dressed gluon propagator Δ. Vertices with blobs represent exact dressed vertex functions Γ and vertices without blobs represent bare vertex functions Γ_o for 3- and 4-gluon vertices.

The color tensor structure of Δ, Δ^{-1}, Π are all taken here to be the trivial δ^{ab} ; a,b = 1...8, although this is not the most general case. The space-time tensor structures are more complicated because TAG is O(3) and translation invariant, but not Lorentz invariant. Consequently vertex functions are functions of spatial and temporal components of 4-vectors separately. For our brief purpose here we need not write all the space-time tensor structures explicitly; see ref. 4.

In TAG the STWT identities are

$$\frac{1}{ig}k^{\sigma}\Gamma_{\mu\nu\sigma}^{abc}(p,q,k)_{p+q=-k} = f^{abm}\left\{ \Delta^{-1}\,{}_{\mu\nu}^{mc}(p) - \Delta^{-1}\,{}_{\mu\nu}^{mc}(q) \right\}_{p+q=-k} \tag{1}$$

and

$$\frac{1}{ig}s^{\rho}\Gamma_{\mu\nu\sigma\rho}^{abcd}(p,q,k,s)_{p+q+k=-s}$$

$$= \left\{ f^{adm}\Gamma_{\mu\nu\sigma}^{mbc}(p,q,k) + f^{bdm}\Gamma_{\mu\nu\sigma}^{amc}(p,q,k) + f^{cdm}\Gamma_{\mu\nu\sigma}^{abm}(p,q,k) \right\}_{p+q+k=-s} , \tag{2}$$

and correspondingly in the zero-momentum limits of interest for our concerns

$$\frac{1}{ig}\Gamma_{\mu\nu\sigma}^{abc}(q,-q,0) = f^{abm}\frac{\partial\Delta^{-1}\,{}_{\mu\nu}^{mc}(q)}{\partial q^{\sigma}} \tag{3}$$

and

$$\frac{1}{ig}\Gamma^{abcd}_{\mu\nu\sigma\rho}(p,q,-p-q,0)$$

$$= \left\{ f^{adm}\frac{\partial\Gamma^{mbc}_{\mu\nu\sigma}(p,q,k)}{\partial p^\rho} + f^{bdm}\frac{\partial\Gamma^{amc}_{\mu\nu\sigma}(p,q,k)}{\partial q^\rho} + f^{cdm}\frac{\partial\Gamma^{abm}_{\mu\nu\sigma}(p,q,k)}{\partial k^\rho} \right\}_{p+q=-k} . \quad (4)$$

Upper indices a,b,c,d = 1,...8 are for color, lower indices $\mu\nu\sigma\rho$ are Lorentz, and p,q,k,s are momentum variables.

The gluon mass is conventionally, but arbitrarily, identified as the infrared limit

$$m^2 = \frac{1}{3}\Pi^\mu_\mu(|\vec{k}|=0,k_o=0) \quad . \quad (5)$$

The transverse and longitudinal masses can be expressed separately as,

$$m_T^2 = \frac{1}{2}(\delta^{jm} - \hat{k}^j\hat{k}^m)\Pi_{jm}(|\vec{k}|=0,k_o=0) \quad , \quad (6)$$

and

$$m_L^2 = -\Pi_{oo}(|\vec{k}|=0,k_o=0)\frac{k^2}{\vec{k}^2} \quad , \quad (7)$$

where the indices j,m = 1,2,3 and $\mu,\nu=0,1,2,3$, and $\hat{k}_j=k_j/|\vec{k}|$. It is the combined average $m^2 = \frac{2}{3}m_T^2 + \frac{1}{3}m_L^2$ that we refer to usually.

With these definitions we are interested in graphs with zero-momentum external lines and so with employment of the zero-momentum differential forms of the STWT identities. The integrand under the the single free loop momentum integral for the sum of the first two 1-loop graphs is

$$\Gamma^{acbd}_{0\ \mu\sigma\nu\rho}(0,q,0,-q)\,\Delta^{cd}_{\sigma\rho}(q) + \Gamma^{cda}_{0\ \sigma\rho\mu}(q,-q,0)\,\Delta^{ce}_{\sigma\eta}(q)\,\Gamma^{meb}_{\zeta\eta\nu}(q,-q,0)\,\Delta^{md}_{\zeta\rho}(q)$$

$$= igf^{bcm}\frac{\partial\Gamma^{cda}_{0\ \sigma\rho\mu}(q,-q,0)}{\partial q^\nu}\,\Delta^{md}_{\sigma\rho}(q) - igf^{bem}\Gamma^{cda}_{0\ \sigma\rho\mu}(q,-q,0)\,\Delta^{cf}_{\sigma\eta}(q)\,\frac{\partial\Delta^{-1}{}^{ef}_{\zeta\eta}}{\partial q^\nu}\,\Delta^{md}_{\zeta\rho}(q)$$

$$= igf^{bcm}\frac{\partial}{\partial q^\nu}\left\{\Gamma^{eda}_{0\ \sigma\rho\mu}(q,-q,0)\,\Delta^{md}_{\sigma\rho}(q)\right\}, \quad (8)$$

which integrates to a surface term that we define to vanish[5,6]. The vanishing of the 1-loop contributions to the effective gluon mass in the TAG is an exact result under the assumptions and definitions used, whose criticism we will mention later.

The 2-loop contributions also simplify and reduce by partial cancellation under the STWT identities to a single 2-loop skeleton graph. The combined integrand under the integrals over the two free loop momenta p and q for the third and fourth graphs is

$$I_{\mu\nu}^{ab}(p,q) \equiv \Gamma_{0\,\mu\rho\sigma\omega}^{aced}(0,-p-q,p,q)\, \Delta_{\sigma\eta}^{er}(p)\, \Delta_{\rho\zeta}^{cs}(-p-q)$$

$$\times\left\{ \frac{1}{3}\Delta_{\omega\gamma}^{du}(q)\, \Gamma_{\eta\zeta\gamma\nu}^{rsub}(p,-p-q,q,0) \;-\; \Delta_{\omega\beta}^{dn}(q)\, \Gamma_{\beta\kappa\nu}^{nmb}(q,-q,0)\, \Delta_{\kappa\gamma}^{mj}(q)\, \Gamma_{\eta\zeta\gamma}^{rsj}(p,-p-q,q) \right\}$$

$$= \Gamma_{0\,\mu\rho\sigma\omega}^{aced}(0,-p-q,p,q)\, \Delta_{\sigma\eta}^{er}(p)\, \Delta_{\rho\zeta}^{cs}(-p-q)$$

$$\times\left\{ \Delta_{\omega\gamma}^{du}(q)\, igf^{bju}\frac{\partial\Gamma_{\eta\zeta\gamma}^{rsj}(p,-p-q,q)}{\partial q^{\nu}} \;+\; igf^{bju}\frac{\partial\Delta_{\omega\gamma}^{du}(q)}{\partial q^{\nu}}\, \Gamma_{\eta\zeta\gamma}^{rsj}(p,-p-q,q) \right\}$$

$$= \Gamma_{0\,\mu\rho\sigma\omega}^{aced}(0,-p-q,p,q)\, \Delta_{\sigma\eta}^{er}(p)\, \Delta_{\rho\zeta}^{cs}(-p-q)\, \frac{\partial}{\partial q^{\nu}}\left\{ \Delta_{\omega\gamma}^{du}(q)\, igf^{bju}\Gamma_{\eta\zeta\gamma}^{rsj}(p,-p-q,q) \right\}. \quad (9)$$

Transferring the partial derivative by a partial integration, employing the STWT Eq.(3), and disposing of the surface term this is equal to the single term

$$I_{\mu\nu}^{ab}(p,q) \qquad\qquad\qquad\qquad\qquad\qquad\qquad\qquad\qquad\qquad\qquad (10)$$
$$= \Gamma_{0\,\mu\rho\sigma\omega}^{aced}(0,-p-q,p,q)\, \Delta_{\omega\beta}^{dn}(q)\, \Gamma_{\beta\kappa\nu}^{njb}(q,-q,0)\, \Delta_{\kappa\gamma}^{ju}(q)\Delta_{\sigma\eta}^{er}(p)\, \Gamma_{\eta\zeta\gamma}^{rsu}(p,-p-q,q)\, \Delta_{\rho\zeta}^{cs}(p+q)\;;$$

which is just the contribution of the fourth graph alone.

The third graph does not contribute in the TAG, leaving just one skeleton graph to calculate. The STWT gauge consistency conditions eliminate the dressed 4-gluon vertex function, and therefore no independent approximation for it needs to be postulated. The longitudinal part of the 3-gluon vertex function is given by the STWT identity in terms of the self-energy, while the transverse part remains unconditioned by these identities. The bare 4-gluon vertex is independent of momenta; which has the consequence that the contributions of the 2-loop graphs to the longitudinal effective mass vanish, and also for the self-energy/polarization tensor at nonvanishing momenta. These results hold for both zero and finite temperature; leaving open the possibility that the effective mass vanishes exactly at zero temperature and becomes finite only at finite temperatures.

A gauge-consistent approximation for the 3-gluon interaction vertex function can be adopted for self-consistent nonperturbative calculations of the gluon mass; for instance that suggested by Baker, Ball and Zachariasen[3] in which it is approximated by only that part which is given by the STWT relations. This type of approach is being pursued, but the 2-loop solution is quite complicated and remains a principal outstanding problem of this field.

There are some caveats to be observed. These results are exact within the context of the assumptions and definitions adopted. However, some of them are somewhat arbitrary. The definition of the gluon mass as the long-wavelength limit of the self-energy/polarization tensor at $q_o = 0$ might be considered inferior to a long-wavelength limit taken at $q_o = m$. But that would be a lot more awkward, and might exclude altogether the direct implementation of the STWT identities. The usual assertion that ghosts need not be considered in computations in the TAG may require deeper substantiation. At certain stages of self-consistent nonperturbative variational treatments the auxiliary source functions[2,4,7] can become implicitly nonzero, thereby possibly re-implicating ghosts to maintain detailed gauge consistency. The vanishing of the partial integration surface terms, before or after renormalization, merits careful attention. However, at worst our results should be entirely valid for the finite-temperature correction parts. Finally, the assumption of color-independent effective gluon masses is not the most general case consistent with the center symmetry of SU(3). Generalizations of this type have been discussed by Polonyi in these proceedings.

References

1. M.C. Ogilvie and J.E. Mandula, *Phys. Lett.* **B 185** (1987) 127; Nucl. Phys. **B** (Proc. Suppl.)**1A** (1987) 117; and in *Hadronic Matter In Collision 1988*, Ed., P. Carruthers and J. Rafelski, (World Scientific, Singapore 1989). See also: P.A. Amundsen and J. Greensite, *Phys. Lett.* **B 173** (1986) 179; and J.H. Hurley and J. Greensite, *Phys. Rev.* **D34** (1986) 513; who also obtained effective gluon masses of about 600 MeV in lattice gauge studies. See also C. Bernard, *Phys. Lett.* **B 108** (1982) 431; and *Nucl. Phys.* **B 219** (1983) 341.

2. J.M. Cornwall, R. Jackiw and E. Tomboulis, *Phys. Rev* **D 10** (1974) 2428.

3. M. Baker, J.S. Ball and F. Zachariasen, *Nucl. Phys.* **B 186** (1981) 531.

4. U. Heinz, K. Kajantie and T. Toimela, *Ann. Phys.* **176** (1987) 218.

5. T. Furusawa and K. Kikkawa, *Phys. Lett.* **B 128** (1983) 218.

6. O.K. Kalashnikov, *Pis'ma Zh. Teor. Fiz.* **39**, No. 7 (1984) 337; BNL preprint; and private communication.

7. E. Shrauner, *Phys. Rev.* **D 16** (1977) 1877; H. Kleinert, *Phys. Lett.* **B 62** (1977) 429; R.W. Haymaker *Phys. Rev.* **D 16** (1977) 1211.

TRANSVERSALLY CONFINED, ACAUSAL GLUONS

H.M. FRIED[*]

Physique Théorique/Institut Nonlinéaire de Nice
Université de Nice, 06108 NICE-Cedex 2, France

and

Physics Department, Brown University
Providence, RI 02912, USA

This brief presentation[1] describes yet another way to approach confinement, analytically, and from first principles. It is based upon a modified version of Halpern's formulation[2] which converts functional integration over gluonic vector potentials to integration over gluonic field strengths; the present discussion deals principally with gluons, with quarks inserted at the end. A mass scale is achieved by dimensional transmutation (DT), which produces a theory which in Gaussian, or Hartree-Fock approximation leads to acausal gluons that effectively propagate within tubes of confined electric or magnetic flux, together with a finite, gauge-invariant condensate, $g^2 < F^2 >$.

One begins with the generating functional

$$z\left\{\mathcal{J}_\mu^a\right\} \sim \int d(A)\, \delta[\mathcal{F}[A]]\, \det\left|\frac{\delta \mathcal{F}}{\delta \omega}\right| \cdot e^{-\frac{i}{4}\int F^2 + i\int \mathcal{J}\cdot A}\,, \tag{1}$$

and for simplicity chooses an (arbitrary) axial gauge, $n_\mu A_\mu^a = 0$. A variant of Halpern's method can be used[3] to rewrite (1) in the form

$$Z\{j\} \sim \int d[\theta_a]\int d[\chi_a]\, e^{-\frac{i}{2}\int \chi^2} \cdot \int d[\psi]\, e^{-i\int \psi^2}.$$
$$\int d[\sigma]\, e^{i\int \sigma\cdot\psi - \frac{1}{2}\mathrm{Tr}\ell n\, K(\sigma) + \frac{i}{2}\int \mathcal{J}\cdot K^{-1}(\sigma)\cdot \mathcal{J}}\,, \tag{2}$$

with

$$\mathcal{J}_\mu^a = J_\mu^a + 2\partial_\mu\,\psi_{\mu\nu}^a\,, \quad J_\mu^a = \mathcal{J}_\mu^a + \theta_a n_\mu - \partial_\mu\chi_a\,,$$

and

$$K(\sigma) = (-\partial^2 - g\sigma\cdot\tau)\,, \quad (\sigma\cdot\tau)_{\mu\nu}^{ab} = \sigma_{\mu\nu}^c\,\tau_{ab}^c\,, \quad \tau_{ab}^c = f_{abc}\,.$$

The reason for this formalism is that $K^{-1}(\sigma)$ is now a "propagator" with a matrix-valued "mass^2", rather than a local operator as in Halpern's formalism. To reproduce Halpern's result, one need only integrate over the Gaussian ψ-dependence; but before that step is performed, one may ask if there exist semi-classical fields $\sigma^a_{o,\mu\nu}(x)$, $\psi^a_{o,\mu\nu}(x)$ about which one may effectively integrate the quantum fluctuations expressed in this language. If those semi-classical fields were essentially constants, one might expect to introduce a mass scale M by DT; but may constant fields really be used?

Suppose that one were given a physically reasonable, semi-classical field $\sigma_o(x)$, where x denotes an appropriate spatial variable, subsequently chosen to be the radius vector of a uniform tube of radius R. What we shall do is replace that $\sigma_o(x)$ by σ_o, its average in the (inner) region where it is nonzero. Corrections to this approximation are given in terms of Fourier modes of momenta $k \gtrsim R^{-1}$, and these will be lumped together with (indistinguishable) quantum fluctuations; in this way, which does not change the essential physics, one can understand the use of a constant field, over a specified finite region, in terms of which one can achieve DT.

We now set up a gap equation by "minimizing" an effective action

$$-i\, S_{\text{eff}}(\sigma\,,\psi) = i \int \sigma \cdot \psi - \frac{1}{2}\, \mathrm{Tr}\, \ell n\, K(\sigma) - i \int \psi^2\,,$$

and search for solutions of

$$\left.\frac{\delta\, S_{\text{eff}}(\sigma\,,\psi)}{\delta\sigma(x)}\right|_{\sigma_0\,,\psi_0} = \left.\frac{\delta\, S_{\text{eff}}(\sigma\,,\psi)}{\delta\,\psi(x)}\right|_{\sigma_o\,,\psi_0} = 0\,.$$

Eventually, one finds a saddle point, rather than a minimum; but since quarks have not yet been introduced, this is only a passing difficulty. One finds

$$\psi^a_{o,\mu\nu} = \frac{1}{2}\, \sigma^a_{o,\mu\nu}\,,$$

and

$$\psi^a_{o,\mu\nu}(x) = \frac{ig}{2}\, tr\left[\tau^a\, <x|\left(\frac{1}{-\partial^2 - g\sigma_o\cdot\tau}\right)_{\nu\mu}|x>\right]$$

or

$$\sigma^a_{o,\mu\nu}(x) = ig^2\, tr\left[\tau^a\, <x|\left(\frac{1}{-\partial^2}\right)(\sigma_o\cdot\tau)\frac{1}{(-\partial^2 - g\sigma_o\cdot\tau)}|x>\right]_{\nu\mu}\,, \qquad (3)$$

upon applying $tr(\tau) = 0$, so that the appearance of a logarithmic divergence becomes clear.

It may now be argued[1] that, for unidirectional fields such as flux in a tube of constant direction, Eq. (3) may be replaced by

$$\sigma^a_{o,\,\mu\nu}(x) = \frac{ig^2}{(2\pi)^4} \int \frac{d^4h}{k^2}\, tr\, \left[\tau^a(\sigma_o(x)\cdot\tau) \left(\frac{1}{k^2 - g\sigma_o(x)\cdot\tau} \right) \right]_{\nu\mu}, \qquad (4)$$

and we shall assume in Eq. (4) that the field $\sigma_o(x)$ is replaced by $\sigma_o\theta(R^2 - x^2)$, so that the only x-dependence resides in the θ-function, which defines a tube pointing in the 3-direction, with axial coordinates x_1 and x_2, and $x = [x_1^2 + x_2^2]^{1/2}$. If the flux in the tube is magnetic one may keep g real; but then the use of an anticausal contour $(k^2 \to \mathbf{k}^2 - k_0^2 + i\varepsilon)$ is required; if the flux is electric, then the continuation $g \to iG$ (as has been made before, in other contexts[4]) is appropriate, retaining the conventional causal contour $(k^2 \to \mathbf{k}^2 - k_0^2 - i\varepsilon)$. For simplicity, the discussion here is presented for a magnetic color field, and the corresponding results for an electric color field are quoted as appropriate.

Causal and anticausal contours yield, in a conventional bosonic problem, just complex conjugates of each other, with exponential fall-off outside, and oscillatory-polynomial fall-off inside the light cone. Causality is really violated in the tachyon case where a propagator's mass^2 becomes negative, in which case the behavior inside and outside the light cone are interchanged; because of the matrix structure of the present "mass^2" terms, with real eigenvalues of equal and opposite signs, this is the situation found here. Hence gluon excitations relative to the σ_o background are acausal. Since gluons are to be confined within hadronization distances, and are not measurable over arbitrary asymptotic distances, one should realize that nothing physically outrageous is being proposed.

A nonzero solution for σ_o may be obtained by multiplying both sides of Eq. (4) by $\sigma^a_{o\,\mu\nu}$ and summing over all indices. For a color magnetic field, we make the choice $\sigma^a_{o\mu\nu} = \xi^a_{\mu\nu34}/\sqrt{2}$, where ξ^a is a constant color vector of magnitude ξ, and define a rescaled set of eigenvalues whose positive members satisfy $\Sigma_{i+}\eta_{i+}^2 = 1$; then, with the definition $|g|\xi e^C = \zeta M^2$, where C is Euler's constant and M is the desired (fixed and finite) mass scale, and with ζ given by $\ln(\zeta) = \Sigma_{i+}\eta_{i+}^2 \ln(1/\eta_{i+})$, one finds from Eq. (4) the familiar relation

$$1 = (g^2/4\pi^2)\, \ln(\Lambda/M) \qquad (5)$$

where Λ is an UV cut off. The "propagator" $K^{-1}(\sigma_o)$ is then given in momentum space by $K^{-1}(\sigma_o) = \left[\left(k^2 - \mu^2\, \epsilon\, \hat{\xi}\cdot\vec{\tau} \right)^{-1} \right]^{ab}_{\mu\nu}$, where $\mu^2 = M^2\, \zeta\, e^C/\sqrt{2}$, and $\hat{\xi}$ is a unit vector in color space. For $SU(2)$, which has been used for all calculations reported here, $\zeta = \sqrt{2}$.

After shifting, according as $\sigma \to \sigma_0 + \sigma_1$, $\psi \to \psi_0 + \psi_1$, we now perform the ψ-integration, with the result (suppressing the subscript of σ_1):

$$Z\{j\} \sim \int d[\theta] \int d[\chi]\, e^{-\frac{i}{2}\int \chi^2} \cdot \int d[\sigma]\, e^{R(\sigma_o,\sigma)-\frac{1}{2}\operatorname{Tr}\ell n\, S(\sigma_o,\sigma)} \cdot$$
$$\cdot\, e^{\frac{i}{2}\int J\cdot K^{-1}(\sigma_o+\sigma)\cdot J+\frac{i}{4}\int L\cdot S^{-1}\cdot L}, \tag{6}$$

where

$$L^a_{\mu\nu} = \sigma^a_{\mu\nu} - \frac{1}{2}\left(J^b_\lambda\, K^{-1}(\sigma_o+\sigma)^{ba}_{\lambda\mu}\,\overleftarrow{\partial}_\nu + \overrightarrow{\partial}_\nu\, K^{-1}(\sigma_o+\sigma)^{ab}_{\mu\lambda}\, J_\lambda \right),$$

$$S^{ab}_{\substack{\mu\nu\\\lambda\sigma}} = \delta_{ab}\,\delta_{\mu\nu}\,\delta_{\lambda\sigma} - \frac{1}{2}\vec{\partial}_\lambda\, K^{-1}(\sigma_o+\sigma)^{ab}_{\mu\nu}\,\overleftarrow{\partial}_\sigma,$$

and where $R(\sigma_o, \sigma)$ denotes only quadratic and higher dependence upon σ. Triple-pair indices here combine in the form $\int \sigma \cdot S \cdot \sigma = \int \sigma^a_{\mu\lambda}\, S^{ab}_{\substack{\mu\nu\\\lambda\sigma}}\, \sigma^b_{\nu\sigma}$.

What is to be done with this rather complicated equation (6)? The most obvious question to ask is if one may take the "naive" limit $g \to 0$ suggested by (5); and the answer is clearly No, because there may well be other logarithmic devergences, or worse, hidden in these functionals which could combine with vanishing factors of g, leaving a finite or divergent remainder. Consider first the determinantal terms $R - (1/2)\operatorname{Tr}\ell n\, S$. Inspection shows that the linear and quadratic σ-dependence contains logarithmic divergences independent of the magnitude of the Fourier components of $\tilde{\sigma}(k)$, and that there are no other obvious divergences in the remaining terms of the expansion. There are no obvious divergences in any of the non-determinantal pieces of the integrand of (6). (More singular divergences can be made to appear after the σ-integration is performed, by using a perturbation expansion of the σ-integrand; but that is a different matter.) One finds, after reference to a Consistency Condition (CC) described briefly below, that the total quadratic and linear dependence here takes the form

$$-\frac{i}{4}\left(\frac{g^2}{4\pi^2}\right)\left(\ell n\left(\frac{1}{M}\right)+\ldots\right)\int d^4x\, \sigma^2(x) \Rightarrow -\frac{i}{4}\int \sigma^2$$
$$i\left(\frac{g^2}{4\pi^2}\right)\left(\ell n\left(\frac{\Lambda}{M}\right)+\ldots\right)\int \sigma^a_{o,\mu\nu}(x)\,\sigma^a_{\mu\nu}(x) \Rightarrow i\int \sigma_o\cdot\sigma. \tag{7}$$

Note that the linear tadpole is of order $1/g$.

Having removed, by DT, the obvious log divergences lurking in the exponential part of the integrand of (6), one may now ask: What would be the result of using (7) and taking the naive limit $g \to 0$ in all the (superficially finite) remainder terms of the integrand of (6). What should happen? The only physics input has been, in effect, an "external field" $\sigma_o(x)$ introduced via DT, and when all further radiative corrections are discarded one must find a result that corresponds to a system containing this external field; that is, when one calculates $< F^a_{\mu\nu}(x) >$, one should reproduce something very much like $\sigma^a_{o,\mu\nu}(x)$, and nothing else. This is the CC referred to above, and in part defines the choice of contours used in evaluating the linear and quadratic σ-dependent integrands.

The result is very encouraging, for all quadratic J (and there j) dependence cancels away, after performing the Gaussian integration which results from the naive limit, leaving only a tadpole result,

$$Z\{j\} \sim e^{\frac{i}{4}\int \partial_\lambda \sigma^a_{o,\mu\lambda}} \left(\frac{1}{\partial^2}\right) j^a_\mu \,, \tag{8}$$

where the axial gauge parameter n_μ has been chosen to lie in some combination of the 3- and 4-directions. A simple calculation then shows that $< F >= -\sigma_o$, a result[5] which suggests that the naive limit of this gluonic system is perfectly colordiaelectromagnetic (PCDEM), since the system generates a color field which opposes the "external" field introduced by DT. Of course this is only valid in the present case when unidirectional tubes of flux have been used; but if other geometries were possible, then (8) would again provide the PCDEM condition. This is very reminiscent of the superconducting models[6] of QCD.

To do better than the naive CC model, one may use a Gaussian, or Hartree-Fock model which retains all quadratic σ- dependence in the exponential of (6), including those pieces which had been dropped in the CC limit $g \to 0$; because of the $O(1/g)$ tadpole part of (7), new terms of $O(g^0)$ will now appear in the Gaussian result, which comes out to be

$$Z\{j\} \sim e^{\frac{i}{2}\int j \cdot \bar{\bar{Q}} \cdot j + i \int j \cdot \bar{\bar{T}}} \,, \tag{9}$$

where

$$\bar{\bar{Q}} = \bar{Q} - \bar{Q} \cdot n \left(n \cdot \bar{Q} \cdot n\right)^{-1} n \cdot \bar{Q} \,,$$

$$\bar{Q} = Q - Q \cdot \overleftarrow{\partial} \left[1 - \vec{\partial} \cdot Q \cdot \overleftarrow{\partial}\right]^{-1} \vec{\partial} \cdot Q \,,$$

$$\bar{\bar{T}} = \frac{1}{2} \left[\bar{\bar{Q}} Q^{-1} T + T Q^{-1} \bar{\bar{Q}}\right] \,,$$

and

$$Q = 3 \hat{K}^{-1} (\tau \cdot \Gamma) \left[1 - \hat{K}^{-1} (\tau \cdot \Gamma)\right] \hat{K}^{-1} \,,$$

$$T = \frac{1}{g} \Gamma \cdot \overleftarrow{\partial} \left[1 - \hat{K}^{-1} (\tau \cdot \Gamma)\right] \hat{K}^{-1} \,,$$

with

$$\hat{K}^{-1} = \left(+\partial^2 - g\,\sigma_o \cdot \tau\right)^{-1} .$$

Here, Γ is a modified σ_o, and acts to produce transverse confinement,

$$\Gamma^a_{\mu\nu}(x) = \tfrac{g}{2}\,\theta(R - x)\,\sigma^a_{o,\,\mu\nu} + \tfrac{g\xi_a}{2\sqrt{2}}\,\theta(x - R)\left(\tfrac{R^2}{x^2}\right)\left[\epsilon_{\mu\nu} - 2\epsilon_{\mu\lambda'}\,\tfrac{x_{\lambda'}\,x_\nu}{x^2}\right] .$$

Certainly, a non-Gaussian σ-approximation should generate non-Gaussian j-dependence in (9), as well as corrections to the Q and T of (9); the latter may be guessed to be given by

$$Q \to Q_E = 3\,\hat{K}^{-1}\,(\tau \cdot \Gamma)\,[\hat{K} + \tau \cdot \Gamma)^{-1},$$

$$T \to T_E = \frac{1}{g}\,\Gamma \cdot \overleftarrow{\partial}\,[\hat{K} + \tau \cdot \Gamma]^{-1},$$

where the subscript E stands for "exact". Whichever forms are used, the numerator factors of $\tau \cdot \Gamma$ produce a transversally confined, acausal, "effective gluon propagator" (EGP).

Finite condensates may be calculated from Eq. (9). If the momentum k is conjugate to the difference $z_1 - z_2$, in $< z_1|Q|z_2 >$, one then finds that $\tilde{Q}(k) \sim (k^2)^{-2}$ for large k^2, so that $< F >$ and $< F^2 >$ are log divergent. However, $g < F >$ and $g^2 < F^2 >$ are finite, and (in the simplest approximation of replacing $\tilde{\tilde{Q}}$ by Q, and neglecting tadpole contributions) are given by

$$g\,\langle F^a_{\mu\nu}(x)\rangle = 3\,\Gamma^a_{\mu\nu}(x),$$

$$g^2\,\langle F^2(x)\rangle = 27\,\mu^4\,\theta(R - x) - \frac{9}{4}\mu^2\left(\frac{R^2}{x^2}\right)^2\theta(x - R).$$

In this computation, where color magnetic flux is assumed inside the tube, one finds a smaller color electric flux outside; if electric flux is assumed inside the tube, then one would find, in this simple computation, magnetic flux outside, a result suggested by the Dual QCD formalism[6].

Finally, quarks may be introduced in a straightforward way; and one finds for the complete generating functional a form which resembles that of QED, except that the latter's photon propagator is replaced by the EGP, and there exists an effective tadpole field built out of T. Now one can see explicit chiral symmetry breaking, in terms of the gluonic scale M; and one must arrange matters such that the total Action, in the presence of all fields generated by DT, is truly a minimum. These questions are presently under study.

References

1. Details may be found in the author's Nice/Physique Théorique preprint (May 1992) of the same title.

2. M.B. Halpern, *Phys. Rev.* **D19** (1979) 517. See also S. Mandelstam, *Ann. Phys. (NY)* **19** (1962) 1 and 25; *Phys. Rev.* **175** (1968) 1580.

3. This method was used in unpublished calculations by Hing-Tong Cho and the author during 1987.

4. D.J. Gross and A. Neveu, *Phys. Rev.* **D10** (1974) 3235.

5. The negative sign relating $< F >$ and $\sigma_o(x)$ was discovered by the author just after the Workshop presentation of this paper; this error has been corrected in these proceedings, as well as in the detailed Nice preprint.

6. M. Baker, J.S. Ball and F. Zachariasen, *Phys. Rev.* **D37** (1988) 1036.

QCD Based Potential Between Heavy Quarks

M. Baker*
Laboratoire de Physique Théorique et Hautes Energies,
Université de Paris XI, Bâtiment 211, F-91405 Orsay Cedex, France

James S. Ball
University of Utah,
Salt Lake City, Utah 84112, U.S.A.

F. Zachariasen
California Instutute of Technology,
Pasadena, California 91125, U.S.A.

ABSTRACT

We calculate the potential between a heavy quark-antiquark pair using dual potentials to describe long distance Yang Mills theory. The resulting potential is essentially that acting between a monopole-antimonopole pair carrying Dirac electric dipole moments and rotating in a relativistic superconductor.

---

*Permanent address: University of Washington, Seattle, WA 98195, USA

To order $1/m^2$ the Hamiltonian V describing the interaction of a quark and an antiquark of masses m_1 and m_2 respectively, separated by a distance R has the general decomposition[1,2]

$$
\begin{aligned}
V &= \;^{\cdot}V_0(R) + \left(\frac{\vec{s}_1 \cdot \vec{L}}{2m_1^2} + \frac{\vec{s}_2 \cdot \vec{L}}{2m_2^2}\right) \frac{1}{R}(V_0'(R) + 2V_1'(R)) \\
&+ \frac{(\vec{s}_1 + \vec{s}_2) \cdot \vec{L}}{m_1 m_2 R} V_2'(R) + \frac{1}{m_1 m_2}\left(\frac{(\vec{R} \cdot \vec{s}_1)(\vec{R} \cdot \vec{s}_2)}{R^2} - \frac{\vec{s}_1 \cdot \vec{s}_2}{3}\right) V_3(R) \\
&+ \frac{\vec{s}_1 \cdot \vec{s}_2}{3m_1 m_2} V_4(R) + \text{ spin independent corrections of order } \frac{1}{m^2}\,.
\end{aligned}
\tag{1}
$$

To higher order in $1/m^2$ the concept of a potential loses its meaning. We use the notation of ref. 1. $V_0(R)$ is the central potential, $V_0'(R) \equiv dV_0/dR$ etc. ($\vec{L}$ is the orbital angular momentum and $\vec{s}_1$ and $\vec{s}_2$ are spin matrices.) The one gluon exchange diagram gives an explicit expression for V which is the Fermi-Breit Hamiltonian for photon exchange, multiplied by a factor 4/3. This one gluon exchange Hamiltonian can also be obtained by calculating the second order classical electromagnetic Lagrangian for charged particles having Dirac magnetic moments and adding the effect of Thomas precession.

In this talk we will show how to calculate V using a Lagrangian[3] describing long distance QCD in terms of dual potentials. Let us briefly review the use of dual potentials to describe long distance QCD. In the case of a source free relativistic medium with dielectric constant $\varepsilon(q)$ and magnetic permeability $\mu(q) = 1/\varepsilon(q)$ Gauss' law and Ampere's law for the electric displacement vector $\vec{D}$ and the magnetic $\vec{H}$ vector are:

$$
\vec{\nabla} \cdot \vec{D} = 0, \qquad \vec{\nabla} \times \vec{H} = \frac{\partial \vec{D}}{\partial t}\,.
\tag{2}
$$

Their solution,

$$
\vec{D} = -\vec{\nabla} \times \vec{C}, \qquad \vec{H} = -\partial_0 \vec{C} - \vec{\nabla} C_0\,,
\tag{3}
$$

define dual potentials C_μ in Abelian gauge theory. The C_μ propagator, $\Delta_C = 1/q^2 \mu(q)$, describes the same physics as the ordinary A_μ propagator, $\Delta_A = 1/q^2 \varepsilon(q)$. In non-Abelian gauge theory dual potentials can be defined[4] but the explicit form of the Lagrangian expressed in terms of dual potentials is not known. We know only that it must be invariant under dual non-Abelian gauge transformations, so that in order to construct this Lagrangian in the long distance regime we must use some information about long distance Yang Mills dynamics.

Our choice was motivated by the work of many authors who have obtained a solution of a truncated set of Dyson equations of Yang Mills theory for the gluon propagator Δ_A which has the behavior $\Delta_A \sim -M^2/q^4$ as $q^2 \to 0$, (M is an undetermined mass scale). As a consequence, $\mu = 1/\varepsilon \sim -M^2/q^2$ as $q^2 \to 0$, and the corresponding C_μ propagator $\Delta_C \sim 1/q^2 - M^2$, i.e. in dual language the truncated Dyson equations give rise to a mass for the dual gluon field C_μ. This mass signals dual superconductivity just as the massive photon in a relativistic superconductor is a signal of ordinary superconductivity. The Lagrangian $\mathcal{L}(C)$, describing long distance Yang Mills theory in terms of dual potentials, must then generate this mass and must be invariant under non-Abelian gauge transformations of the octet C_μ^a of dual potentials which couple[4] via the dual coupling constant $g = 2\pi/e$, where e is the ordinary Yang Mills coupling constant, i.e. $\alpha_s \equiv e^2/4\pi$.

To construct $\mathcal{L}(C)$ we first define non-Abelian generalizations of $\vec{D}$ and $\vec{H}$:

$$\vec{D} = -\vec{\nabla} \times \vec{C} - i\frac{g}{2}[\vec{C} \times \vec{C}], \quad \vec{H} = -\partial_0 \vec{C}_0 - ig[\vec{C},C_0],\tag{4}$$

where $C_\mu \equiv \sum_a C_\mu^a\, \lambda_a/2$ and $\lambda_a/2$ are matrices generating $SU(3)$. To obtain massive dual gluons we couple C_μ to three scalar octets B_1^a, B_2^a, B_3^a or equivalently to three matrices $B_i \equiv \sum_a B_i^a\, \lambda_a/2$. These additional degrees of freedom arise automatically as a consequence of the non-local form of $\mu(q)$. We use the vector notation $\vec{B} = (B_1, B_2, B_3)$ so that

$$2 \text{ trace } \vec{B} \cdot \vec{B} = \sum_i 2 \text{ trace } B_i^2 = \sum_{ia}(B_i^a)^2 \equiv \frac{\tilde{F}^2}{2}.\tag{5}$$

The Lagrangian $\mathcal{L}(C)$ is then given by

$$\mathcal{L} = 2 \text{ trace } \left\{\frac{1}{2}(\vec{H}^2 - \vec{D}^2) + \frac{1}{2}(\mathcal{D}_\mu \vec{B})^2\right\} - W(\vec{B}),\tag{6}$$

where

$$\mathcal{D}_\mu \vec{B} = \partial_\mu \vec{B} - ig[C_\mu,\, \vec{B}],\tag{7}$$

and

$$W(\vec{B}) = 3\mu^2 \text{ trace } \vec{B}^2 + \frac{3\lambda}{4}W_4(\vec{B})\tag{8}$$

is the counter-term needed for renormalization. It accounts for the $\vec{B}$ self-energy graphs and the $\vec{B}$, $\vec{B}$ scattering graphs induced by the $\vec{B}$, $\vec{C}$ interaction. The explicit form of the quartic term $W_4(\vec{B})$ is given in ref. 3. We take $\lambda > 0$ and $\mu^2 < 0$ so that W has a minimum at a non-vanishing value $\vec{B}_0$ of $\vec{B}$ which has the color structure

$$\vec{B}_0 = B_0(\lambda_7,\, -\lambda_5,\, \lambda_2).\tag{9}$$

The dual gauge symmetry is then completely broken and all the dual gluons become massive. Their masses are of order

$$M_g^2 = 4g^2 \text{ trace } \vec{B}_0^2 \equiv g^2 \tilde{F}_0^2.\tag{10}$$

Eight of the "Higgs" fields B_i^a remain massless and become the longitudinal components of the dual potentials while the remaining ones are massive. Using the $SU(3)$ trace

92

anomaly and the fact that the value of W at $\vec{B} = \vec{B}_0$ is the vacuum energy density we obtain the following expression[3] for the gluon condensate G_2:

$$G_2 = \frac{32\lambda}{99} \left(\tilde{F}_0^2 \right)^2 . \tag{11}$$

The classical equations of motion generated by $\mathcal{L}$ have solutions[5] corresponding to tubes of color electric flux along the z axis for which $\vec{C}$ lies in the ϕ direction and is proportional to the hypercharge matrix $Y = \lambda_8/\sqrt{3}$. With this ansatz $\vec{D}$ assumes the Abelian form, eq. (3). At large distances ρ from the flux tube $\vec{C} \rightarrow -\hat{e}_\phi Y/g\rho$ and $\vec{B}$ approaches a gauge transformation of its vacuum value $\vec{B}_0$. The electric flux Φ contained in the flux tube is then

$$\Phi = \int \vec{D} \cdot \vec{da} = - \oint \vec{C} \cdot \vec{d\ell} = \frac{2\pi}{g} Y = eY . \tag{12}$$

The flux tube then contains a unit $e/3$ of flux Φ. Its energy per unit length, i.e. the string tension κ, is of order $\tilde{F}_0^2$ and its radius R_{FT} is of order $1/\sqrt{\lambda \tilde{F}_0^2}$.

This flux tube describes the limit of infinite quark separation. For finite separation $R = R_1 + R_2$ with a heavy quark and antiquark at $\vec{x}_1 = R_1 \hat{e}_z$ and $\vec{x}_2 = -R_2 \hat{e}_z$ respectively, C_μ depends upon z and has the form:

$$\vec{C} = C(\rho, z) \hat{e}_\phi Y , \quad C_0 = 0 . \tag{13}$$

The quark charge density $\rho(\vec{x})$ must also lie in the Y direction in order to absorb the flux of $\vec{D}$ so that

$$\vec{\nabla} \cdot \vec{D} = \rho , \tag{14}$$

where

$$\rho(\vec{x}) = e\delta(x)\delta(y)[\delta(z - R_1) - \delta(z + R_2)]Y . \tag{15}$$

Hence $\vec{D}$ can no longer be written in the form (3). Following the procedure by which Dirac coupled magnetic monopoles to ordinary potentials, we couple quarks to dual potentials by replacing eq. (3) by[6]

$$\vec{D} = -\vec{\nabla} \times \vec{C} + \vec{D}_s , \tag{16}$$

where $\vec{D}_s \equiv \vec{D}(\text{STRING})$ is given by

$$\vec{D}_s = e\delta(x)\delta(y)[\Theta(z - R_1) - \Theta(z + R_2)]\hat{e}_z Y . \tag{17}$$

Since $\vec{\nabla} \cdot \vec{D}_s = \rho$, eq. (16) satisfies Gauss' law, eq. (14). Inserting (16) into the $\vec{D}^2$ term in eq. (6) we obtain the Lagrangian coupling dual potentials to static quarks*. The

---

*S. Maedan, Y. Matsubara and T. Suzuki,[7] using different reasoning have arrived at a Lagrangian resembling $\mathcal{L}$ from which they have calculated the central potential $V_0(R)$.

equations of motion obtained from the Lagrangian close under the following ansatz[6] for $\vec{B}$:

$$\vec{B} = (B(\rho,\ z)\lambda_7,\ -B(\rho,\ z)\lambda_5,\ B_3(\rho,\ z)\lambda_2). \tag{18}$$

Eqs (6), (13), (16), (17) and (18) then yield a set of non-linear equations for C, B, and B_3 in which the distance R between the quarks appears as a parameter.

We have solved these equations[6][†] numerically for a range of values of R and found a solution for each R which we denote as $\vec{C} = \vec{C}^{ST}$, $\vec{B} = \vec{B}^{ST}$. ($ST \equiv$ STATIC). The energy $V_0(R)$ of the corresponding static configuration, is

$$V_0(R) = -\int d\vec{x}\ \mathcal{L}(\vec{C}^{ST},\ \vec{B}^{ST}). \tag{19}$$

The solutions have the following features: for $R \to 0$, $\vec{C}^{ST} \to \vec{C}_D$ the usual Dirac monopole potential and eq. (16) yields a pure dipole field for $\vec{D}^{ST}$. The field $\vec{D}^{ST}$ evolves from a squashed dipole distribution at small R to a flux tube distribution for large separations. For $R \to 0$ the central potential $V_0(R)$, eq. (19), approaches a Coulomb potential and as $R \to \infty$ it becomes linear.

Next we calculate the spin dependent potential[9] assuming that the quark and antiquark have Dirac magnetic moments and that they are rotating with angular velocity $\vec{\omega} = \omega \hat{e}_y$. At time $t = 0$ they are at positions $\vec{x}_1 = R_1 \hat{e}_z$ and $\vec{x}_2 = -R_2 \hat{e}_z$, and their velocities at any time t are $\vec{v}_1(t) = \vec{\omega} \times \vec{x}_1(t)$ and $\vec{v}_2(t) = \vec{\omega} \times \vec{x}_2(t)$ respectively. They produce a magnetization density $\vec{M}(\vec{x},\ t)$:

$$\vec{M}(\vec{x},\ t) = \left[\frac{e\vec{\sigma}_1}{2m_1}\delta\left(\vec{x} - \vec{x}_1(t)\right) - \frac{e\vec{\sigma}_2}{2m_2}\delta\left(\vec{x} - \vec{x}_2(t)\right) \right] Y, \tag{20}$$

where $\vec{\sigma}_1$ and $\vec{\sigma}_2$ are Pauli matrices. The moving magnetization produces an electric polarization $\vec{P}(\vec{x},\ t)$:

$$\vec{P}(\vec{x},\ t) = \vec{v} \times \vec{M}(\vec{x},\ t), \tag{21}$$

where $\vec{v} = \vec{\omega} \times \vec{x}$. Gauss' law then becomes

$$\vec{\nabla} \cdot \vec{D} = \rho(\vec{x},\ t) - \vec{\nabla} \cdot \vec{P}(\vec{x},\ t), \tag{22}$$

where

$$\rho(\vec{x},\ t) = e[\delta\left(\vec{x} - \vec{x}_1(t)\right) - \delta\left(\vec{x} - \vec{x}_2(t)\right)]Y. \tag{23}$$

The solution of eq. (22) is

$$\vec{D} = \vec{D}_s(\vec{x},\ t) - \vec{P}(\vec{x},\ t) - \vec{\nabla} \times \vec{C}, \tag{24}$$

where $\vec{D}_s(\vec{x},\ t)$ is a rotating string; i.e., $\vec{D}_s$ satisfies the equation

$$\frac{\partial \vec{D}_s}{\partial t} = -(\vec{v} \cdot \nabla)\vec{D}_s + \vec{\omega} \times \vec{D}_s, \tag{25}$$

[†]See also ref. 8.

and at $t = 0$ is given by eq. (17). Eq. (24) replaces (16) for $\vec{D}$ in $\mathcal{L}$.

Furthermore the quarks generate a current density $\vec{J}(\vec{x}, t)$:

$$\vec{J} = \rho\vec{v} + \vec{\nabla} \times \vec{M} + \frac{\partial \vec{P}}{\partial t}. \tag{26}$$

Inserting (24) and (26) into Ampere's law,

$$\vec{\nabla} \times \vec{H} = \frac{\partial \vec{D}}{\partial t} + \vec{J}, \tag{27}$$

we find that the general solution for $\vec{H}$ is

$$\vec{H} = \vec{M}(\vec{x}, t) + \vec{v} \times \vec{D}_s(\vec{x}, t) - \frac{\partial \vec{C}}{\partial t} - \vec{\nabla} C_0. \tag{28}$$

Eq. (28) then replaces eq. (3) for $\vec{H}$ in $\mathcal{L}$. Inserting eq. (24) and eq. (28) into eq. (6) gives the final expression

$$\mathcal{L} = 2 \operatorname{trace} \left\{ \frac{1}{2}(\vec{M} + \vec{v} \times \vec{D}_s - \partial\vec{C}/\partial t - \vec{\nabla} C_0)^2 \right. \tag{29}$$

$$\left. -\frac{1}{2}(\vec{D}_s - \vec{P} - \vec{\nabla} \times \vec{C})^2 + \frac{1}{2}(\mathcal{D}_\mu \vec{B})^2 \right\} - W(\vec{B}) \tag{30}$$

for the Lagrangian $\mathcal{L}$ as a function of C_μ and $\vec{B}$ in presence of sources determined by $\vec{D}_s$, $\vec{M}$ and $\vec{v}$.

To solve the equations of motion generated by $\mathcal{L}$ we use a rotating coordinate system with variables $\vec{C}'$ and $\vec{B}'$. Then

$$\vec{C}(\vec{x}, t) = \vec{C}'(\vec{x} - \vec{x}_1(t), \vec{x} - \vec{x}_2(t), t), \tag{31}$$

$$\vec{B}(\vec{x}, t) = \vec{B}'(\vec{x} - \vec{x}_1(t), \vec{x} - \vec{x}_2(t), t), \tag{32}$$

and

$$\frac{\partial \vec{C}}{\partial t} = \frac{\partial \vec{C}'}{\partial t} - (\vec{v} \cdot \vec{\nabla})\vec{C}' + \vec{\omega} \times \vec{C}', \tag{33}$$

$$\frac{\partial \vec{B}}{\partial t} = \frac{\partial \vec{B}'}{\partial t} - (\vec{v} \cdot \vec{\nabla})\vec{B}'. \tag{34}$$

To calculate the energy to second order in $1/m^2$ we insert eqs. (31) – (34) into $\mathcal{L}$ and make the following replacements:

$$\vec{C}' \to \vec{C}^{ST}, \; \vec{B}' \to \vec{B}^{ST}, \; \frac{\partial \vec{C}'}{\partial t} \to 0, \; \frac{\partial \vec{B}'}{\partial t} \to 0. \tag{35}$$

The corrections to eq.(35) can be shown to be of order v^2 and since $\vec{C}^{ST}$ and $\vec{B}^{ST}$ are stationary points of the action we can use eq. (35) to calculate the v^2 contribution to the energy. The replacements (35) in $\mathcal{L}$ yield a Lagrangian:

$$\mathcal{L}^{ST}(C_0,\ \vec{v},\ \vec{M})\,, \tag{36}$$

where the explicit form of $\mathcal{L}^{ST}$ is determined by the static field configuration $\vec{C}^{ST}$, $\vec{B}^{ST}$. The equation of motion

$$\frac{\delta \mathcal{L}^{ST}}{\delta C_0} = 0\,, \tag{37}$$

determines C_0, and the energy V is given by

$$V = -\int d\vec{x}\,\mathcal{L}^{ST}(C_0,\ \vec{v},\ \vec{M}) + V\,(\text{Thomas})\,, \tag{38}$$

where

$$V\,(\text{Thomas}) = -\left(\frac{\vec{s}_1 \cdot \vec{L}}{2m_1^2} + \frac{\vec{s}_2 \cdot \vec{L}}{2m_2^2}\right)\frac{V_0'}{R}\,, \tag{39}$$

is the contribution of Thomas precession.

Substitution of the numerical solution of (37) for C_0 into (38) yields expressions for the potentials $V_i(R)$ defined in (1). (We have not yet completed the calculation of the spin independent corrections of order $1/m^2$.) We find the following results:

a) The tensor and spin-spin potentials $V_3(R)$ and $V_4(R)$ resemble one gluon exchange, but they vanish exponentially at large distances as a consequence of confinement.

b) The spin orbit potential $V_2'(R)$ has the qualitative behavior of one gluon exchange.

c) The spin orbit potential $V_1'(R)$, which vanishes in the one gluon approximation, is long range and we find

$$V_1'(R) = -V_0'(R)\ \text{ for large } R\,. \tag{40}$$

The relation $V_1'(R) = -V_0'(R)$ is valid for all R for a potential produced by scalar particle exchange. Hence the long range part of the spin orbit potential mimics scalar particle exchange although its physical origin has nothing to do with one particle exchange. Indeed the potential V, eq. (38), is essentially the same as the potential between a monopole-antimonopole pair carrying Dirac electric dipole moments and rotating in a relativistic superconductor. It provides a concrete realization of the idea of Mandelstam and 't Hooft[10,11] that dual superconductivity is the physical mechanism for confinement.

The Lagrangian $\mathcal{L}(C)$ contains the two parameters $\tilde{F}_0^2$ and λ in addition to α_s. We have used our potential to fit the 16 known energy levels of $b\bar{b}$ and $c\bar{c}$ systems.[12] With $\alpha_s = .36$ we find,

$$\lambda = 1.74\,, \quad \hat{F}_0^2 = (437 MeV)^2\,. \tag{41}$$

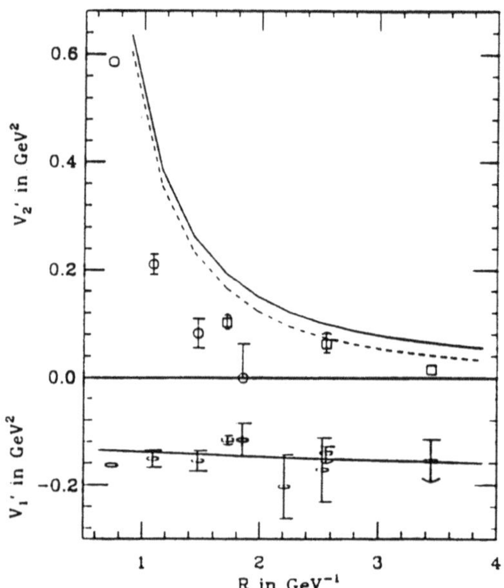

Figure 1. Comparison of calculated V_1' and V_2' (solid curves) with lattice gauge theory data.[14] Dotted curve is perturbative theory.

Using these values of λ and $\tilde{F}_0^2$ we find a flux tube radius $R_{FT} \approx .5f$ and a gluon condensate $G_2 \approx (380\,MeV)^4$. This value for G_2 is on the high side of original estimates but falls within the range allowed by recent determinations.[13]

In the figure we compare our results for the spin orbit potentials V_1' and V_2' with the results obtained from Monte Carlo lattice gauge theory calculations.[14] Because of the complications involved in obtaining the lattice data, these lattice results should be regarded as representing only the general trends of $V_1'(R)$ and $V_2'(R)$.

In summary we have seen how the dynamics of long distance Yang Mills theory can lead to a concrete realization of the dual superconducting picture of confinement which is compatible with lattice gauge theory calculations and with experimental data on heavy quark systems.

Acknowledgements

M.B. would like to thank Michel Fontannaz and the members of the theory group of the Laboratoire de Physique Théorique et Hautes Energies at Orsay for their warm hospitality and helpful conversations during the course of this work. The work of M.B. was supported in part by the US Dept. of Energy under Contract No. DOE/ER/40614. The work of J.S.B. was supported in part by the US National Science Grant No. PHY 9008482. The work of F.Z. was supported in part by the US Department of Energy under Contract No. DEAC_03_81ER40050.

References

1. W. Lucha, F. Schoberl and D. Gromes, **Phys. Rep. 200** (1990) 127.
2. E. Eichten and F. Feinberg, **Phys. Rev. D23** (1981) 2724.
3. M. Baker, James S. Ball and F. Zachariasen, **Phys. Rep. 209** (1991) 73.
4. S. Mandelstam, **Phys. Rev. D19** (1979) 2391.
5. M. Baker, James S. Ball and F. Zachariasen, **Phys. Rev. D41** (1990) 2612.
6. M. Baker, James S. Ball and F. Zachariasen, **Phys. Rev. D44** (1991) 3328.
7. S. Maedan and T. Suzuki, **Prog. Theor. Phys. 81** (1989) 229.
8. S. Maedan, Y. Matsubara and T. Suzuki, **Prog. Theor. Phys. 84** (1990) 130.
9. M. Baker, James S. Ball and F. Zachariasen, **Phys. Rev. D44** (1991) 3949.
10. S. Mandelstam, **Phys. Rep. 23C**, (1976) 245.
11. G. 't Hooft, **High Energy Physics, Proceedings of the European Physical International Conference on High Energy Physics**, 1979, edited by A. Zichichi (Editrice Compositori, Bologna 1976) p. 1225.
12. M. Baker, James S. Ball and F. Zachariasen, **Phys. Rev. D45** (1992) 910.
13. A. Pich and E. de Rafael, **Nucl. Phys. 358** (1991) 311.
14. A. Huntley and C. Michael, **Nucl Phys. B** (1987) 211.

MACROSCOPIC MODEL OF THE COLOUR CONFINEMENT

JIŘÍ HOŠEK

Department of Theoretical Nuclear Physics, Nuclear Physics Institute
250 68 Řež near Prague, Czechoslovakia

ABSTRACT

We demonstrate that the medium described Lorentz - covariantly by a condensing colourless spin - zero field, and by a colour - octet spin - one ghost exhibits the confining properties ascribed to the nonperturbative QCD vacuum: It behaves like a relativistic type - II dual colour superconductor.

1. Introduction and Motivation

The short - distance behaviour of QCD is well understood due to its distinguished perturbative property of asymptotic freedom. Microscopic origin of the large - distance colour confinement generally ascribed to the highly nontrivial properties of the nonperturbative QCD ground state is, however, not understood at all.

An alluring analogy was pointed out[1] between the nonperturbative QCD vacuum medium which should confine the chromoelectric charges, and between the superconducting medium known to confine the magnetic charges. The superconductor behaves like a perfect diamagnet characterized by vanishing magnetic permeability μ. Correspondingly, the confining QCD vacuum should behave like a perfect colour dielectric characterized by vanishing dielectric function ϵ. Since, moreover, the QCD vacuum must look the same in all Lorentz frames which implies $\epsilon.\mu = 1$, it should simultaneously behave like a perfect colour paramagnet characterized by $\mu \to +\infty$. In order to employ an analogy which does not falter too much we should have in mind for comparison a Lorentz - covariant superconductor. Fortunately, it is at hand. It is described by the electrically charged condensing scalar Higgs field.

The key question to be asked and answered is this: Which fields the gluon field A_a^μ should move in and interact with in order to generate the dielectric function ϵ vanishing at the vanishing momenta? Finding the appropriate fields we will briefly discuss further physical properties of such an interacting field system.

2. London Limit
2.1. Calculating the Dielectric Function

In the London approximation dealing with a linearized Lagrangian one possible answer to the question formulated above is disarmingly simple[2]: The massless gluon field A_a^μ should interact via a vector - meson - dominance term with a massive colour

- octet spin $S = 1$ ghost Φ_a^μ, and with a colourless constant scalar field Φ_0. The London - like Lagrangian is

$$
\begin{aligned}
\mathcal{L}_L &= -\tfrac{1}{4}(\partial_\mu A_{a\mu} - \partial_\nu A_{a\mu})^2 + \tfrac{1}{4}(\partial_\mu \Phi_{a\nu} - \partial_\nu \Phi_{a\mu})^2 - \tfrac{1}{2}\Phi_0^2 \Phi_{a\mu}\Phi_a^\mu - \\
&\quad - \tfrac{1}{2}\Phi_0^2 A_{a\mu}A_a^\mu + \Phi_0^2 A_{a\mu}\Phi_a^\mu.
\end{aligned}
\tag{1}
$$

Using the ordinary Feynman rules it is easy to calculate from (1) the gluon polarization tensor $i\Pi_{ab\mu\nu}(k) \equiv -i(k^2 g_{\mu\nu} - k_\mu k_\nu)\delta_{ab}\Pi(k^2, \Phi_0)$ with the result $\Pi(k^2, \Phi_0) = \Phi_0^2/(k^2 - \Phi_0^2)$. The dielectric function $\epsilon(k^2, \Phi_0)$ is defined by the relation between the bare and the full gluon propagator which translates into $\epsilon(k^2, \Phi_0) \equiv 1 + \Pi(k^2, \Phi_0)$. Hence, the dielectric function belonging to (1) is

$$
\epsilon(k^2, \Phi_0) = \frac{k^2}{k^2 - \Phi_0^2}.
\tag{2}
$$

It vanishes at the vanishing momenta, and in the static limit it yields the potential $V(\vec{r}) = -1/|\vec{r}| + \tfrac{1}{2}\Phi_0^2|\vec{r}|$. The opposite charges immersed in such a prefectly dia-electric medium described by Φ_a^μ and Φ_0 will try to avoid to be separated from each other by all possible means. In our case described below their chromoelectric field will destroy the condensate Φ_0 between them.

It is known that the $S = 1$ object can also be described by appropriately constrained antisymmetric tensor field. It is easily demonstrated[3] that the London Lagrangian

$$
\begin{aligned}
\mathcal{L}_L &= -\tfrac{1}{4}(\partial_\mu A_{a\nu} - \partial_\nu A_{a\mu})^2 + \tfrac{1}{2}\partial^\mu \Phi_{a\mu\nu}\partial_\rho \Phi_a^{\rho\nu} - \tfrac{1}{2}\Phi_0^2 \Phi_{a\mu\nu}\Phi_a^{\mu\nu} \\
&\quad + \tfrac{1}{2}\Phi_0 \Phi_{a\mu\nu}(\partial^\mu A_a^\nu - \partial^\nu A_a^\mu)
\end{aligned}
\tag{3}
$$

dealing with the order parameter $\Phi_a^{\mu\nu}$ instead of Φ_a^μ also yields the dielectric function (2). We find (3) advantageous. First, it enables straight-forward non-Abelian gene-ralization. Second, it exhibits explicitly the chromoelectric (nonminimal) coupling, since only the components Φ_a^{0n} are the propagating fields.

2.2. Taming the Ghosts

It should be emphasized that for obtaining the desired result (2) we had to pay a price: In both Lagrangians (1) and (3) the $S = 1$ order parameters Φ_a^μ and $\Phi_a^{\mu\nu}$ enter with the "wrong" sign of their kinetic term. That is why we call them the ghosts. Should we be afraid of them? Presumably not. In a simplified case of two bilinearly coupled scalar fields of which one is a ghost Narnhofer and Thirring[4] have demonstrated that the addition of interaction with an appropriate external source or with the harmonic oscillator yields reasonable dynamics. Verification of this property in the realistic case starting from (3) is our programme for the nearest future.

2.3. Preserving the Ghosts

We believe that in phenomenological field systems of the London - Ginzburg - Landau type the ghosts can be preserved on physical grounds. Their "wrong" sign kinetic term certainly lowers the energy of the system in comparison with the no-field perturbative configuration. This is analogous to the generally accepted way of using the "wrong" sign mass term of some scalar order parameters which causes the vacuum condensation i.e., the nonzero vacuum expectation values.

In QCD an introduction of the order parameters of both types associated with potential instabilities of the perturbative QCD ground state is rather suggestive[5]. First, the colourless scalar order parameter Φ can be associated with the gluon-gluon most attractive channel in $S = 0$, colour $C = 1$ (singlet) state. Being real, scalar and colourless, it can develop spontaneously the VEV Φ_0. Second, another gluon-gluon attractive channel exists in $S = 1$ $C = 8$ (octet) configuration. Obviously, the order parameter $\Phi_a^{\mu\nu}$ associated to it cannot condense in physically acceptable ground state. Yet its very presence can be energetically advantageous provided it is a ghost.

3. The Ginzburg - Landau Extension

The gauge - invariant Lagrangian of interacting fields which would yield (3) upon linearization is not unique. Simplest candidate has the form[3]

$$
\begin{aligned}
\mathcal{L}_{GL} = & -\tfrac{1}{4}F_{a\mu\nu}F_a^{\mu\nu} + \tfrac{1}{2}(D^\mu\Phi_{\mu\nu})_a(D_\rho\Phi^{\rho\nu})_a - \tfrac{1}{4}\Phi^2\Phi_{a\mu\nu}\Phi_a^{\mu\nu} - \tfrac{1}{2}\Phi\Phi_{a\mu\nu}F_a^{\mu\nu} - \\
& - \tfrac{1}{4}\xi(\Phi_{a\mu\nu}\Phi_a^{\mu\nu})^2 + \tfrac{1}{2}(\partial_\mu\Phi)\partial^\mu\Phi - \tfrac{1}{4}\lambda(\Phi^2 - \Phi_0^2)^2 - \tfrac{1}{4}\lambda\Phi_0^4.
\end{aligned}
\tag{4}
$$

Of course, all interaction terms must be confronted with, and eventually adapted to, the experimental data.

Strong support in favour of (4) is that it exhibits the property of *disappearance of the condensate* Φ_0 *in a strong chromoelectric field* $E_a^n \equiv F_a^{0n}$. The argument proceeds as follows. First, minimization of the Hamiltonian corresponding to (4) implies that the quantity $\Phi_{a\,0n}\Phi_a^{0n}$ tends to be different from zero. Second, the part of energy dependent upon Φ acquires the form

$$
E(\Phi, J) = \left[-\tfrac{1}{2}(\lambda\Phi_0^2)\Phi^2 + \tfrac{1}{4}\lambda\Phi^4 + J.\Phi \right].\Omega
\tag{5}
$$

to which we apply the argument of T. D. Lee[6]. For simplicity we assume that $J \equiv \tfrac{1}{2}\Phi_{a0n}E_a^n$ is constant inside a large volume Ω, and zero outside. This assumption justifies neglecting the surface energy. From (5) we get that for $J \geq J_{cr}$ $\Phi_0 = 0$, whereas for $J < J_{cr}$ $\Phi_0 \neq 0$.

This phenomenon is essential for spontaneous colourless bag formation around the chromoelectric charges carried, e.g., by static quarks. Detailed calculations similar to those available in the case of magnetic confinement[7] will amount to numerically solving the GL equations following from (4).

4. Other Manifestations of the Confining Medium

At present, with rather detailed lattice QCD data available, just to describe colour confinement in a GL - like manner is certainly not enough. Fortunately, its present operational formulation offers several specific manifestations which can be experimentally tested.

(i) In general, *surface energy* associated with the interface between the confining phase with $\Phi_0 =$ const and the deconfined phase with $\Phi_0 = 0$ cannot be neglected. We think that the experimental data suggest that the confining QCD vacuum medium is a dual colour superconductor of type-II characterized by negative surface energy. First, there are no hadrons in nonminimal colourless configurations. Second, there seems to exist a universal short-range repulsion between the hadronic bags. Referring to Abrikosov[8] we argue[3] that the Lagrangian (4) can yield, by construction, the negative surface energy.

(ii) In strong chromoelectric field where Φ_0 disappears i.e. inside hadrons or in the quark-gluon plasma produced in heavy-ion collisions, the gluons should not behave quite perturbatively. In particular, the term

$$\mathcal{L}_{int}^L = \tfrac{1}{2}(D^\mu \Phi_{\mu\nu})_a (D_\rho \Phi^{\rho\nu})_a$$

should manifest itself by specific thermodynamic behaviour of the plasma phase.

(iii) The fields Φ and $\Phi_a^{\mu\nu}$ arranged in colourless configurations should transmit specific interactions between the hadronic bubbles which do not correspond to the hadron exchanges.

(iv) The massless quarks should automatically be confined in colourless bags together with the gluon field of which they are a source. Consequently, the chirally invariant Lagrangian

$$\mathcal{L} = \mathcal{L}_{GL} + \bar{\psi} i \!\!\not{D} \psi$$

seems to be a particular realization of the possibility[9] that the spontaneous breakdown of chiral symmetry of QCD i.e., the spontaneous appearance of the massless pion bags is a consequence of the colour confinement. It is due to the spontaneous emergence of the confining surface forcing the massless quarks to flip their helicity.

5. References

1. J. Kogut and L. Susskind, *Phys. Rev.* **D9** (1974) 3501; Y. Nambu, *Phys. Rev.* **D10** (1974) 4262; S. Mandelstam, *Phys. Lett.* **53B** 5313 (1974) 476; M. Baker, J. S. Ball and F. Zachariasen, *Phys. Rep.* **209** (1991) 73.

2. J. Hošek, *Phys Lett.* **226B** (1989) 377; H. Narnhofer and W. Thirring, in *Springer Tracts in Modern Physics Vol. 119* (Springer - Verlag, Berlin, 1990) p.1.

3. J. Hošek, *Phys. Rev.* **D** (1992) to be published.

4. H. Narnhofer and W. Thirring, *Phys. Lett.* **76B** 1978) 428.

5. T. H. Hanson, K. Johnson and C. Petersson, *Phys. Rev.* **D26** (1982) 2069.

6. T. D. Lee, *Particle Physics and Introduction to Field Theory* (Harwood Academic Publishers, New York, 1981), p. 381.

7. J. S. Ball and A. Caticha, *Phys. Rev.* **D37** (1988) 524.

8. A. A. Abrikosov, *Zh. Eksp. Teor. Fiz.* **32** (1957) 1442.

9. A. Casher, *Phys. Lett.* **83B** (1979) 395.

CORRELATION FUNCTIONS OF THE QCD VACUUM AND INSTANTONS

Edward SHURYAK

Physics Department, State University of New York
Stony Brook NY 11794, USA

ABSTRACT

Short-range interaction between light quarks is much more complicated than it was previously believed. We extract some point-to-point correlation functions from experimental data and compare them to the results of calculations based on 'random instanton approximation'. We also briefly discuss the present status of the theory of interacting instantons.

1. Introduction

The structure of the QCD vacuum is, in a way, the central problem of strong interaction physics. Its general discussion can be found in reviews [1, 2]. 'Elementary particles', or hadrons, are but collective excitations of this complicated matter. Respectively, like for phonons in condense matter, one cannot really understand them without first understanding the structure of matter itself.

Hadronic spectrum and excitation cross sections can be converted into more fundamental quantities, the correlation functions, which provide more direct information about short-range structure of matter itself. Say, neutron scattering on solids and liquids have produced enough examples of the kind. Recent review [17] compile available information on such correlator functions, being extracted directly from experimental data. With such phenomenological input, one can discuss models of the vacuum structure and results of lattice simulations with greater confidence.

It turns out that it is extremely important to study *point-to-point* correlation functions rather than 3d-plain-to-plain one, traditionally used by lattice people. The reason is short-range forces between quarks and anti-quarks are much more complicated than previously believed, and those can only be studied if the distance between the points is sufficiently small. In other words: studying *virtual* hadrons, the wave packets of small size, one can learn many things which it is difficult to get out of properties of lightest hadronic states alone[1].

During the last decade it was becoming more and more clear, that 'instanton vacuum' can qualitatively reproduce many features of these functions. The present

---

[1]One can use here analogy with nuclear physics. In 30's, when only general properties of nuclei were known, the NN interaction were believed to be a simple attractive central forces. Only much later, due to precise NN scattering experiments, its full complexity was revealed.

talk contains a report of some preliminary results of much more accurate data [18] on the subject, which aim at *quantitative* comparison with data.

The talk is structured as follows: after some brief introduction into the instanton physics, we discuss a series of examples, comparing the calculated correlation functions for the most important mesonic and baryonic channels to phenomenology. We use here only the simplest possible model of the instanton ensemble, called 'random instanton approximation' (RIA), and add remark about current status of the theory of interacting instantons later.

2. Introductory instanton physics and approximations

Tunneling phenomena in gauge theories were discovered in 70's. This *one-instanton era* is marked by the discovery of the instanton solution [3], its physical interpretation as tunneling, semiclassical integration and relation with chiral anomalies [4]. Their first applications to QCD problems [5] based on *dilute gas approximation*, attracted a lot of attention in late 70's. However, as no explanation for diluteness of the instanton ensemble and to their semiclassical nature were suggested from first principles, pessimism has prevailed and most people has left the field.

However, phenomenological studies of possible instanton effects [6, 17] have shown, that instanton-induced effects definitely can explain many puzzles of the hadronic world, and should not be forgotten. The so called 'instanton liquid' model [7] was suggested, with two free parameters, the main radius and distance between the instantons. In this talk we essentially use it again, demonstrating that it does work, at new and more detailed level.

Attempts to describe *interacting* tunneling events were initiated by Dyakonov and Petrov [8], who used *variational approach.* and have reproduced the 'instanton liquid' parameters surprisingly well. Further numerical studies of this problem were initiated by one of us in [20], which get rid of many approximations and eventually included fermionic effects to all orders in *'t Hooft effective Lagrangian* [4]. We return to their discussion at the end of the talk.

The main reason why instantons are believed to be so important in physics of light fermions is related to the so called 'zero modes', solutions of the Dirac equation

$$D_\mu \gamma_\mu \phi_0(x) = 0 \tag{1}$$

where D is a covariant derivative containing the instanton field. Formally, importance of small eigenvalues of D follows from the (Euclidean) definition of the quark propagator

$$S = -\frac{1}{iD_\mu \gamma_\mu + im} \tag{2}$$

at $m \to 0$. Their existence explains chiral anomaly: while tunneling, quarks with one chirality 'dive into Dirac sea', and with another 'emerge' from it. Also, quark condensate is known to be proportional to density of 'nearly-zero' modes.

Instead of going into discussion of all these complicated phenomena, we now concentrate on evaluation of a quark propagator in the multi-instanton field configuration. Trying to understand spectrum and eigenfunctions of the Dirac operator, one can use the following analogy. 'Zero modes' can be viewed as quark bound states to instantons (a 'potential wells'), and at finite density of such wells they are collectivized (as electrons in condenced matter) and form the so called 'zero mode zone'. In simple physical language, quarks can easily jump from one instanton to another, and thus travel very far. If they can go infinitely far in this way, modes with infinitely small eigenvalues appear, and chiral symmetry becomes spontaneously broken even for massless quarks.

Significant efforts were done in the past to work out some set of approximations providing quark propagators in the 'instanton vacuum'. Our expression looks as follows:

$$S(x,y) = \Sigma_{ZMZ}\frac{\phi_\lambda(x)\phi_\lambda^+(y)}{\lambda - im} + iS_{NZM}(x,y))$$
(3)

Here the first term is the sum over all states belonging to 'zero mode zone' (ZMZ), or those being linear combinations of zero modes of individual instantons. The non-zero modes (analogs of 'scattering states') are taken into account by two last terms. For *single* instanton and for *massless* quarks the analytic expression for fermion propagator is known from explicit solution of the Dirac equation. Generalizing it for *many* instantons, we use an approximation summing all deviations for instantons and antiinstantons:

$$S_{NZM} = S_0 + \Sigma_I(S_I - S_0)$$
(4)

valid for sufficiently dilute system. The last step is generalization of this expression to the case of *non-zero quark mass*: this is needed for discussion of effects related to strange quarks. The resulting expression is lengthy, and we do not give it here.

For reasons to be explained below, we are going to use a simplified instanton ensemble. The so called *random instanton approximation* (RIA) is based on the following assumptions: (i) all instantons have the same size ρ_0; (ii) they have *random* positions and orientations; (iii) instanton and anti-instanton densities are equal to $n_0/2$. Thus, there is one 'diluteness' parameter $f = \pi^2/2n_0\rho_0^4$ which describes the model, apart from the overall scale given e.g. by the distance R defined by $n_0 = 1/R^1$. We are going to show, that already this simple model leads to reasonable description of many correlation functions. For definiteness, we use below R=1 fm (corresponding to right gluon condensate) and $\rho_0 = 1/2$ fm, without attempting to fit them to the correlators. For random model, results are not very sensitive to them, apart of obvious

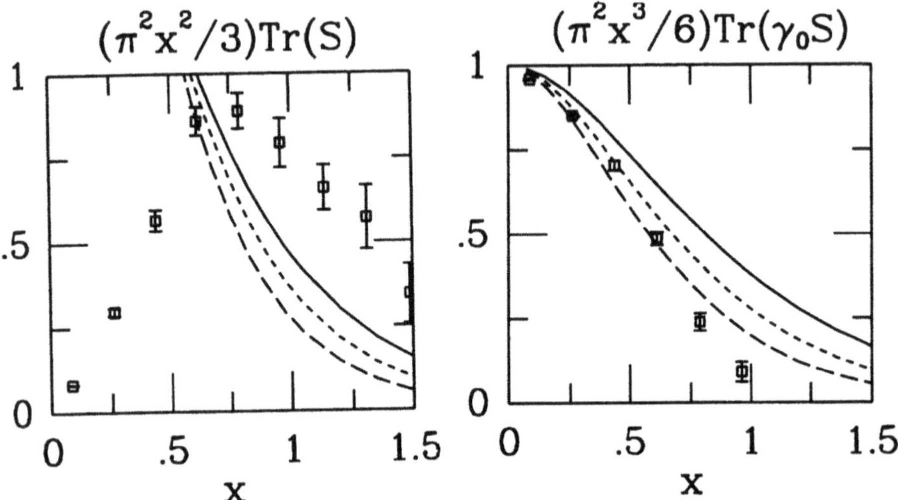

Figure 1: Chirality-flipping (a) and chirality conserving (b) parts of the quark propagator, normalized as indicated in the figure. Three curves correspond to 'constituent quark model' with masses 500,600,700 MeV (solid, dashed, long-dashed, respectively). Here and below dots are our results.

change in overall scale.

3. Correlators in random instanton approximation

We start with showing in Fig.1 our measurements of the quark propagators[2]. At small x part (b) is normalized to a propagator of massless quarks, so it starts at 1, while the part (a) is normalized so that it should be equal to quark mass. Part (b) can be approximately represented by a 'constituent quark model' (curves), although with a surprisingly large quark mass. The chirality-flipping part (a) does not agree with this model at all. One may doubt whether it is possible at all to produce hadrons with reasonable properties from such quarks at all, but let us look at those correlators.

Let us now consider correlators, starting with few general remarks. One can classify correlation functions considering quark paths, recognizing two different types of diagrams: the *one-loop* ones, in which both quark and anti-quark travel from 0 to x, and the *two-loop* ones, in which they return back to where they started. Isospin $I=1$ correlators (e.g. of $\bar{u}d$ currents) get contribution only from one-loop diagrams, and therefore most of lattice work deal with this case. Although we also mainly deal with $I=1$ mesons, we have also performed measurements of some important two-loop

---

[2]We remind that those quantities are gauge dependent and shown only for illustration.

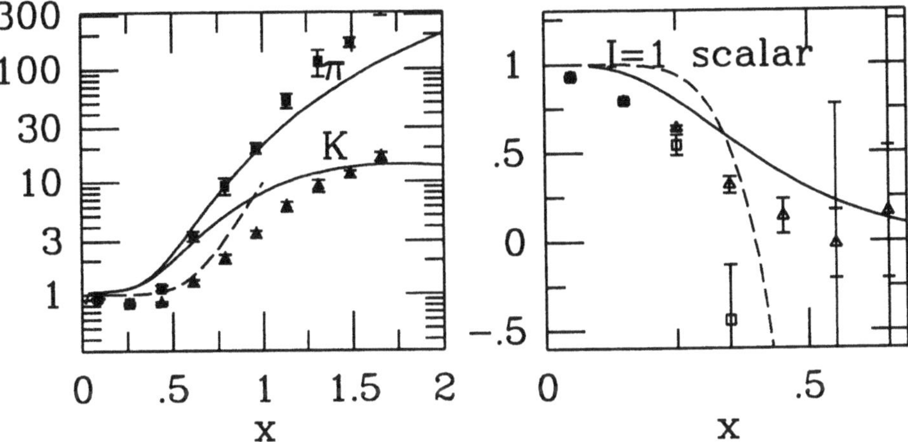

Figure 2: Pseudoscalar (a) and scalar (b) correlators, normalized to the free quark loop. The curves are explained in the text. Squares and triangles correspond to $\bar{u}d$ and $\bar{u}s$ flavor structure of the currents.

diagrams as well, see below.

Our second comment emphasizes some observations related to chiral properties fermionic zero modes. They have chirality, directly related to the topological charge of the gauge field: there is only a *left-handed* solution for the instantons and a *right-handed* one for the anti-instanton. Therefore, the first order corrections in 't Hooft effective interaction are: (i) present in the scalar and pseudoscalar correlators, but absent in the vector and axial ones; (ii) they have the opposite sign for the scalar and pseudoscalar channels; (iii) and, since it has $\bar{u}u\bar{d}d$ flavor structure, the have the opposite sign for the isospin 1 and 0 channels. All three points are phenomenologically welcomed, and this is by itself a very strong hint, suggesting that instanton-induced effects are crucial for short-range correlators.

The pion and kaon solid curves in (a) correspond to phenomenology [17]: notice the scale: the ratio to free quark propagator is very large, because pion is so light! On the contrary, scalar correlator (b) goes down. We do not have phenomenological input in this case, and the solid curve in (b) represent a 'maximally decreasing' guess, without any resonance and with free quark threshold pushed up to $E_{th} = 2$ GeV. And still, the RIA points decrease even stronger. It should be eventually wrong: the correlators cannot become negative!

Here and below the dashed curves stand for 'vacuum dominance approximation', by which we mean the that quark propagator is taken as free one plus the quark

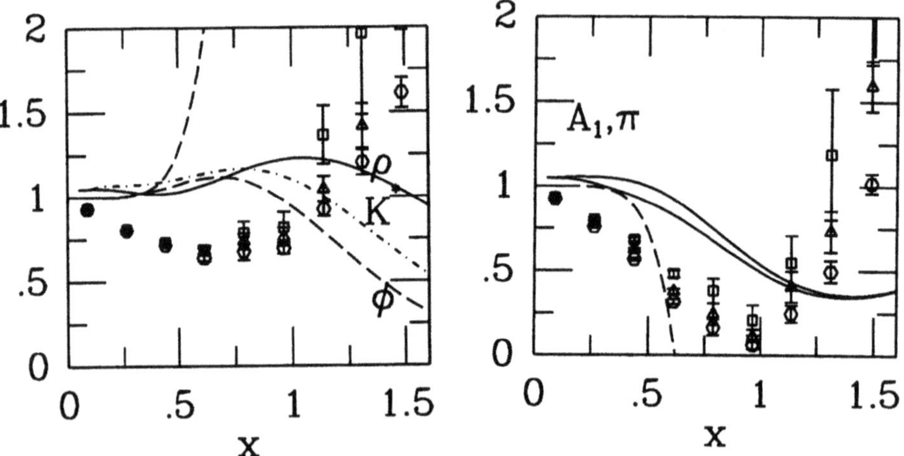

Figure 3: Vector (a) and axial (b) correlation functions normalized to the free quark loop. Three sets of points correspond to $\bar{u}d$, $\bar{u}s$ and $\bar{s}s$ (without annihilation) flavor structure of the currents.

condensate. One can see, that this approximation always predict the right sign of the deviations, but not the correct magnitude. The next pair of channels is shown in Fig.3: those are vector and axials. Three ρ, K^*, ϕ solid curves in (a) and region between the two A_1, π curves in (b) correspond strictly to phenomenology [17]. Actually, those correlation functions have the smallest ambiguities because we have rather good data for electromagnetic annihilation into hadrons and weak τ lepton decays. The most interesting feature of these data is 'superduality' in vectors: all three flavor channels are very close to 1 till rather large distances, about 1 fm. It is reproduced by the model, although deviation is about 20-30 percent[3]. Axial-vector splitting is also reproduced reasonably well.

In Fig.4 we show two examples of channels, involving two-loop (or annihilation) diagrams. Splitting of η, η' channels is the famous Weinberg U(1) problem. To reproduce their splitting, especially at small distances like .3-.4 fm, is a challenge to lattice calculations. One can see that RIA 'overshoot' at this point, making so strong splitting that η' one becomes negative.

The I=0 scalar behaves in a way opposite to I=1 scalar: in this case the correlation function is very large (relative to free quark one). Physically it happens because there exists a disconnected part of the correlator $< \bar{q}q >^2$ (shown by solid line at (b)): our data reproduce it very well.

---

[3]Note that the first perturbative correction is $(1 + \alpha_s/\pi)$, which explains about 10 percent of it.

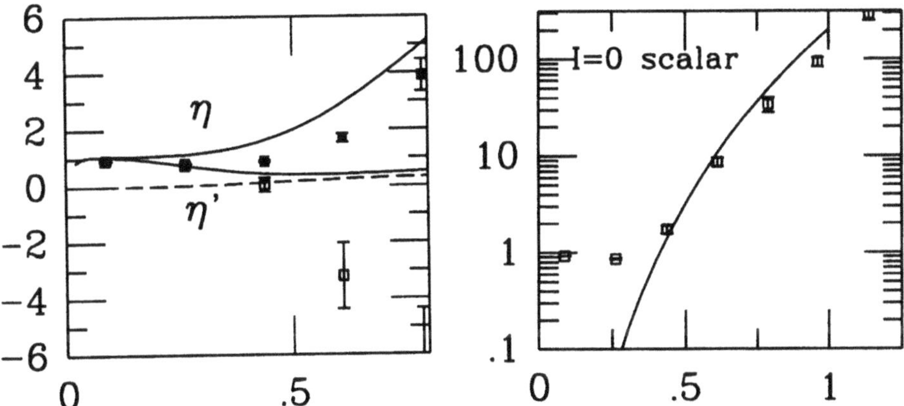

Figure 4: SU(3) octet and singlet pseudoscalars (a) and I=0 scalar (b) correlation functions normalized to the free quark loop. The η, η' solid curves in (a) correspond to phenomenology, and points are our data. Solid curve in (b) is explained in the see text.

Taken together, four channels π, η',I=1,-1 scalars behave at small distances exactly like they should if the main non-perturbative correction is the one-instanton 't Hooft interaction. But at larger distances multi-instanton (and other effects) should correct them: we have seen that in RIA it does not happen.

Finally let us show some results for baryonic systems. In this case we do not have phenomenological input, thus we use predictions based on QCD sum rules. The solid curves marked BI and FZOZ are predictions of ref. [9] and [21], respectively: they were recalculated in x space in [17]. Strong increase of the correlation functions is not really reproduced in RIA till x $\sim$ 1 fm, and after that the data are too noisy. At the same time, these data still imply existence of strong attraction between 'constituent quarks': if they would travel independently, correlator would decay roughly as cube of the propagator shown in Fig.1(b) rather than being flat, as observed.

4. Studies of interacting instantons

Our discussion above have ignored correlation between tunneling events (instantons). However, at least the very phenomenon studied above, namely quark 'jumping' from one instanton to another, produce strong correlation between them. Another obvious source of interaction is non-linear gluonic Lagrangian: a superposition of instanton field always have action different from a sum of two independent ones. In-

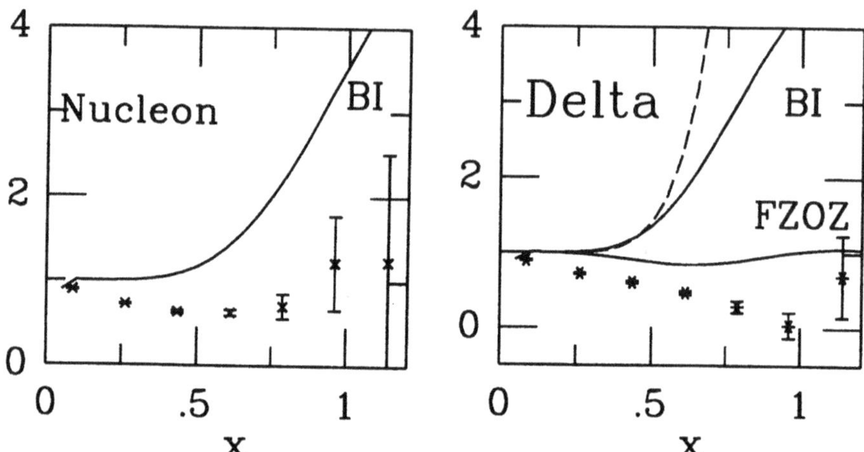

Figure 5: The nucleon (a) and delta (b) correlation functions normalized to the free quark ones. The solid curves marked BI and FZOZ are explained in the text.

teracting instantons can be described by a statistical system with a partition function

$$Z = \int \Pi_i[d\Omega_i exp(-S_i)]exp(S_{int})\Pi_{f=1,N_f}[det(i\hat{D} + im_f)] \tag{5}$$

where by $d\Omega_i$ we have denoted the measure in space of *collective coordinates* of the i-th instanton[4]. The action of an individual instanton is $S_i = 8\pi^2/g^2(\rho_i)$. The next term S_{int} describes the classical (gluonic) interaction and the *fermionic determinant* describes quark-induced interactions.

This formulation leads to a problem similar to those traditionally studied in statistical mechanics [5]. At fixed value of the collective coordinates, evaluation of the fermionic determinant resembles a problem of quantum chemistry: in both cases one has to determine spectrum of fermionic operator in some multi-centered background field.

Treating instantons as atoms (and quarks as electrons) one is lead to the following natural questions: *Is the instanton ensemble in solid, liquid or gas phase? Is the chiral symmetry broken or not?* The answer, obtained in [8, 22, 19], reads as follows: for densities roughly n=.01-64. Λ_{PV}^4 it is liquid, with 'freezing' at the upper end of this

---

[4]There are 12 of them in QCD, including the size ρ, 4-d coordinates of the center and color rotation angles.

[5]It is somewhat more complicated than traditional atomic systems because of fermionic determinant: however, it is still orders of magnitude simpler than full lattice gauge theory

interval. The physical density is not exactly known, but it is roughly in the middle of the interval n $\sim 1\Lambda_{PV}^4$. Chiral symmetry is broken in this phase, and the obtained $< \bar{\psi}\psi >$ has reasonable magnitude.

The main new element which has emerged during the last few years is that (rather arbitrary) trial functions for gauge field configurations used previously, are substituted by the so called *streamline configurations*, corresponding to the bottom of the $I\bar{I}$ valley. Those configurations can be found numerically, 'descending' down the valley [23], which produce the solution to the 'streamline' equation [10]. For gauge theories that was recently found by one of us [24], using conformal symmetry of the classical Yang-Mills theory. It was also found, that the specific ansatz proposed by Yung [10] give very accurate description of the action of these configurations.

However, naive application of the 'streamline-based' gluonic and fermionic interaction have not improved the results obtained previously. In fact, it was found [18] to result in 'over-correlated' liquid leading to very small quark condensate and wrong correlators. This interaction is too attractive.

These results have taught us a lesson, that *a problem of close instanton - anti-instanton pairs* need much closer attention. Interplay between perturbative and non-perturbative (semiclassical) contributions should be understood much better. Configurations with small separation and particular relative orientation have small action, and therefore these are not semiclassical. As it was emphasized by Yung [10], those fluctuations were already included in the usual perturbation theory (which, by definition, includes *all* small fluctuations of the fields) and they should not be included twice.

While this challenging problem still waits for its theoretical solution, one may introduce an artificial repulsive core, and study how large it should be to reproduce the correlators. It is indeed possible to do[6].

5. Conclusions and discussion

Summarizing this talk, we can conclude that even in the simplest 'random instanton approximation' one gets correlation functions which are very similar to what is observed in nature.

We do get chiral symmetry breaking, manifested e.g. in splitting of vector and axial correlators of proper magnitude. Thus, this approximation is at least doing as well as QCD sum rules, based on operator product expansion and 'vacuum dominance' [11].

Second, we got much stronger splittings between pseudoscalar and scalar channels, have seen practical resolution of Weinberg U(1) problem for η' channel (where OPE-based expressions fail).

Third, although quark propagators qualitatively correspond to 'constituent quark'

---

[6]In fact, the resulting correlators are even in much better agreement with data than those shown above: in particular, negative values of scalar and η' correlators disappear.

regime, a quark effective mass is surprisingly large, about 1/2 GeV. However, those are compensated by strong short-range interaction, and produce reasonable masses and other parameters for mesons and baryons.

A number of other instanton-induced effects were suggested in literature, and we are looking forward to study them. Among those are: (i) Spin splittings of baryons [12, 13], (ii) NN short-range potential and the absence of a well-bound dihyperon H [14], (iii) Suppression of the singlet axial charge of the proton (known as the "*spin crisis*") [15], (iv) Restoration of chiral symmetry at large temperatures, by pairing of instantons into 'molecules' [16].

At the same time, many important questions remain open. We have mentioned above the unclear situation with correlation between instantons. Previous trial functions had lead to moderate correlations, which *improved* the results, as compared to data. However, new *streamline-based* interaction leads to over-correlated ensemble, leading to definitely wrong correlators.

Concluding the discussion, let me comment, that we deal here with properties of the QCD vacuum which is much more complicated than, say, nuclear matter. Thus, it might be very naive to expect that straightforward application of semiclassical ideas will lead us directly from the QCD Lagrangian to correct correlators. However, comparison with data and lattice results presumably shows that we are on the right track.

Finally, first lattice measurements of these correlation functions were recently reported [25]. Their results are in rough agreement with both phenomenology and instanton model. More detailed comparison is on the way, so at the moment the field is becoming very interesting.

6. Acknowledgements

First of all, I should thank my collaborator Jac Verbaarschot, with whom I have worked out the reported correlators in the instanton vacuum. I am also indebted to organizers of this interesting conference, especially to Berndt Muller, for invitation. The reported work was partially supported by the US DOE grant DE-FG-88ER40388.

7. References

1. E. Shuryak, *Physics Rep.* **115**, 151 (1984).

2. E. Shuryak, 1988, *The QCD vacuum, Hadrons and the Superdense matter*, World Scientific, Singapore.

3. A. A. Belavin, A. M. Polyakov, A. A. Schwartz, and Y. S. Tyupkin, *Phys. Lett.* **59B**, 85 (1975).

4. G. 't Hooft, *Phys. Rev.* **14D**, 3432 (1976).

5. C. G. Callan, R. Dashen, and D. J. Gross, *Phys. Rev.* **D17**, 2717 (1978).

6. B. V. Geshkenbein and B. L. Ioffe, *Nucl.Phys.* **B166**, 340 (1980).

7. E. Shuryak, *Nucl. Phys.* **B203**, 116, 140, 237 (1982).

8. D. I. Diakonov and V. Y. Petrov, *Nucl. Phys.* **B245**, 259 (1984).

9. V. M. Belyaev and B. L. Ioffe, *Sov.Phys.-JETP* **83**, 976 (1982).

10. A.Yung, *Nucl.Phys.* **B297**, 47 (1988).

11. M. A. Shifman, A. I. Vainshtein, and V. I. Zakharov, *Nucl. Phys.* **B147**, 385,448,519 (1979).

12. R. G. Betman and L. V. Laperashvili, *Sov.J. of Nucl.Phys.* **41**, 295 (1985).

13. E. Shuryak and J. L. Rosner, *Phys. Lett.* **B218**, 72 (1989).

14. S. Takeuchi and M. Oka, *Phys.Rev.Lett.* **66**, 1271 (1991).

15. S. Forte and E. Shuryak, *Nucl. Phys.* **B357**, 153 (1991).

16. E. M. Ilgenfritz and E. Shuryak, *Nucl. Phys.* **B319**, 511 (1989).

17. E.Shuryak, *Correlation functions in the QCD vacuum*,SUNY-NTG-91/45, *Rev.Mod.Phys.*, in press.

18. E.Shuryak and J.J.M.Verbaarschot, *QCD correlation functions and instantons*, to be published.

19. E. Shuryak and J.J.M.Verbaarschot, *Nucl. Phys.* **B341**, 1 (1990).

20. E. Shuryak, *Nucl. Phys.* **B302**, 559,574,599,621 (1988).

21. G. Farrar, H. Zhoang, A. A. Ogloblin, and I. R. Zhitnitsky, *Nucl.Phys.* **B311**, 585 (1981).

22. E. Shuryak, *Nucl. Phys.* **B319**, 521 (1989).

23. E. Shuryak, *Nucl. Phys.* **B302**, 621 (1988).

24. J.Verbaarschot, *Nucl.Phys.* **B362**, 33 (1991).

25. M.-C. Chu, J. Grandy, S. Huang, and J. Negele, 1992, *Lattice calculation of QCD vacuum correlation functions*, CTP 2113, *Phys.Rev.Lett.*, in press.

SCHWINGER DYSON EQUATIONS: DYNAMICAL CHIRAL SYMMETRY BREAKING AND CONFINEMENT

CRAIG D. ROBERTS

Physics Division, Bldg. 203, Argonne National Laboratory
Argonne, IL 60439-4843, USA

ABSTRACT

A representative but not exhaustive review of the Schwinger-Dyson equation (SDE) approach to the nonperturbative study of QCD is presented. The main focus is the SDE for the quark self energy but studies of the gluon propagator and quark-gluon vertex are also discussed insofar as they are important to the quark SDE. The scope of this article is the application of these equations to the study of dynamical chiral symmetry breaking, quark confinement and the phenomenology of the spectrum and dynamics of QCD.

1. Introduction

The fact that, from the field equations of a quantum field theory, one can derive a system of coupled integral equations relating the Green's functions for the theory has been know for quite some time[1] and a general introductory review of the complex of Schwinger-Dyson equations (SDE's) can be found in Ref. [2]. There have been spasmodic attempts to employ this infinite tower of equations in the study of field theories but there are perhaps two main impediments: 1) the fact that one must truncate the system to make any progress whatsoever and the concommitant uncertainty that this introduces regarding the verity of the solution of the truncated equations to the field theory; and 2) the problem of the renormalisation of these equations.

In this contribution I will present an incomplete but representative summary of recent applications of these equations to the study of quantum chromodynamics (QCD). In the main I will concentrate on studies using the SDE for the fermion self energy with the major focus being what these studies can reveal about dynamical chiral symmetry breaking (DχSB) and quark confinement. This preoccupation will not, however, prevent a description of some interesting results obtained from the study of the gauge boson SDE's since, for example, the gauge boson propagator is an important element in the SDE for the fermion self energy.

This article is structured as follows. Section 2 contains a general discussion of the SDE for the fermion self energy and the constraints that can be placed on the gluon propagator. In Sec. 3 a discussion of the fermion–gauge-boson vertex and the constraints one may place upon it is presented. A general discussion of the phenomenology of DχSB is presented in Sec. 4; the SDE approach in both QED and QCD is discussed. There is also a discussion of the broader applicability of the SDE approach to the phenomenology of QCD. In Sec. 5 a discussion of the application of the SDE approach to the determination of the analytic structure of the quark propagator, its implications for quark confinement and the validity of the Wick Rotation is presented through the medium of a simple but instructive model. The article is summarised in Sec. 6.

2. Schwinger-Dyson Equation for the Fermion Self Energy

Of all the equations in the complex this equation is the easiest to write down. It describes the dressing acquired by a fermion propagating in the field generated by its own charge. In Euclidean metric [†] the equation takes the form:

$$\Sigma(p) = m + \frac{4}{3}g^2 \int \frac{d^4q}{(2\pi)^4} \gamma_\mu S(q) D_{\mu\nu}(p-q) \Gamma_\nu(q,p) \,, \tag{1}$$

where m is the fermion bare mass (and may be a matrix in flavour space in the appropriate situation). In covariant gauges one has the following representations of the dressed fermion propagator:

$$S(p) \;=\; \frac{1}{i\gamma \cdot p A(p^2) + B(p^2)} = \frac{Z(p^2)}{i\gamma \cdot p + M(p^2)} \tag{2}$$

$$\;=\; -i\gamma \cdot p \, \sigma_V(p^2) + \sigma_S(p^2) \tag{3}$$

and dressed gauge boson propagator:

$$D_{\mu\nu}(q) = \left\{ \delta_{\mu\nu} - \frac{q_\mu q_\nu}{q^2} \right\} \frac{1}{q^2[1 + \Pi(q^2)]} + \xi \frac{q_\mu q_\nu}{q^4} \tag{4}$$

where $\Pi(q^2)$ is called the polarisation scalar and represents the vacuum polarisation contribution to the gauge boson propagator. In Eq. (1), $\Gamma_\mu(p,q)$ is the dressed fermion–gauge-boson vertex and it is clear that that one must know the form of this and $D_{\mu\nu}(q)$ before one can proceed to obtain a solution of this equation.

2.1a *Constraints on the UV Form of the Gauge Boson Propagator*

The first approximation made in connection with the gluon propagator is to assume that

$$\frac{g^2}{4\pi} \frac{1}{[1 + \Pi(q^2)]} = \alpha(q^2) \tag{5}$$

[†]The convention used herein is that all Euclidean Dirac matrices are Hermitian and obey the anticommutation relations $\{\gamma_\mu, \gamma_\nu\} = 2\delta_{\mu\nu}$ with a positive semidefinite metric: $a \cdot b = \sum_{i=1}^{4} a_i b_i \geq 0 \,\forall\, a, b \in \mathcal{R}^4$.

where $\alpha(q^2)$ is the QCD running coupling constant. This relation is true in Abelian gauge theories but in non Abelian theories it amounts to neglecting ghost contributions to the gluon vacuum polarisation.[3] If one accepts this, however, then constraining the form of the gluon propagator is reduced to the problem of determining the QCD running coupling constant. In this case one has the constraint of asymptotic freedom.

As is well known, in the deep spacelike region the QCD running coupling constant is, at one loop in the modified minimal subtraction scheme ($\overline{\text{MS}}$):

$$\alpha(q^2) = \frac{\lambda \pi}{\ln\left(q^2/\Lambda^2_{\overline{\text{MS}}}\right)} \tag{6}$$

with $\lambda = 12/[33 - 2N_f]$. Here $\Lambda^{n_f=4}_{\overline{\text{MS}}} = 200^{+150}_{-80}$ MeV is the renormalisation group invariant QCD mass scale.[4] (The subscript $\overline{\text{MS}}$ will be suppressed in the following.) However, even allowing for higher loop corrections (the value of Λ is actually extracted from a fit to the two loop expression) one cannot rely on this formula for $q^2 \lesssim 25$ GeV^2 and hence a good deal of uncertainty remains in the intermediate and infrared q^2 regimes. This admits a phenomenological element into studies of the fermion SDE in QCD with the form of the gluon propagator in this region simply being modelled according to whatever criteria given authors give greatest weight to.

2.1b *Gluon Propagator in the infrared: $q^2 \simeq 0$*

There is an expectation that the structure of the gluon propagator at $q^2 \simeq 0$ has important implications for quark confinement: in an imprecise way one might say that the behaviour of the $q - \bar{q}$ interaction in this region determines the long range properties of the $q - \bar{q}$ potential and hence incorporates the physics of confinement. In the context of the fermion SDE, a sufficient condition for confinement; i.e., the absence of free quarks as asymptotic states in QCD, is the absence of a singularity in the propagator at timelike momenta.[†] This then allows a constraint on the infrared behaviour of the gluon propagator: it should entail the absence of a singularity in the soloution of Eq. (1) at timelike momenta. However, as will become clear, this is not a very tight constraint.

In QCD there are seven superficially divergent proper vertices (quark, gluon and ghost propagators; quark-gluon, three-gluon, four-gluon and gluon-ghost). The set of SDE's that couple these vertices is the starting point for the studies of Ref. [7] (which formalises the ideas presented in Ref. [8]). Therein an approach to closing this system of equations is presented and argued to be systematic. In a subsequent article[9] the

---

[†] For a discussion of this see Ref. [5]. This is not a necessary condition, however. As an example one need only consider QCD$_2$ where, at leading order in a $1/N_c$ expansion, confinement is ensured in a conspiratorial fashion: the quark propagator has a mass pole but the vertex in the Bethe-Salpeter equation has zeros at the momenta which correspond to one of the quarks going on shell and hence there is no scattering out of the bound state.

purely gluonic sector of this modified system of SDE's is studied (neglecting coupling to the four-gluon vertex) and it is argued that, in Landau gauge, a gluon propagator of the form:

$$D_{\mu\nu}(q) = \left\{ \delta_{\mu\nu} - \frac{q_\mu q_\nu}{q^2} \right\} \frac{q^2}{q^4 + b^4} \,, \tag{7}$$

with b a constant, is likely to be a reasonable approximation to the solution. It is suggested in these articles that a quark propagator which is consistent with the other vertices would be of the form:

$$S(p) = \frac{i\gamma \cdot p + c_0}{(i\gamma \cdot p - c_{+1})(i\gamma \cdot p + c_{-1})} \tag{8}$$

where the c's are complex constants with non zero imaginary parts. One important observation is that the gluon propagator is zero at $q^2 = 0$ and another is that neither of these propagators has a singularity at timelike momenta and thus both admit the interpretation that they describe confined particles.

It should also be noted that neither of these propagators has a spectral (or Lehmann) representation. It follows, therefore, that the gauge technique[10] could not yield these functions as solutions of the complex of SDE's.

In isolation the result of Eq. (7) might be disregarded given the many uncertainties involved in its derivation, however, a similar result has been obtained in a completely different manner in Ref. [11]. It is argued therein that, in order to completely eliminate Gribov copies and fix Landau gauge uniquely in the lattice formulation of QCD, one must introduce new ghost fields in addition to those associated with the Fadde'ev-Popov determinant. Analysing the action thus obtained in the strong coupling limit it is found that, for $q^2 \simeq 0$, the gluon propagator is given by Eq. (7) .

A summary of other studies of the SDE for the gauge boson sector can be found in Ref. [12]. Even a cursory glance at the literature reveals that most[3,13–16] obtain the following behaviour for the gluon propagator:

$$\frac{1}{q^4} \text{ for } q^2 \simeq 0 \; ; \tag{9}$$

a result that rather extremely contradicts Eq. (7). [†] Some of these studies[3,13–15] use Landau gauge while others[16] use axial gauge but all simplify the system of equations they study by eliminating the four-gluon vertex. In addition, the Landau gauge studies neglect ghost contributions which are argued to be quantitatively small.

In axial gauge the gluon propagator can be written:

$$D_{\mu\nu}(q, \gamma) \;\; = \;\; F_1(q, \gamma) M_{\mu\nu}(q, n) + F_2(q, \gamma) N_{\mu\nu}(q, n) \tag{10}$$

[†]It is interesting to note that a fermion propagator with, in Eq. (2), $M(p^2) \equiv 0$ and $Z(p^2 = 0) = 0$, can be obtained as a self-consistent solution of Eq. (1) when the form of the gluon propagator in the infrared is given by Eq. (9).[15]

118

with

$$M_{\mu\nu}(q,n) = g_{\mu\nu} - \frac{q_\mu n_\nu + q_\nu n_\mu}{q\cdot n} + n^2\frac{q_\mu q_\nu}{(q\cdot n)^2}\;, \tag{11}$$

$$N_{\mu\nu}(q,n) = g_{\mu\nu} - \frac{n_\mu n_\nu}{n^2} \tag{12}$$

and where $\gamma = [q\cdot n]^2/[q^2 n^2]$ is the "gauge parameter". In the studies of Ref. [16], F_2 is assumed to be zero (this corresponds to the assumption that the dressed propagator has the same tensor structure as the bare propagator) and a solution with

$$F_1(q^2,\gamma) \propto \frac{1}{q^4} \tag{13}$$

is found. There is some doubt whether this result survives a more thorough treatment of the gluon equations, however, because it has been argued[17] that the net coefficient of the $\delta_{\mu\nu}$ term in the spectral representation of the axial gauge gluon propagator cannot be more singular than q^{-2}. This suggests a cancellation between the F_1 and F_2 terms in a more thorough treatment.

Given this objection one might choose to concentrate on the Landau gauge analyses which circumvent the restrictions of Ref. [17] because of the presence of ghosts and most obviously contradict Eq. (7). No attempt has been made to reconcile this difference but it is clear that both the "improved" equations studied in Refs. [7,9] and the approach used to find a solution (which involves rational polynomial Ansätze for the vertices) are quite distinct from the more "conventional" approach of Ref. [15].

In connection with this singular form for the propagator in the infared it should first be pointed out that it implies area law behaviour for the Wilson Loop[18] which many would regard as a signal of confinement. (This may be understood, in a simple minded fashion, through the fact that q^{-4} yields a linear potential in four dimensions.) It is, however, very difficult to solve the SDE for the fermion self energy with a gluon propagator that is so singular in the infrared (although it can be done - see below). This fact prompts the introduction of a phenomenological form for the infrared behaviour of the gluon propagator[19] that may be interpreted as a "regularisation" of q^{-4} ; i.e.,

$$\frac{\alpha(q^2)}{q^2} \propto \delta^4(q)\;. \tag{14}$$

This is a regularisation in the sense that it is integrable on $q^2 \in [0,\epsilon]$ whereas q^{-4} is not.

A final remark on the structure of the gluon propagator is in order. One can couple the fermion SDE with the ladder Bethe-Salpeter equation to obtain a description of the pseudoscalar mass spectrum. A good fit is found[20] with a gluon propagator obtained from the piecewise sum of the delta function in the infrared, the two loop renormalisation group result in the ultraviolet and a term:

$$\sim \left(1 - \frac{2}{3}\frac{q^2}{q_0^2}\right)\exp\left(-\frac{q^2}{q_0^2}\right)\;, \tag{15}$$

which contributes in the intermediate q^2 range. The term in Eq. (15) is found to be quite important in obtaining a uniformly good fit to light and heavy mesons: π, κ, D, D_s, η_c, B, B_s, B_c, η_b.

3. Fermion-Gauge Boson Vertex

Another important element of the Eq. (1) is the fermion–gauge-boson vertex. Over many years most studies of this equation have simply used

$$\Gamma_\mu(p,q) = \gamma_\mu \tag{16}$$

which yields what is often called the "rainbow" approximation to Eq. (1). (This corresponds to ladder approximation in the Bethe-Salpeter equation.) It is obvious that, even in QED$_4$, Eq. (16) is inadequate. The whole point of solving Eq. (1) is to obtain the dressed fermion propagator and, of course,

$$i(k-p)_\mu \gamma_\mu \neq A(k^2)i\gamma \cdot k + B(k^2) - A(p^2)i\gamma \cdot p - B(p^2) \ . \tag{17}$$

Hence the Ward-Takahashi identity is violated immediately and it is consequently impossible to obtain gauge covariant solutions and gauge invariant physical observables.

This leads to a consideration of the most rigorous constraints one can place on Γ_μ without actually solving the SDE for this function. This SDE involves the complete kernel of the Bethe-Salpeter equation which, in principal, cannot be written in a closed form. This makes it extremely difficult even to formulate a sensible approximation-to/truncation-of this SDE and consequently the most efficacious procedure at present is to formulate an Ansatz for the vertex which does not manifestly violate any of the physical constraints that the equation might impose.

3.1 *Constraints on the Vertex*

A) The first constraint is that the vertex must satisfy the Ward-Takahashi identity:

$$i(p-q)_\mu \Gamma_\mu(p,q) = S^{-1}(p) - S^{-1}(q) \ . \tag{18}$$

Of course, in QCD one has the Slavnov-Taylor identity which involves the ghost self energy and the ghost-ghost-quark-quark scattering kernel[21] but these contributions are, in practice, neglected in all of the covariant gauge studies of Eq. (1) to date.

B) One can "solve" Eq. (18) in a number of ways and this leads to the second constraint; i.e., that the vertex should be free of kinematic singularities. It has been shown[22] that this is the case at $O(\alpha^2)$ in perturbation theory and the expectation is that this feature should survive at arbitrary order. With this in mind the following

120

solution of the Ward-Takahashi identity was proposed:[22]

$$\Gamma^0_\mu(p,q) \;=\; \frac{[A(p^2)+A(q^2)]}{2}\,\gamma_\mu$$
$$+\; \frac{(p+q)_\mu}{p^2-q^2}\left\{\left[A(p^2)-A(q^2)\right]\frac{[\gamma\cdot p+\gamma\cdot q]}{2}-i\left[B(p^2)-B(q^2)\right]\right\}. \quad (19)$$

C) Another simple and obvious constraint is that the vertex Ansatz should transform under C, P and T in the same way as the bare vertex which requires, for example, that

$$\left[\Gamma_\mu(q,p)\right]^{\mathrm T} = -\mathcal{C}^\dagger\Gamma_\mu(p,q)\mathcal{C} \quad (20)$$

where "T" denotes matrix transpose and $\mathcal{C}=\gamma_2\gamma_4$ is the charge conjugation matrix.

D) It is also reasonable to require that the vertex Ansatz should reduce to the bare vertex in the free field case; i.e., if one substitutes the free fermion propagator in the expression that specifies the vertex Ansatz then one should recover the bare vertex. This is not guaranteed, especially if one chooses to construct an Ansatz for the non-amputated vertex: $\Lambda_\mu(p,q) = S(p)\Gamma_\mu(p,q)S(q)$.

E) An important constraint is that the vertex should ensure multiplicative renormalisability of Eq. (1).[23] In general, a solution of Eq. (18) can be written in the form:

$$\Gamma_\mu(p,q) = \Gamma^0_\mu(p,q) + \sum_{i=1}^{8} f^i(p^2,q^2,p\cdot q)\, T^i_\mu(p,q)\,, \quad (21)$$

where T^i_μ are the eight tranverse tensors of Eq. (3.4) in ref. [22] of which those with $i=1,2,3$ are symmetric under $p\leftrightarrow q$ and the remainder are antisymmetric. The requirement **C)** above implies that all of the f^i are symmetric except for f^6 which is antisymmetric. The simplest form of the vertex allowed by **A)**-**C)** and multiplicative renormalisability has[23]

$$f^6 \neq 0 \quad \text{and} \quad f^i \equiv 0\ \forall\, i \neq 6\,. \quad (22)$$

F) An important observation regarding the Ansatz for the vertex is that the transverse part is crucial to ensuring gauge covariance of Eq. (1).[34] Simply ensuring that Eq. (18) is satisfied does not entail the gauge covariance of particle propagators.[35] This is only guaranteed if the vertex and propagators obey the Landau-Khalatnikov (LK) transformation laws[36] which describe how these elements must respond to a change in the gauge parameter. These transformation laws and their importance in SDE studies are discussed in detail in Refs. [37,38]. Here it is important only to state that these transformation laws augment the specification of the vertex Ansatz. One is only required to specify the Ansatz in a given gauge and then assert that the form of the Ansatz in any other gauge is simply obtained by applying the appropriate LK transformation. Following this procedure the gauge covariance of particle propagators is assured.

4. Phenomenology of DχSB

A very important property of the QCD vacuum is DχSB. The implications this has for the low energy hadronic spectrum can be studied in a variety of ways; successful examples of which are the Nambu–Jona-Lasinio model approach[24], chiral perturbation theory[25] and the Global Colour-symmetry Model.[26–33] The dynamics of DχSB is an important area of application of the SDE approach. In this connection the study of QED$_3$ is useful because this field theory has a number of things in common with QCD; for example, it is confining, at least in quenched approximation, and it is the infrared singularity in the kernel of the SDE that is responsible for that. It therefore provides a simplified environment in which to study the implications of the interplay between the singularities in the fermion–gauge-boson vertex and those in the gauge-boson propagator; an interplay which, in QCD, is likely to be crucial in determining the analytic properties of the quark propagator. I will not discuss these studies here since there is an independent report on them in these proceedings[38] but simply remark that Refs. [34,37–39] address a number of current issues.

4.1 DχSB in QED$_4$

I will begin with studies of DχSB in QED$_4$. If the fermions have zero bare mass then it is obvious that there is a solution with $M(p^2) \equiv 0$. This corresponds to a Wigner-Weyl realisation of chiral symmetry.[40] There is also a non-trivial solution and, in Landau gauge and using Eq. (16), one finds easily[40,41] that this solution has $Z(p^2) \equiv 1$ and $M(x = p^2)$ given by the solution of the non-linear differential equation:

$$xM''(x) + 2M'(x) + \frac{3\alpha}{4\pi}\frac{M(x)}{x + B^2(x)} = 0 \tag{23}$$

with the boundary conditions: $M(0) = 1$ and $M(\Lambda_{UV} = \infty) = 0$.

Equation (1) takes such a simple form in this case because in four dimensions:

$$\frac{1}{2\pi^2}\int d\Omega \frac{\alpha}{(p-q)^2} = \theta(p^2 - q^2)\frac{\alpha}{p^2} + \theta(q^2 - p^2)\frac{\alpha}{q^2} , \tag{24}$$

with $\int d\Omega = \int_0^\pi d\beta \sin^2\beta \int_o^\pi d\theta \sin\theta \int_0^{2\pi} d\phi$, which is the kernel in the equation for $M(p^2)$, and the angular integral in the equation for $Z(p^2)$ is identically zero in Landau gauge. Of course, this is not true if one does not use Eq. (16).

In studying Eq. (23) one finds that as long as $\Lambda_{UV} = \infty$ then, for all values of the coupling α, there is a solution which satisfies the boundary conditions[†] and corresponds to a Nambu-Goldstone realisation of chiral symmetry. Further, stability

---

[†]This is reported in Ref. [41] - a fact which is often neglected.

122

analysis[40] in the ladder approximation establishes this solution to be dynamically favoured; i.e., chiral symmetry is dynamically broken in QED$_4$ for all α in this approximation.

However, if one chooses to work with $\Lambda_{UV} < \infty$ then there is no solution of Eq. (23) that satisfies the UV boundary condition unless $\alpha > \pi/3$. The stability analysis is the same and this leads to the statment that QED$_4$ has DχSB for $\alpha > \alpha_c = \pi/3$; i.e., there is a critical coupling for DχSB in QED$_4$.

It should be noted that a finite UV cutoff is not necessary in the rainbow approximation, massless QED$_4$ SDE; the integral is finite because of the asymptotic properties of the solution. (When the bare mass is non-zero this is no longer true.) Of course, the spectrum of QED$_4$ is such that one does not expect DχSB and one might speculate that a finite UV cutoff is actually necessary because QED$_4$ is really only an effective theory valid up to some (large) momentum scale. Before such speculation is necessary, however, studies that go beyond the rainbow approximation may provide some interesting results.

4.2 Modelling DχSB in QCD

Early studies of DχSB in QCD using Eq. (1) with a momentum dependent coupling can be found in Refs. [42–44]. It was found[42,44] that one can obtain a mass function whose UV behaviour agrees with that obtained from an analysis of QCD using the renormalisation group and operator product expansion[45]:

$$M(p^2)\Big|_{p^2 \to \infty} \to -\frac{4\pi^2\lambda}{3}\frac{\left(\ln\left[\mu^2/\Lambda^2\right]\right)^\lambda \langle\bar{q}q\rangle_\mu}{p^2 \left(\ln\left[p^2/\Lambda^2\right]\right)^{1-\lambda}} \tag{25}$$

by making the approximation

$$\frac{1}{2\pi^2}\int d\Omega \frac{\alpha((p-q)^2)}{(p-q)^2} \simeq \theta(p^2 - q^2)\frac{\alpha(p^2)}{p^2} + \theta(q^2 - p^2)\frac{\alpha(q^2)}{q^2}, \tag{26}$$

which is obviously based on the QED$_4$ result: Eq. (24). Many of the more recent studies have also used this approximation.

In one such study[46] the following choice of the running coupling constant was made:

$$\alpha(q^2) = \frac{\lambda\pi}{\ln\left(1 + \epsilon + q^2/\Lambda^2\right)} \,. \tag{27}$$

It will be observed that, for $\epsilon \neq 0$, this form does not manifest absolute confinement in a sense associated with potential models since $\alpha(0) < \infty$. This study went beyond rainbow approximation using a vertex which satisfied **A)** and **D)** above but which violated **B)** and **C)**. It also used the approximation of Eq. (26) which in Landau gauge, as in QED$_4$, ensures that $Z(p^2) \equiv 1$ and that M is obtained as a solution of a differential equation:[47]

$$0 = b''(x) + \beta(x)b'(x) + \lambda\gamma(x)\frac{b(x)}{x + b^2(x)} \tag{28}$$

with

$$\beta(x) = \frac{2x^2 + x(x + 2\xi)\ln\xi + 2\xi^2(\ln\xi)^2}{x\xi\ln\xi(x + \xi\ln\xi)} \quad \text{and} \quad \gamma(x) = \frac{x + \xi\ln\xi}{x\xi(\ln\xi)^2} \qquad (29)$$

where $b(x) = M(\Lambda^2 x)/\Lambda$ and $\xi = 1 + \epsilon + x$.

The solution of this equation has the UV asymptotic form of Eq. (25) but it is the IR behaviour that is important in a study of DχSB since $M(p^2 = 0)$ can serve as an order parameter equivalent to $\langle\bar{q}q\rangle$. A study of this model[47] leads to the conclusion that one only has DχSB for $\alpha(0) > \alpha_c(0) = 0.78$.

A question naturally arises concerning the validity of the approximation of Eq. (26). This has been addressed in Refs. [47,48]. When the angular integral is evaluated numerically and used to solve the integral equations for Z and M directly one finds[47] that $Z(p^2) \not\equiv 1$ and that, although the UV behaviour of M is unchanged, the IR behaviour is changed quantitatively. This leads to an upward shift in $\alpha_c \rightarrow 0.89$; an increase of 14%. So, in the study of the critical coupling for DχSB the approximation of Eq. (26) is qualtiatively correct but quantitatively unreliable.

In this connection the "small" shift in α_c is misleading. As discussed in Ref. [47], as one increases $\alpha(0)$ the approximation of Eq. (26) gets progressively worse and so, in studies that attempt to incorporate confinement, which usually involve $\alpha(0) = \infty$, this approximation can be expected to be very poor indeed. The DχSB study, attempting as it does to find the minimal value of $\alpha(0)$ for which $\langle\bar{q}q\rangle$ is non-zero, naturally limits itself to that region of couplings in which Eq. (26) is most reliable.

The $\epsilon = 0$ limit of the model defined by Eq. (27) has been studied.[49] In this limit the propagator has a strong, nonintegrable q^{-4} singularity and one must choose the fermion–gauge-boson vertex with extreme care in order to ensure that $M(p^2)$ remains finite. The vertex used is not related to that discussed above but the regularisation of the singularity through the choice of the vertex leads to an outcome that is qualitatively similar to that of Refs. [46,47].

A study of DχSB using the $\delta^4(q)$ infrared form for the gluon propagator has also been carried out.[50] Using Landau gauge with

$$\frac{\alpha(q^2)}{q^2} = C\Lambda^2\delta^4(q) + \frac{\lambda\pi}{q^2\ln\left(1 + \epsilon + \frac{q^2}{\Lambda^2}\right)}, \qquad (30)$$

with C a dimensionless constant, and a light cone singular vertex (which violates **B**) above):

$$\begin{aligned}
\Gamma_\mu(p,q) &= \frac{[A(p^2) + A(q^2)]}{2}\gamma_\mu \\
&+ \frac{(p-q)_\mu}{(p-q)^2}\left\{[A(p^2) - A(q^2)]\frac{[\gamma\cdot p + \gamma\cdot q]}{2} - i[B(p^2) - B(q^2)]\right\}. \quad (31)
\end{aligned}$$

it was shown that a good fit to f_π and $\langle\bar{q}q\rangle$ could be obtained with $C = 500\text{-}600$. The results were insensitive to ϵ and N_f. In a subsequent study,[51] which used the light cone regular vertex of Eq. (19), it was found that, keeping all other parameters unchanged, there was an increase in $\langle \bar{q}q \rangle$.

4.3 *Hadron Phenomenology: SDE and BSE*

One can develop a very successful phenomenology of QCD based on the SDE for the fermion self energy, the Bethe-Salpeter equation for fermion-antifermion and fermion-fermion bound states and the relativistic Fadde'ev equation for three body bound states.[20,26-33] To illustrate this I have included Table. 1 in which a representative collection of the results of these studies is presented. These results were obtained with the propagator of Eq. (30) (with $\Lambda = 0.200$ GeV, $\epsilon = 2.0$, $C = (3\pi)^3$) and the vertex of Eq. (16). They are insensitive to ϵ and hence C is the only parameter. Evidently, this one-parameter model can provide a very good description of the static and dynamic properties of hadrons.

Table 1: In this table an illustrative set of calculations of physical quantities obtained in the SDE approach to QCD phenomenology is presented. The propagator of Eq. (30) ($\Lambda = 0.200$ GeV, $\epsilon = 2.0$, and $C = (3\pi)^3$) and ladder approximation were used in the solution of the SDE and Bethe-Salpeter equations. The superscripts indicate the reference that a given result was taken from.

	Calculated (GeV) Massless u,d	Experiment (where applicable)
$m_\pi{}^{26}$	0	(0 if quarks massless)
$f_\pi{}^{26}$	0.076	0.093
$\frac{1}{r_\pi}{}^{27}$	0.290	0.303
$m_{f_0}{}^{26}$	0.811	0.975
$m_\omega{}^{28-30}$	0.745	0.783
$m_\omega - m_\rho{}^{28-30}$	0.053	0.013
$\Gamma_\rho{}^{28-30}$	0.232	0.154
$m_{f_1}{}^{28-30}$	1.310	1.283
$m_{f_1} - m_\omega{}^{28-30}$	0.565	0.500
$m_{0^+}^{qq\,30,31}$	0.607	
$m_{1^+}^{qq\,30,31}$	1.170	
$m_N{}^{32}$	$1.20 \sim 1.30$	0.939

It has been remarked[52] that such a simple model of the interaction between quarks lacks sufficient strength in the intermediate q^2 region to provide as good a description of mesonic correlation functions as the instanton liquid model. This may be the case but one cannot be certain since no attempts have been made to calculate these correlation functions in this model. Where comparable calculations do exist this model does at least as well as the instanton liquid model. It may also be remarked that there are fewer paramters in this model and that, in contrast to the instanton liquid model, it incorporates quark confinement in the sense that the quark propagator, when correctly continued to the complex p^2 plane, has no singularity on the timelike p^2 axis.

Finally, it should be recognised that Eq. (30) is simply a single illustrative example of a particularly simple kernel in the combined SDE-BSE approach. Some studies[20] have found that extra strength in the intermediate q^2 region is useful in fitting other data in the meson spectrum, however, this is merely a fine tuning of the particular model of the quark-quark interaction. Steps continue to be made toward a determination of the best phenomenological form for the quark-quark interaction and the quark-gluon vertex but progress beyond phenomenology lies in a reliable calculation of these two quantities.

5. Analytic Structure of the Fermion Propagator

5.1 *Simple Model of Fermion Confinement*

A particularly useful illustrative model SDE for the fermion self energy is obtained with the choice:[19,53]

$$\frac{\alpha(q^2)}{q^2} = 2\pi^3 D \delta^4(q) , \tag{32}$$

where D is a parameter that sets the mass scale. This is an infrared dominant model and hence is useful for studying confinement in the SDE approach.

When the bare vertex, Eq. (16), is used in Eq. (1) this equation reduces to a pair of coupled algebraic equations:

$$Z = \frac{M}{2M - m} , \tag{33}$$

$$0 = 2M^4 - 3mM^3 + M^2(2p^2 - D + m^2) - 3mp^2M + p^2m^2 . \tag{34}$$

For fermions with zero bare mass these equations decouple and have the dynamical chiral symmetry breaking solution $(s = p^2)$:[19]

$$M(s) = \begin{cases} \sqrt{\frac{1}{2}D - s} , & s < \frac{1}{2}D \\ 0 , & s > \frac{1}{2}D \end{cases} \tag{35}$$

$$Z(s) = \begin{cases} \frac{1}{2} , & s < \frac{1}{2}D \\ \frac{s}{2D}\left(\sqrt{1 + \frac{4D}{s}} - 1\right) , & s > \frac{1}{2}D \end{cases} \tag{36}$$

In order to determine whether this non-trivial solution is favoured over the chirally symmetric $M = 0$ solution one evaluates the difference between the CJT Effective

Action[54] evaluated at these extremals:[55]

$$\beta = V[M = 0] - V[M = \text{Eq. (35)}] = N_c N_f \frac{D^2}{32\pi^2} \left(4 \ln 2 - \frac{11}{4} \right) > 0 \ . \tag{37}$$

It follows that one has DχSB in this model when the bare vertex is used.

The DχSB solution is also a confining solution since the fermion propagator constructed from these functions has no singularity on the timelike p^2 axis (although it does have a square-root branch point at spacelike p^2). When the quarks have non zero bare mass this feature survives. (The branch point becomes a pair of complex conjugate branch points with spacelike real part.) In this case one has:

$$M(s) = m \left(1 + \frac{D}{s} \right) \quad \text{and} \quad Z(s) = 1 - \frac{D}{s} \ ; \tag{38}$$

in the spacelike UV limit ($s \to \infty$) while in the timelike UV limit ($s \to -\infty$):

$$M(s) = M_0(s) + m \frac{3D}{8} \frac{1}{M_0^2(s)} \quad \text{and} \quad Z(s) = Z_0(s) + m \frac{1}{4} \frac{1}{M_0(s)} \ , \tag{39}$$

where M_0 and Z_0 are the $m = 0$ solutions of Eqs. (35) and (36).

If, instead of the bare vertex, one that is consistent with all of the constraints described in Sec. 3 is used then one obtains the following pair of coupled differential equations:[53]

$$\mu'(x) = \frac{2\mu}{x} + 2 \frac{x + \mu^2(x)}{x\zeta(x)} (\zeta(x)\, \overline{m} - \mu(x)) \ , \tag{40}$$

$$\zeta'(x) = \frac{\mu(x)\mu'(x) - 1}{x + \mu^2(x)} \zeta(x) + 2 \left(1 - \zeta(x) \right) \ , \tag{41}$$

where $x = p^2/(2D), \overline{m} = m/\sqrt{2D}, M(s) = \sqrt{2D}\mu(x)$ and $Z(s) = \zeta(x)$, with the boundary conditions: $\mu(x \to +\infty) = \overline{m}$ and $\zeta(x \to +\infty) = 1$. In this case it is still a simple matter to study the solution in the complex p^2 plane which is the utility of this model.

Writing $\overline{\sigma}_V(x) = \zeta(x)/(x + \mu^2(x))$ and $\overline{\sigma}_S(x) = \overline{\sigma}_V(x)\mu(x)$ these differential equations simplify considerably:

$$\overline{\sigma}_S'(x) = 2(\overline{m}\, \overline{\sigma}_V(x) - \overline{\sigma}_S(x)) \ , \tag{42}$$

$$\overline{\sigma}_V'(x) = -\frac{2}{x} (\overline{\sigma}_V(x)(x + 1) + \overline{m}\, \overline{\sigma}_S(x) - 1) \ . \tag{43}$$

In the case of quarks with zero bare mass Eqs. (42) and (43) decouple, whereas those for μ and ζ do not, and one has a solution:

$$\overline{\sigma}_V(x) = \frac{2x - 1 + e^{-2x}}{2x^2} \ , \quad \overline{\sigma}_S(x) = 0 \ , \tag{44}$$

which corresponds to a Wigner-Weyl mode realisation of chiral symmetry and also a solution:

$$\overline{\sigma}_V(x) = \frac{2x - 1 + e^{-2x}}{2x^2} \;, \quad \overline{\sigma}_S(x) = e^{-2x} \;, \tag{45}$$

which corresponds to a Nambu-Goldstone mode. The cancellation of zeros between the numerator and denominator of $\overline{\sigma}_V$ is not a trivial matter, as will be seen below.

In this case the difference between the CJT effective action evaluated at its extremals is

$$V[\text{Eq. (44)}] - V[\text{Eq. (45)}] = -2N_cN_f \int \frac{d^4q}{(2\pi)^4} \ln\left(1 + \frac{\mu^2(q^2)}{q^2}\right) < 0 \tag{46}$$

and hence, in the dressed vertex version of the model, there is no DχSB. One notes then that dressing the vertex has dramatically changed this important feature of the model. [†]

When the fermions are massive one obtains the following differential equation:

$$\frac{d^2}{dx^2}\overline{\sigma}_S(x) + 4\left(1 + \frac{1}{2x}\right)\frac{d}{dx}\overline{\sigma}_S(x) + 4\left(1 + \frac{1+\overline{m}^2}{x}\right)\overline{\sigma}_S(x) \;\; = \;\; \frac{4\overline{m}}{x} \tag{47}$$

with $\overline{m}\,\overline{\sigma}_V(y) = \{\overline{\sigma}_S(y) + \overline{\sigma}_S'(y)/[4y]\}$ $(x = y^2)$. The solution of Eq. (47) is:

$$\overline{\sigma}_S(y) = \frac{\overline{m}^2}{y}\int_0^\infty d\xi\,\xi K_1(\overline{m}\xi)J_1(y\xi)\exp\left(-\frac{\xi^2}{8}\right) \tag{48}$$

which, with a little thought, is seen to be an <u>entire</u> functon in the complex $x = p^2$ plane. (It follows that in this case the propagator does not have a spectral [or Lehmann] representation and again could not be obtained using the approach of Refs. [10].) As discussed above, this is a sufficient condition for confinement since it ensures the absence of asymptotic quark states in the spectrum of the model.

It is clear that changing the structure of the vertex has had a dramatic qualitative and quantitative effect on the solution of the model. For example, the behaviour of the solution in the deep timelike region in the present case is ($E = \sqrt{-p^2}$):

$$\overline{\sigma}_S(E) \;\; = \;\; \sqrt{\frac{2\pi\overline{m}^3}{E^3}}\exp\left(2(E - \overline{m})^2\right) \;, \tag{49}$$

$$\overline{\sigma}_V(E) \;\; = \;\; \sqrt{\frac{8\pi\overline{m}}{E^3}}\exp\left(2(E - \overline{m})^2\right)\left(1 - \frac{\overline{m}}{2E} - \frac{3}{16E^2}\right) \;. \tag{50}$$

which is to be compared with Eq. (39). Further, the branch points present in the bare vertex case have disappeared.

[†]It should be remembered that if an asymptotic freedom tail is added to Eq. (32) the model does have DχSB.[51] It is a general result that when the equations for $\overline{\sigma}_V$ and $\overline{\sigma}_S$ decouple in a given model then this model will not manifest DχSB.[56]

Expressed in terms of the functions μ and ζ that are most often used to represent the quark propagator one finds that a pole is avoided only because of a cancellation between zeros in the numerator and denominator. In terms of the positive definite function $K(x) = 1 + \overline{\sigma}_S'(x)/(2\overline{\sigma}_S(x))$, one can write:

$$\mu(x) \; = \; \frac{\overline{\sigma}_S(x)}{\overline{\sigma}_V(x)} \equiv \frac{\overline{m}}{K(x)} \, , \tag{51}$$

$$\zeta(x) \; = \; \frac{1}{\overline{m}} \left(x + \mu^2(x) \right) \overline{\sigma}_S(x) K(x) \, . \tag{52}$$

A study of the denominator shows that for all $\overline{m}$ there is an x_c such that:

$$\left(x_c + \mu^2(x_c) \right) = 0 \tag{53}$$

and hence a pole is avoided only because $\zeta(x)$ and $x + \mu^2(x)$ have coincident zeros. This is an unexpected and surprising feature of the model.

Models with similarities to this have been studied in axial gauges where the fermion propagator can be written as

$$S(p) = i\gamma \cdot p \, \sigma_V(p) + \sigma_S(p) + \left[i\gamma \cdot p \, \omega_V(p) + \omega_S(p) \right] i\gamma \cdot n \, . \tag{54}$$

One such study[57] concentrated on a gluon propagator with the q^{-4} infrared singularity and found that the quark mass shell singularity disappears and that DχSB is a possible but not necessary outcome in this model. Another[58] used the $\delta^4(q)$ singularity and found the fermion propagator to be an entire function of $(p_\mu - n_\mu n \cdot p/n^2)^2$ and $\sqrt{(n \cdot p)^2 - m^2}$. In this case the fermion propagator does have a branch point at $(n \cdot p)^2 = m^2$ but, nevertheless, the propagator obtained is not one that can be associated with a free particle.

5.2 *Wick Rotation*

In perturbative analyses of quantum field theory, where the singularity structure of the elements of Feynman diagrams (propagators, for example) is known, it is always possible to rotate the k_0 integration into one over k_4, and vice versa, and obtain no additional terms in the process. Effectively, this encourages the belief that one may make the transition between Minkowski and Euclidean space simply by employing the transcription: $d^4k_M \to i d^4k_E$, $k_M^2 \to -k_E^2$ and $\not{k}_M \to -i\gamma_E \cdot k_E$, etc. One might refer to this procedure as the *naive* Wick Rotation.

An interesting point that arises in connection with the studies of the SDE for the fermion self energy described herein, and others, is that in all cases where the structure of the solution in the complex plane is known[53,58–61] one finds that it is not possible to proceed from Euclidean to Minkowski space via a *naive* Wick rotation. An actual rotation of the contour will at least yield non-trivial pole contributions, which means that a given Euclidean space SDE will differ in form from its actual

Minkowski space counterpart. In fact, in many cases the contour simply cannot be rotated; for example, in the dressed vertex model discussed above the rotation of the contour is completely precluded because of the behaviour of the solution on the boundary at infinity. In this case the Wick rotation cannot be performed. (The general implications of this are discussed in detail Refs. [5,60].)

It is conceiveable that this is a property of any and all *model* SDE (although the *complete* SDE may permit such a simple transcription) and so one must make a decision: whether to formulate a model in Euclidean or Minkowski space? The point of view adopted in Refs. [5,60] is that the natural choice is the Euclidean space formulation. A complete discussion of the reasoning will not be presented here but it will be recognised that all of the information that one has about QCD is known, strictly, only at spacelike momentum transfer. The spacelike region is therefore the region where the tightest constraints can be placed on any given model. It is important to emphasise that the invalidity of the *naive* Wick Rotation entails that two equations, one in Minkowski space and the other in Euclidean, that are related by this straightforward transcription, will have solutions that bear no relation to each other.[5]

As a final comment it should be noted that formulating a model in Euclidean space in no way precludes the calculation of physical observables. For example, many of the techniques employed in lattice gauge theory can be used unchanged in the SDE framework, as may be seen in Ref. [29]. In broad terms, one defines the model and evaluates a given amplitude, which is a well defined function of the external momenta with definite analytic properties in the complex plane. It is a simple matter then to extract the physical observables via analytic continuation (which may be done explicitly or implicitly[5]).

6. Summary

The SDE for the fermion self energy allows a gauge invariant study of $D\chi SB$ in the continuum. At the present time this study remains somewhat phenomenological in QCD since the exact form of the gluon propagator and quark-gluon vertex is unknown except in the deep spacelike region. Given an Ansatz for each of these quantities, however, one may use this SDE to obtain information about the analytic structure of the quark propagator and make inferences about the dynamics of confinement.

The fermion propagator is crucial element in the calculation of S-matrix elements and hence a sensible and physically reasonable model quark propagator is very useful in the development of a covariant, Feynman diagram based, confining hadronic phenomenology. This is true of all propagators and vertices and makes their study important in its own right. In this connection it is interesting to note that the models described herein lead to or incorporate propagators and vertices that do not have a Lehmann representation.

A good deal of progress has been made in the general approach to the nonperturbative solution of a field theory using the complex of SDE's - I have tried to summarise this herein. Much remains to be done, of course, but with the current surge of interest there is reason to be optimistic about an increased rate of progress. In the context of the phenomenology of QCD there are three current problems: the structure of the gluon propagator in the IR and intermediate q^2 region; the structure of the quark-gluon vertex in the same region; and an increased understanding of the quantitative and qualitative importance of ghost fields in covariant gauges. It is conceivable that, with some effort, lattice gauge calculations might provide some information about the first two of these problems but the calculations that exist at present[62] are inadequate and poorly understood.

Acknowledgements

I would like to thank and congratulate Berndt Müller and Herb Fried, the organisers of this meeting, which was timely and stimulating. I would also like to thank the conference secretary, Susan Bell, for helping the workshop to run so smoothly. This work was supported by the Department of Energy, Nuclear Physics Division under contract number W-31-109-ENG-38. Additional financial assistance from the National Science Foundation, which aided my participation, is also gratefully acknowledged.

References

1. F. J. Dyson, *Phys. Rev.* **75** (1949) 1736; J. S. Schwinger, *Proc. Nat. Acad. Sc.* **37** (1951) 452,455.

2. C. Itzykson and J. -B. Zuber, *Quantum Field Theory* (McGraw-Hill, New York, 1980), Chap. 10.

3. U. Bar-Gadda, *Nucl. Phys.* **B163** (1980) 312.

4. Particle Data Group, *Review of Particle Properties, Phys. Lett.* **B239** (1990).

5. C. D. Roberts, A. G. Williams and G. Krein, *Int. J. Mod. Phys.*, **A7** (1992) 5607.

6. M. B. Einhorn, *Phys. Rev.* **D14** (1976).

7. U. Häbel, *et al*, *Z. Phys.* **A336** (1990) 423.

8. M. Stingl, *Phys. Rev.* **D34** (1986) 3863.

9. U. Häbel, *et al*, *Z. Phys.* **A336** (1990) 435.

10. A. Salam, *Phys. Rev.* **130** (1963) 1287; A. Salam and R. Delbourgo, *Phys. Rev.* **135** (1964) B1398; J. Strathdee, *Phys. Rev.* **135** (1964) B1428; R. Delbourgo, *Nuovo Cimento* **A49** (1979) 484.

11. D. Zwanziger, *Nucl. Phys.* **B364** (1991) 127.

12. A. Hädicke, *Int. J. Mod. Phys.* **A6** (1991) 3321.

13. S. Mandelstam, *Phys. Rev.* **D20** (1979) 3223.

14. N. Brown and M. R. Pennington, *Phys. Lett.* **B202** (1988) 257.

15. N. Brown and M. R. Pennington, *Phys. Rev.* **D38** (1988) 2266.

16. M. Baker, J. S. Ball and F. Zachariasen, *Nucl. Phys.* **B186** (1981) 531,560; **B226** (1983) 455.

17. G. B. West, *Phys. Rev.* **D27** (1983) 1878.

18. G. B. West, *Phys. Lett.* **B115** (1982) 468.

19. H. J. Munczek and A. M. Nemirovsky, *Phys. Rev.* **D28** (1983) 181.

20. H. J. Munczek and P. Jain, *Phys. Rev.* **D46** (1992) 438.

21. W. Marciano and H. Pagels, *Phys. Rep.* **36** (1978) 137.

22. J. S. Ball and T. -W. Chiu, *Phys. Rev.* **D22** (1980) 2542.

23. D. C. Curtis and M. R. Pennington, *Phys. Rev.* **D42** (1990) 4165.

24. S. P. Klevansky, *Rev. Mod. Phys.* **64** (1992) 649 and references therein.

25. See, for example, J. Gasser and H. Leutwyler, *Ann. Phys.* **158** (1984) 142 and *Nucl. Phys.* **B250** (1985) 465.

26. C. D. Roberts, *Hadron Dynamics from QCD, PhD Thesis - Flinders University of South Australia* (1987), unpublished; C. D. Roberts, R. T. Cahill and J. Praschifka, *Ann. Phys. (N.Y.)* **188** (1988) 20.

27. C. D. Roberts, in preparation.

28. J. Praschifka, C. D. Roberts and R. T. Cahill, *Phys. Rev.* **D36** (1987) 209; C. D. Roberts, R. T. Cahill and J. Praschifka, *Int. J. Mod. Phys.* **4** (1989) 719; C. D. Roberts, J. Praschifka and R. T. Cahill, *Int. J. Mod. Phys.* **A4** (1989) 1681.

29. L. C. L. Hollenberg, C. D. Roberts and B. H. J. McKellar, U. Melbourne and ANL preprint (1992), unpublished.

30. J. Praschifka, R. T. Cahill and C. D. Roberts, *Int. J. Mod. Phys.* **A4** (1989) 4929.

31. R. T. Cahill, C. D. Roberts and J. Praschifka, *Phys. Rev.* **D36** (1987) 2804.

32. J. Burden, R. T. Cahill and J. Praschifka, *Aust. J. Phys* **42** (1989) 147; R. T. Cahill, private communication.

33. D. W. McKay and H. J. Munczek, *Phys. Rev.* **D32** (1985) 266; R. T. Cahill and C. D. Roberts, *Phys. Rev.* **D32** (1985) 2419; D. W. McKay, H. J. Munczek and B. -L. Young, *Phys.Rev.* **D37** (1988) 195; Y. Dai, C. Huang and D. Liu, *Phys. Rev.* **D43** (1991) 1717; K. -I Aoki, T. Kugo and M. G. Mitchard, *Phys. Lett.* **B266** (1991) 467; R. T. Cahill, *Nucl. Phys.* **A543** (1992) 63.

34. C. J. Burden and C. D. Roberts, *Phys. Rev.* **D44** (1991) 540.

35. H. A. Slim, *Nucl. Phys.* **B177** (1981) 172.

36. L. D. Landau and I. M. Khalatnikov, *Sov. Phys. JETP* **2** (1956) 69; B. Zumino, *J. Math. Phys.* **1** (1960) 1.

37. C. J. Burden, J. Praschifka and C. D. Roberts, *Phys. Rev.* **D** (1992), in press.

38. C. J. Burden, *How can QED_3 help us understand QCD_4?*, these proceedings.

39. M. R. Pennington and D. Walsh, *Phys. Lett.* **B253** (1991) 246.

40. C. D. Roberts and R. T. Cahill, *Phys. Rev.* **D33** (1986) 1755.

41. R. Fukuda and T. Kugo, *Nucl. Phys.* **B117** (1976) 250.

42. K. Higashijima, *Phys. Lett.* **B124** (1983) 257.

43. P. I. Fomin, V. P. Gusynin, V. A. Miransky and Yu. A. Sitenko, *Riv. Nuov. Cim.* **6** (1983) 1.

44. K. Higashijima, *Phys. Rev.* **D29** (1984) 1228.

45. K. Lane, *Phys. Rev.* **D10** (1974) 2605; H. D. Politzer, *Nucl. Phys.* **117** (1976) 397; H. Pagels, *Phys. Rev.* **D19** (1979) 3080.

46. D. Atkinson and P. W. Johnson, *Phys. Rev.* **D37** (1988) 2296.

47. C. D. Roberts and B. H. J. McKellar, *Phys. Rev.* **D41** (1990) 672.

48. H. J. Munczek and D. W. McKay, *Phys. Rev.* **D42** (1990) 3548.

49. L. v. Smekal, P. Amunsden and R. Alkofer, *Nucl. Phys.* **A529** (1991) 663.

50. A. G. Williams, G. Krein and C. D. Roberts, *Ann. Phys. (N.Y.)* **210** (1991) 464.

51. F. T. Hawes and A. G. Williams, *Phys. Lett.* **B268** (1991) 271.

52. E. V. Shuryak, these proceedings.

53. C. J. Burden, C. D. Roberts and A. G. Williams, *Phys. Lett.* **B285** (1992) 347.

54. J. M. Cornwall, R. Jackiw and E. Tomboulis, Phys. Rev. D **10** 2428 (1974).

55. K. Stam, Phys. Lett. B **152** 238 (1985).

56. C. J. Burden and C. D. Roberts, in preparation.

57. J. S. Ball and F. Zachariasen, *Phys. Lett.* **B106** (1981) 133.

58. H. J. Munczek, *Phys. Lett.* **B175** (1986) 215.

59. D. Atkinson and D. W. E. Blatt, *Nucl. Phys.* **B151** (1979) 342.

60. S. J. Stainsby and R. T. Cahill, *Phys. Lett.* **A146** (1990) 467.

61. P. Maris and H. Holties, *Int. J. Mod. Phys.* **A7** (1992) 5369.

62. P. Coddington, A. Hey, J. Mandula and M. Ogilvie, *Phys. Lett.* **B197** (1987) 191;
J. E. Mandula and M. Ogilvie, *Phys. Lett.* **B185**, (1987) 127; *Nucl. Phys.* **B1A**
Proc. Suppl. (1987) 117; *Phys. Lett.* **B201** (1988) 117; R. Gupta *et al*, *Phys. Rev.*
D36 (1987) 2813.

NONPERTURBATIVE SOLUTIONS TO THE
DYSON–SCHWINGER EQUATIONS OF PURE QCD

Jörg Ahlbach, Andreas Streibl, Martin Schaden*
*Physik Department (T30) der Technischen Universität München
James-Franck Straße, D–8000 München, Fed. Rep. Germany*

ABSTRACT

The selfconsistent reproduction of nonperturbative power corrections to vertices
is reviewed. We use the operator product expansion to obtain an ansatz for
the nonperturbative vertices of pure QCD. It is shown that the Dyson-Schwinger
equation for the ghost selfenergy is selfconsistently solved by them if the logarithmic renormalization group improvement is taken into account. We find that
Ward-identities are satisfied, but that dimension four operators which are not
BRS-invariant contribute to the operator product expansion of the vertices. This
result is interpreted as an indication that BRS-symmetry may be spontaneously
broken.

1. Introduction

Only fairly recently a systematic method for obtaining nonperturbative solutions to the Dyson-Schwinger equations of QCD was proposed[1]. It hinges on the
definition that nonperturbative solutions should have a nonanalytic dependence on
the coupling constant g^2 around $g^2 = 0$, that cannot be found in renormalization
group improved perturbation theory. It is assumed, that this nonanalyticity is
introduced only through the fundamental scale

$$\Lambda = \mu \exp\left(-\int_{g_0}^{g} \beta^{-1}(\xi) d\xi\right) \sim C(g_0)\mu \exp(-8\pi^2/11g^2)\,, \tag{1}$$

where for pure QCD

$$\beta(g) = -11g^3/16\pi^2 + \ldots \tag{2}$$

is the β-function of the Renormalization Group Equation[2] (RGE) and μ is the
renormalization point corresponding to the coupling g^2 such that Λ is held fixed.
In the following we will restrict ourselves to pure QCD in covariant gauges and in
particular in most cases to the Landau gauge.

The RGE for the vertices $\Gamma_{n_A,n_G}(p_i, k_j; \mu, g)$ with n_A gluonic and n_G ghost
fields having momenta p_i and k_j respectively, imply that $\tilde{\Gamma}$ defined by,

$$\Gamma_{n_A,n_G}(p_i, k_j; \mu, g) = \tilde{\Gamma}_{n_A,n_G}(p_i, k_j; \Lambda)e^{\int_{g_0}^{g}(n_A\gamma^A(\xi)+n_G\gamma^G(\xi))\beta^{-1}(\xi)d\xi}\,, \tag{3}$$

---

*Work supported by: Gesellschaft für Schwerionenforschung, Darmstadt
E-mail: mschaden@physik.tu-muenchen.de

with the gluon and ghost anomalous dimensions γ^A and γ^G, satisfies a homogeneous RGE. $\tilde{\Gamma}$ therefore only depends on the coupling and the renormalization point in the RG-invariant combination of Λ.

The *ansatz* of Ref.1 for these (homogeneous) vertices is a power expansion in Λ,

$$\tilde{\Gamma}(p_i, k_j; \Lambda) = \sum_{r=0}^{\infty} \Lambda^r C^{(r)}(p_i, k_j; \Lambda), \tag{4}$$

where the coefficients $C^{(r)}$ may still depend logarithmically on the scale. This ansatz was originally motivated by asymptotic freedom[1], but gains even more credibility with the observation that it is just an Operator Product Expansion[2] (OPE) for the vertices of the theory. Note that only the coefficient $C^{(0)}$ is obtained in RG-improved perturbation theory.

Assuming that the anomalous dimensions and the β-function are analytic at $g^2 = 0$ and have expansions of the form

$$\beta(g) = -\beta_0 \frac{g^3}{16\pi^2}(1 + \beta_1 g^2 + \ldots); \quad \beta_0 = \frac{11N - 2N_F}{3}$$

$$\gamma^A(g) = -\gamma_0^A \frac{g^2}{16\pi^2}(1 + \gamma_1^A g^2 + \ldots); \quad \gamma_0^A = \frac{(13 - 3\xi)N - 4N_F}{6} \tag{5}$$

$$\gamma^G(g) = -\gamma_0^G \frac{g^2}{16\pi^2}(1 + \gamma_1^G g^2 + \ldots); \quad \gamma_0^G = \frac{(3 - \xi)N}{4}.$$

The corresponding ansatz for $\Gamma_{n_A, n_G}(p_i, k_j; g, \Lambda)$ with Eq.(3) becomes

$$\Gamma_{n_A, n_G}(p_i, k_j; g, \Lambda) = g^{\frac{(n_A \gamma_0^A + n_G \gamma_0^G)}{\beta_0}} \sum_{n=0}^{\infty} g^{2n} \sum_{r=0}^{\infty} \Lambda^r C_{n_A, n_G}^{(n,r)}(p_i, k_j; \Lambda), \tag{6}$$

where we have absorbed g-independent constants in the coefficient-functions. Except for the (logarithmic) dependence of these functions on the scale and the inclusion of the overall (nonanalytic) coupling constant dependence in front, which are dictated by the renormalization group, this ansatz is that of ref.1. The coefficient functions in Eq.(6) can be systematically found in the framework of the OPE, leaving only a few parameters (condensates) in each dimension undetermined. The ansatz Eq.(6) would nevertheless be practically useless, could we not truncate the summations severely, and still obtain (approximately) selfconsistent solutions to the DS-equations.

The scheme proposed in Ref.1 is to try to obtain selfconsistent solutions of the DS-equations order by order in the coupling g^2, i.e. a weak coupling expansion at fixed Λ – or equivalently, for a large renormalization scale μ. One should thus obtain approximately correct vertices for large momenta. This limit is interesting, because certain power corrections not present in a perturbative calculation were

136

seen to survive[1]. To obtain selfconsistency for the $C^{(0,r)}_{n_A,n_G}$ in this sense, one can reduce the infinite set of DS-equations to those for the five primitively divergent vertices

$$\Gamma \in \{\Gamma_{VV}, \Gamma_{GG}, \Gamma_{GVG}, \Gamma_{VVV}, \Gamma_{VVVV}\}, \tag{7}$$

since these coefficients vanish for higher vertices (we have here used the notation V=gluon, G=ghost).

Condensates with dimension $r \leq 4$ may still lead to a logarithmic ultraviolet divergence in one loop. In addition to the perturbative divergences, the power dependence on Λ in the propagators will in lowest order also lead to (ultraviolet) logarithmically divergent integrals of the form

$$-ig^2 \int \frac{d^4k(k^2)^{2s}}{(k^4 + a\Lambda^4)^{s+1}}(g^2\ln(\frac{k^2}{\Lambda^2}))^\alpha = \frac{\pi^2(g^2\ln\frac{\mu^2}{\Lambda^2} + g^2\ln\frac{M^2}{\mu^2})^{1+\alpha}}{1+\alpha} + g^2 \times \text{finite terms}$$

$$= \frac{\pi^2}{1+\alpha}\left(\frac{16\pi^2}{\beta_0}\right)^{1+\alpha} + O(g^2),$$

$$\tag{8}$$

where M is a large cutoff mass and the power α of the logarithmic dependence on the loop momentum is from the anomalous dimension of the integrand (the logarithmic divergence in Eq.(8) is independent of $s \geq 0$). In the second line of Eq.(8) we have used that

$$g^2\ln(\frac{\mu^2}{\Lambda^2}) = \frac{16\pi^2}{\beta_0} + O(g^2), \tag{9}$$

and expanded in g^2. Such logarithmically divergent integrals therefore produce a term of order g^0. The divergence is of higher order in g^2 and will only contribute to coefficients $C^{n \geq 1}$ (leading among other things to a renormalization of the condensates). The way the extra factor of g^2 from the loop is cancelled by the dependence of the divergent integral on Λ (called "nonperturbative logs" in Ref.1) is essentially the same as if one neglected anomalous dimensions (i.e. setting $\alpha = 0$)[1]. The integral Eq.(8) shows however, that an extra factor $1/(1+\alpha)$ appears as a result of the anomalous dimensions, and the corresponding logarithmic deviation of the vertex functions from a power behaviour at large momenta. This factor will be essential for a quantitative reproduction of the ansatz Eq.(6) in orders of the coupling g^2. Fortunately, for selfconsistency in lowest order of g^2 we do not need to know more about the logarithmic dependence of the vertex function on the momenta than how it scales, since that already determines the divergent part of the integral. This anomalous scaling behaviour can be obtained from the renormalization group (to one loop).

Since the superficial degree of (ultraviolet) divergence in QCD is at most two, the coefficient functions with $r > 4$ will always result in (ultraviolet) convergent loop integrals in the DS-equations. These will therefore lead to an extra factor of g^2 and give a contribution to terms one order higher in the coupling constant expansion[1]. The infinite sum over condensates of dimension r in Eq.(6) can thus in lowest order g^2 be restricted to those with $r \leq 4$[1], all higher dimensional condensates appearing only with coefficients $C^{(n,r)}$, $n \geq 1$.

The above analysis also shows that only logarithmic divergences can lead to inverse powers of the coupling, g. Linear and higher divergences would not have this feature and one should take care in the ansatz Eq.(6) that they do not appear. They could arise from the coefficient function $C^{0,2}$, which essentially is a mass insertion (in pure QCD without quarks the $C^{r=1}$ and $C^{r=3}$ coefficients that correspond to dimension one and three condensates vanish anyhow). The only globally gauge invariant condensate of dimension two (in the absence of scalar fields) is $<: g^2 \mathrm{tr} A_\mu A^\mu :>$, which we will therefore assume to vanish in lowest order. (One might have to introduce such a condensate in higher orders of the coupling to cancel quadratic divergences due to (convergent) terms coming from lower orders).

Our discussion so far concerned the selfconsistent reproduction of the ansatz Eq.(6) via logarithmic ultraviolet divergences. The ansatz also might lead to infrared divergences, because of the inverse powers of the momenta in the vertices. Although these actually lead to a better infrared behaviour of the propagators, we cannot exclude such divergences from occurring in loops. They have nothing to do with the small coupling behaviour of the theory and are certainly beyond the scope of the ansatz. By using cumulants[1] instead of inverse powers in p^2 (e.g. expanding in $(p^2 + \Lambda^2)^{-1}$) one can in principle avoid such infrared divergences without changing the ultraviolet behaviour of the vertices. Note however, that such infrared regulators would appear as condensate contributions of dimension six and higher in a large momentum expansion if we do not allow for dimension two condensates. From the above, they are then of order g^2 and cannot be selfconsistently determined in lowest order. The ultraviolet divergences are unaffected by such a modification of the ansatz and we will only extract them.

With these considerations, we still have to obtain the coefficients $C^{(0,0)}$ and $C^{(0,4)}$ for the five primitively divergent vertex functions of pure QCD and show that they selfconsistently satisfy the DS-equations to leading order in g^2.

2. The coefficient functions $C^{0,0}$ and $C^{0,4}$ from the OPE

Even if we neglect the logarithmic scale dependence, the dependence of the higher vertex functions on more than one momentum allows for a large amount of freedom in the homogeneous coefficient functions. Any unguided ansatz is liable to fail[1]. With the above considerations, the two-point vertices for the ghost and gluon propagators are however, already given in terms of two parameters (up to logarithmic corrections). The ghost selfenergy has the form

$$\Pi(k^2) = \frac{c_G}{k^2}, \tag{10}$$

138

and the corresponding ghost propagator is

$$D(k^2) = \frac{k^2}{k^4 - c_G} \, . \tag{11}$$

The gluon polarization by the Slavnov-Taylor[3] identity has to be transverse, and is thus

$$\Pi_{\mu\nu}(p) = g^{\perp}_{\mu\nu}(p)\frac{c_V}{p^2} \, , \tag{12}$$

and the propagator in a covariant gauge becomes

$$D_{\mu\nu}(p) = -g^{\perp}_{\mu\nu}(p)\frac{p^2}{p^4 - c_V} - \xi\frac{p_\mu p_\nu}{p^4} \, . \tag{13}$$

The parameters c_G and c_V here defined are renormalization group invariants (in lowest order g^2) and therefore proportional to Λ^4. They can (and will) however, be gauge dependent.

To determine the momentum structure of higher vertices, we took advantage of the similarity of the ansatz Eq.(6) with an OPE. Only relatively few parameters, the condensates up to dimension four will remain undetermined in the ansatz. In covariant gauges these have to be Lorentz- (in Euclidean space O(4)-) invariants and if colour symmetry is unbroken also invariant under (global) colour transformations. We however, will not demand their invariance under (global) Becchi-Rouet-Stora[4] (BRS) transformations – which in pure QCD would leave us only with a nonvanishing gluon condensate, $<: g^2 G_{\mu\nu}G^{\mu\nu} :>$. Although physical correlation functions (which are BRS invariant) can only depend on BRS-invariant combinations of condensates, this need not (and will not) be the case for the vertex functions we are considering. Their OPE only has to satisfy the Slavnov-Taylor Identities[2,3] (STI) in every order. Together with Bose-symmetry and the fact that divergences of currents have vanishing expectation values these restrictions still allow for a set of 13 condensates of dimension two and four in a pure SU(N) gauge theory

$$<: g^2 \text{tr} A_\mu A^\mu :> \qquad\qquad <: g^2 \bar{\eta}_a \eta_a :>$$

$$<: g^2 \text{tr}(\partial_\mu A^\mu)^2 :> \qquad\qquad <: g^3 (\partial_\mu \bar{\eta}_a) A^\mu_b d^{abc} \eta_c :>$$

$$<: g^2 \bar{\eta}_a \Box \eta_a :> \qquad\qquad <: g^3 (\partial_\mu \bar{\eta}_a) A^\mu_b f^{abc} \eta_c :>$$

$$<: g^2 \text{tr}(\partial_\mu A_\nu - \partial_\nu A_\mu)^2 :> \qquad\qquad <: ig^3 \text{tr}(\partial^\mu A^\nu)[A_\mu, A_\nu] :>$$

$$<: g^4 \text{tr}[A_\mu, A_\nu][A^\mu, A^\nu] :> \qquad\qquad <: g^3 \text{tr}(\partial_\mu A^\mu) A_\nu A^\nu :>$$

$$<: g^4 \text{tr}\left(A_\mu A_\nu A^\mu A^\nu + 2A_\mu A^\mu A_\nu A^\nu + \tfrac{6}{N} A_\mu A_\nu \text{tr}(A^\mu A^\nu) + \tfrac{3}{N} A_\mu A^\mu \text{tr}(A_\nu A^\nu)\right) :>$$

$$<: g^4 \text{tr}\left(A_\mu A_\nu \text{tr}(A^\mu A^\nu) - A_\mu A^\mu \text{tr}(A_\nu A^\nu)\right) :>$$

$$<: g^4 \mathrm{tr}\left(A_\mu A_\nu \mathrm{tr}(A^\mu A^\nu) + A_\mu A^\mu \mathrm{tr}(A_\nu A^\nu)\right) :> \qquad (14)$$

The derivative coupling between ghosts and gluons in covariant gauges excludes any contribution of $<: g^2 \bar\eta_a \eta_a :>$ to the OPE of the vertex functions. From the STI for the gluon propagator, which asserts that its longitudinal part is just the perturbative one, the condensate $<: g^2 \mathrm{tr}(\partial_\mu A^\mu)^2 :>$ must vanish. We found[5] that a nonvanishing value of $<: g^3 (\partial_\mu \bar\eta_a) A_b^\mu d^{abc} \eta_c :>$ would contribute to the purely longitudinal part of the three gluon vertex function, and therefore imply the anomalous breakdown of the corresponding STI. This is in accord with recent two-loop calculations in chiral gauge theories[6] (where the STI is anomalously broken by precisely such a counterterm). The other two condensates involving ghost fields can be related to gluonic ones through the equations of motion, which imply the two additional constraints

$$<: \mathrm{tr}\left((\partial_\mu A_\nu - \partial_\nu A_\mu)^2 - 2ig(3\partial^\mu A^\nu - ig[A^\mu, A^\nu])[A_\mu, A_\nu]\right) - g(\partial_\mu \bar\eta_a) A_b^\mu f^{abc} \eta_c :>$$
$$=<: g(\partial_\mu \bar\eta_a) A_b^\mu f^{abc} \eta_c - \bar\eta_a \Box \eta_a :>= 0$$
$$(15)$$

These can also be found by demanding that simple STI be satisfied, i.e. that the longitudinal parts of the two and three gluon vertex functions vanish[7,8]. We can thus express the OPE of the vertex functions in terms of the remaining seven dimension four condensates as parameters (assuming $<: g^2 \mathrm{tr} A_\mu A^\mu :>$ to vanish for the reasons stated above, although this condensate did not immediately violate any STI[9]). In terms of these condensates the coefficients c_G and c_V of equations (10), (11), (12) and (13) are[8]

$$c_G = \frac{N}{24(N^2 - 1)}\{(114 + 54\xi) <: ig^3 \mathrm{tr}(\partial^\mu A^\nu)[A_\mu, A_\nu] :> +2\times$$

$$<: g^4 \mathrm{tr}\left(A_\mu A_\nu A^\mu A^\nu + 2A_\mu A^\mu A_\nu A^\nu + \tfrac{6}{N} A_\mu A_\nu \mathrm{tr}(A^\mu A^\nu) + \tfrac{3}{N} A_\mu A^\mu \mathrm{tr}(A_\nu A^\nu)\right) :>$$

$$+ (36 + 18\xi) <: g^4 \mathrm{tr}[A_\mu, A_\nu][A^\mu, A^\nu] :> -(20 + 9\xi) <: g^2 \mathrm{tr}(\partial_\mu A_\nu - \partial_\nu A_\mu)^2 :>\},$$

$$c_V = \frac{N}{36(N^2 - 1)}\{(68 + 3\xi) <: g^2 \mathrm{tr}(\partial_\mu A_\nu - \partial_\nu A_\mu)^2 :> +(5 + 3\xi)\times$$

$$<: g^4 \mathrm{tr}\left(A_\mu A_\nu A^\mu A^\nu + 2A_\mu A^\mu A_\nu A^\nu + \tfrac{6}{N} A_\mu A_\nu \mathrm{tr}(A^\mu A^\nu) + \tfrac{3}{N} A_\mu A^\mu \mathrm{tr}(A_\nu A^\nu)\right) :>$$

$$- (216 - 36\xi) <: ig^3 \mathrm{tr}(\partial^\mu A^\nu)[A_\mu, A_\nu] :> -(44 - 15\xi) <: g^4 \mathrm{tr}[A_\mu, A_\nu][A^\mu, A^\nu] :>\}$$
$$(16)$$

(In c_G we have corrected the wrong sign of the $<: g^3 (\partial_\mu \bar\eta_a) A_b^\mu f^{abc} \eta_c :>$ contribution to the ghost selfenergy of Ref.8) The OPE of the longitudinal part of the gluon polarization was shown to vanish[7,8] with the constraints Eq.(15) imposed by the equations of motion. Contrary to the OPE of gauge invariant correlation functions, the various gluon condensates in Eq.(16) do not appear in the combination of $<: g^2 G_{\mu\nu} G^{\mu\nu} :>$, and their coefficients are explicitly gauge parameter dependent. Furthermore the OPE of these vertex functions also has a contribution from a four gluon operator which is not a part of the Lagrangian. Similar results pertain for the OPE of the three point vertex functions Γ_{VVV} and Γ_{GVG}, which have also been calculated[5,10]. (Due to their more complicated momentum dependence, the OPE for these vertex functions is rather involved and too lengthy to be reproduced here.)

The dependence of the OPE of vertex functions on operators that are not BRS-invariant can also be verified for the quark operators[7,11]. In the light of the above, it should not come as a surprise that early attempts to determine the coefficient of $<: g^2 G_{\mu\nu} G^{\mu\nu} :>$ in the OPE of these vertex functions came to contradictory conclusions, depending on whether the coefficient was determined from a two[12] or a four[13] gluon operator, e.g. whether the coefficient of $<: g^2 \mathrm{tr}(\partial_\mu A_\nu - \partial_\nu A_\mu)^2 :>$ or $<: g^4 \mathrm{tr}[A_\mu, A_\nu][A^\mu, A^\nu] :>$ was obtained. That the OPE of (gauge dependent) vertex functions should depend on BRS-variant condensates with gauge dependent coefficients is not really a problem, as long as these condensates combine to gauge-invariants in physical quantities constructed from these vertices. This should be guaranteed, if the vertices satisfy all the STI. For the gluon propagator[8] and the quark propagator[11] the STI were explicitly verified.

To see this mechanism at work, we have also calculated the $<: g^2 \mathrm{tr}(\partial_\mu A_\nu - \partial_\nu A_\mu)^2 :>$ contribution to the (gauge invariant) vacuum energy density using the OPE of the vertices[14] and found that it is not gauge-parameter dependent and that the coefficient is precisely that of the scale anomaly (to lowest order in the coupling). The gauge parameter dependence also vanished in a calculation of the $<: g^2 \mathrm{tr}(\partial_\mu A_\nu - \partial_\nu A_\mu)^2 :>$-contribution[5,15] to the (gauge invariant) effective Cornwall[16] propagator.

3. Selfconsistency of the DS-equation for the ghost selfenergy

Before drawing possible conclusions from the OPE of vertex functions on the structure of the ground state, we will show how the simplest DS-equation of the purely gluonic theory, the one for the ghost vertex Γ_{GG}, is indeed selfconsistently reproduced by the ansatz Eq.(6) in lowest order g^2. This DS-equation is also essential for the STI of the theory and thus for the gauge invariance of physical correlation functions.

The RGE to one loop implies that the coefficient functions $C^{(0,r)}_{n_A, n_G, O_i}$ to eigenoperators O_i of the anomalous dimension matrix will have a large momentum behaviour

$$C^{(0,r)}_{n_A, n_G, O_i}(p_i, k_j; \Lambda) = \text{hom. func.} \times \left(\frac{\beta_0}{16\pi^2} \ln \frac{s^2}{\Lambda^2} \right)^{(n_A \gamma_0^A + n_G \gamma_0^G - \gamma_0^i)/2\beta_0 - (n_A + n_G - 2)/2},$$

$$(17)$$

where γ_0^i is the eigenvalue of the anomalous dimension matrix for the operator O_i and $s^2 = \sum_i p_i^2 + \sum_j k_j^2$ is a (symmetric) reference momentum. The last factor in the exponent of Eq.(17) takes account of the dependence on the running coupling constant of the three and four point functions. Instead of determining the anomalous dimension matrix for the condensates of dimension four listed in Eq.(14) and finding its eigenvalues, we insert this ansatz in the DS-equations for the vertex functions after having found the homogeneous dependence of the coefficients from the "naive" OPE. We then determine the eigenoperators and their anomalous dimensions by demanding that the DS-equations be satisfied to lowest order in g^2. The advantage of this procedure is that we do not have to obtain the whole anomalous dimension matrix before selfconsistently solving a particular DS-equation – only those eigenoperators contributing enter the problem.

The DS-equation for the ghost selfenergy reads

$$\Gamma_{GG}(k; \Lambda, g) = \tilde{Z} k^2 - ig N \int \frac{d^4 p}{(2\pi)^4} p_\nu \Gamma_\mu^{GVG}(p, k; \Lambda, g) D_{VV}^{\mu\nu}(k-p; \Lambda, g) D_{GG}(p; \Lambda, g),$$

(18)

where D_{VV} and D_{GG} are the gluon and ghost propagators and colour indices have been suppressed.

We then inserted the ansatz Eq.(6) for the relevant vertex functions with the coefficients we have obtained[10] from the OPE. The perturbative coefficients $C^{(0,0)}$ satisfy the DS-equation for the ghost vertex function Γ_{GG} to order g^2 and only the divergent part of the loop integral is important for the selfconsistent reproduction to that order by the mechanism we described above. This DS-equation then leads to the requirement that

$$g^{\frac{2\gamma_0^G}{\beta_0}} \frac{c_G}{k^2} \left(\frac{\beta_0}{16\pi^2} \ln\left(\frac{k^2}{\Lambda^2} \right) \right)^{\frac{2\gamma_0^G - \gamma_0^{cG}}{2\beta_0}} =$$

$$= g^{1 - \frac{\gamma_0^A}{\beta_0}} \frac{c_G N \xi}{2k^2} \left(-i \int_{UV} \frac{d^4 p}{(2\pi)^4} \frac{p^4 \left(\frac{\beta_0}{16\pi^2} \ln\left(\frac{p^2}{\Lambda^2} \right) \right)^{-\frac{\gamma_0^A + \gamma_0^{cG} + \beta_0}{2\beta_0}}}{(p^4 - c_G)(p^4 - c_V)} + O(g^2) \right),$$

(19)

where the index UV of the integral indicates, that we have only written the ultraviolet divergent part of the loop integral. Note that the same combination c_G of condensates appears on both sides for the divergent contribution, although the vertex function Γ_{GVG} depends on other condensates as well[10]. This is a check, that the homogeneous parts of the coefficient functions are correct. Because of the proportionality of the integral to the gauge parameter ξ we cannot really calculate in Landau gauge. Allowing for arbitrary covariant gauges would however lead to a more complicated renormalization group analysis. If we only keep terms to order ξ we do not yet have to take the running of the gauge into account.

Because the right hand side of Eq.(19) shows no logarithmic dependence on the external momentum, the logarithmic dependence on the left must also vanish and we thus find for the anomalous dimension

$$\gamma_0^{cG} = 2\gamma_0^{G}. \tag{20}$$

We therefore finally have to verify that

$$1 = -i\frac{N\xi}{2}g^2 \int\limits_{UV} \frac{d^4p}{(2\pi)^4} \frac{p^4\left(\frac{g^2\beta_0}{16\pi^2}\ln\left(\frac{p^2}{\Lambda^2}\right)\right)^{-\frac{\gamma_0^A+2\gamma_0^G+\beta_0}{2\beta_0}}}{(p^4-c_G)(p^4-c_V)} + O(g^2). \tag{21}$$

Using Eq.(8) and Eq.(5) this is seen to be the case, because

$$N\xi = \beta_0 - \gamma_0^A - 2\gamma_0^G. \tag{22}$$

The selfconsistency is also obtained in Landau gauge, because the divergent part of the integral Eq.(21) is itself proportional to $1/\xi$. The inclusion of anomalous dimensions is seen to be crucial for selfconsistency – they also cancel the N_F-dependence of the beta function, which would be impossible to achieve otherwise, since quarks do not contribute to this DS-equation to one loop.

4. Conclusions

We have shown the selfconsistent reproduction of the ansatz Eq.(6) for the vertex functions in lowest order g^2 with nontrivial coefficients $C^{(0,4)}$ in the DS-equation for the ghost selfenergy. We used the techniques of the OPE to determine the coefficient functions of the ansatz Eq.(6) for the vertex functions and found that they depend on condensates that are not BRS-invariant. The DS-equation for the ghost selfenergy has nontrivial solutions, if the logarithmic scale dependence of the coefficient functions is included. We have thus found a slightly more general ansatz than that of Ref.1, which has the correct asymptotic momentum behaviour of the OPE and at the same time only depends on a few parameters (at most seven independent condensates in pure QCD). This approach also has the advantage that the nonperturbative vertex functions can be constructed with the "naive" OPE before actually verifying that they solve the DS-equations to lowest order g^2 (up to the anomalous dimensions of the operators). We do not yet know the OPE of the four gluon vertex function and therefore could only verify the DS-equation for the ghost selfenergy (which does not depend on this vertex).

The use of the "naive" OPE in the construction of the homogeneous parts of the coefficient functions automatically gives their Lorentz and colour structure. It can be quite complicated for the three and four point vertices[5,10], where several independent external momenta are involved. The STI we have studied hold if the constraints Eq.(15) from the equations of motion are satisfied and $<: g^2\text{tr}(\partial_\mu A^\mu)^2 :>$

and $<: g^3(\partial_\mu \bar{\eta}_a) A_b^\mu d^{abc} \eta_c :>$ vanish. The latter condensate would lead to a violation of the STI for the three gluon vertex[5]. More complicated STI for higher vertices which have not been considered might lead to additional constraints on the four "exotic" condensates in Eq.(14), which are not expectation values of operators already present in the Lagrangian. The DS-equation for the ghost selfenergy however, does not constrain the 'exotic'
$$<: g^4 \text{tr} \left(A_\mu A_\nu A^\mu A^\nu + 2A_\mu A^\mu A_\nu A^\nu + \tfrac{6}{N} A_\mu A_\nu \text{tr}(A^\mu A^\nu) + \tfrac{3}{N} A_\mu A^\mu \text{tr}(A_\nu A^\nu) \right) :>$$
condensate entering it.

The fundamental vertices of the theory were seen to depend on condensates that are not BRS-invariant with coefficient functions that are explicitly gauge parameter dependent. In the cases we have studied[5,14,15], physical quantities constructed from the vertex functions do not show any gauge parameter dependence and the expected coefficients for the gauge-invariant condensates $<: g^2 G_{\mu\nu} G^{\mu\nu} :>$ and $<: m\bar{\psi}\psi :>$ in the effective potential were obtained[14]. We however, cannot demand vanishing expectation values for all the BRS-dependent operators occuring in the OPE of the vertex functions and still have physical condensates.

The situation is reminiscent of that in models which show spontaneous breakdown of a global symmetry, such as chiral symmetry in the linear σ-model: the OPE of chirally invariant correlation functions only depends on condensates such as $<: \sigma^2 + \vec{\pi}\vec{\pi} :>$, which are also chirally invariant; chirally variant correlation functions however, depend on $<: \sigma :>$, $<: \sigma^2 :>$ and $<: \vec{\pi}\vec{\pi} :>$ etc. separately. The ground state one selects has vanishing $<: \vec{\pi}\vec{\pi} :>$. This physically appropriate choice can however, only be verified in quantities that are not symmetric, such as the fermion mass.

In our case, the BRS-variant vertex functions also have a BRS-dependent OPE. Physical (BRS-invariant) quantities will however, only be sensitive to the BRS-invariant condensates, such as $<: g^2 G_{\mu\nu} G^{\mu\nu} :>$ or $<: m\bar{\psi}\psi :>$. For this to be possible, one must have some freedom in the construction of the physical ground state $|\Omega >$, although one demands[17] that $Q_{BRS}|\Omega >= 0$ holds (with the BRS-charge Q_{BRS}). In contrast to a bosonic symmetry, this condition however, does not uniquely specify the state, if other (unphysical) degenerate states $|G_i >$ exist, with $Q_{BRS}|G_i > \neq 0$. In that case, any state

$$|\Omega(\vec{a}) >= |\Omega > + a^i Q_{BRS}|G_i >, \tag{23}$$

will also satisfy $Q_{BRS}|\Omega(\vec{a}) >= 0$ because $Q_{BRS}^2 = 0$ and could serve as a ground state. BRS-invariant quantities will not depend on a particular choice, but the OPE of the vertex functions would. This scenario would imply that scalar zero modes with fermionic statistics and ghost number ± 1 have to exist in the *unphysical* sector of the Hilbert space – the $|G_i >$'s. We could then interpret the result that the OPE of the vertex functions depends on BRS-noninvariant condensates as an indication that such zero modes have to exist.

As originally pointed out by Gribov[18] and recently verified by Zwanziger[19] in an extended calculation, the restriction of the gauge field integration in the path integral to fields within the primary Gribov horizon leads to an effective gluon propagator of precisely the form Eq.(13). The BRS-dependent condensates would then be an indication of the Gribov horizon in covariant gauges, i.e. for zero-modes of the ghosts. The unphysical scalar zero modes with fermionic statistics could even be a more general phenomenon: one also finds them in a completely different nonperturbative approach, which does not fix a covariant gauge. In the field strength approach[20] to QCD they appear as kink-like solutions of zero action (the flips). We would also like to point out that the effective gluon propagator Eq.(13) has an analytic structure in the complex k_0 plane, which is acausal[21] and might reflect the confinement of these modes[1,22].

This interrelation of widely different approaches to a nonperturbative description of the QCD-vacuum we find exciting and hope that further clues to the infrared properties of the theory will begin to emerge.

5. Acknowledgements

We are indebted to Prof. K. Dietrich for his support and continued interest in this work. M.S. would like to thank the organizers for their kind hospitality and the very good atmosphere of this workshop and the participants for numerous enlightening discussions.

6. References

1. U. Häbel, R. Könning, H.-G. Reusch, M. Stingl und S. Wigard, *Zeit. Phys.* **A336** (1990) 423;435.
2. For a review see P. Pascual, R. Tarrach, *QCD: Renormalization for the Practitioner* (Springer Verlag 1984).
3. A.A. Slavnov, *Theor. Math. Phys.* **10** (1972) 99.
   J.C. Taylor, *Nucl. Phys.* **B33** (1971) 436.
4. C. Becchi, A. Rouet, R. Stora, *Phys. Lett.* **B52** (1974) 344; *Commun. Math. Phys.* **42** (1975) 127.
5. A. Streibl, Diplomawork TU-München 1992: *Die Operator-Produkt-Entwicklung für Vertexfunktionen und nicht-störungstheoretische Lösungen der Dyson-Schwinger-Gleichungen der QCD.*
6. G.C. Rossi, R. Sarno, R. Sisto, *I.N.F.N. preprint* **n.888** and **Rom2/F n.22** of May 1992.
7. V.P. Spiridinov, *Mod. Phys. Lett.* **A5** (1990) 653; **A6** (1991) 277(E).
8. J. Ahlbach, M.J. Lavelle, M. Schaden, A. Streibl, *Phys. Lett.* **B275** (1992) 124.
9. M.J. Lavelle, M. Schaden, *Phys. Lett.* **B208** (1988) 297. See however Refs.7,8.
10. J. Ahlbach, Diplomawork TU-München 1992: *Untersuchung der nicht-stö-*

rungstheoretischen Anteile der Vertexfunktionen der QCD mit Hilfe der Operator-Produkt-Entwicklung.

11. M.J. Lavelle, M. Oleszczuk, *Zeit. Phys.* **C51** (1991) 615; *Phys. Lett.* **B275** (1992) 133.
12. F.J. Ynduráin, *Quantum Chromodynamics* (Springer Verlag 1983).
13. T.I. Larsson, *Phys. Rev.* **D32** (1985) 956.
14. M.J. Lavelle, M. Schaden, *Phys. Lett.* **B246** (1990) 4870.
15. M.J. Lavelle, *Phys. Rev.* **D44** (1991) R26.
16. J.M. Cornwall, *Phys. Rev.* **D26** (1982) 1453.
    J. Papavassiliou, *Phys. Rev.* **D41** (1990) 3179.
17. T. Kugo, I. Ojima, *Prog. Theor. Phys.* **60** (1978) 1869; **61** (1979) 294;644.
18. V.N. Gribov, *Nucl. Phys.* **B139** (1978) 1.
19. D. Zwanziger, *Nucl. Phys.* **B323** (1989) 513.
20. P.A. Amundsen, M. Schaden, *Phys. Lett.* **B252** (1990) 265. See also P.A. Amundsen's talk at this workshop.
21. H.M. Fried, talk given at this workshop and *INLN preprint* **92.21** from May 1992.
22. Talk presented by V.N. Gribov at this workshop.

CHIRAL SYMMETRY, BROKEN SCALE INVARIANCE AND THE NUCLEAR EQUATION OF STATE

SERGE RUDAZ

School of Physics and Astronomy, University of Minnesota
Minneapolis, Minnesota 55455 U.S.A.

ABSTRACT

An effective Lagrangian with broken chiral and scale invariance, generalizing the linear sigma model, is suggested that leads to a realistic decription of the saturation properties of symmetric nuclear matter while maintaining the successful low-energy theorems for the interactions of pions and nucleons at low energy.

1. Introduction

The linear sigma model[1] provides a simple field-theoretic realization of the spontaneous breaking of chiral invariance that is a fundamental property of low-energy QCD, and allows for the easy derivation of a number of phenomenologically successful low-energy theorems for pion-pion and pion-nucleon interactions. It is therefore somewhat surprising that even when supplemented with the short-range repulsion arising from the exchange of the ω isoscalar vector meson between nucleons, this model, treated in the mean-field approximation fails to reproduce nuclear matter saturation[2]: the difficulty is readily traced to presence of large attractive three-nucleon forces due to the presence of the σ^3 term in the Lagrangian.

This suggests that an important piece of physics is missing from the pictures. In this talk, I will suggest that this missing ingredient is the classical scale invariance of QCD with massless quarks, broken explicitly by quantum effects (the so-called "trace anomaly"[3]). A new explicit realization of the trace anomaly in the form of an effective Lagrangian for a scalar dilaton or glueball within the framework of the linear sigma model will be put forward: the resulting augmented model allows for a successful description of the saturation properties of symmetric nuclear matter, while at the same time maintaining the successes of the old linear sigma model, given a set of parameters some of whose values are constrained by other physics considerations.

2. The Trace Anomaly and a Chiral Effective Lagrangian

Following Schechter[4] and Migdal and Shifman[5], the trace anomaly of massless QCD, namely

$$\theta_\mu^\mu (x) = \frac{\beta(g)}{2g} \, F_{\mu\nu}^a (x) \, F^{\mu\nu a} (x) \tag{1}$$

is reproduced in an effective Lagrangian framework in the simplest possible fashion, namely through the introduction a single scalar glueball or dilaton field $\phi(x)$. Given a Lagrangian of the form

$$L_{eff} (N, \sigma, \vec{\pi}, \phi, \omega^\mu) = L_0 (N, \sigma, \vec{\pi}, \phi, \omega^\mu) - V_G (\phi, \sigma, \vec{\pi}) \tag{2}$$

consisting of a chiral and scale-invariant part L_0 and an explicitly scale-breaking but still chiral invariant V_G, one can calculate the divergence of the scale current as given by the trace of the "improved" energy-momentum tensor[6]

$$\theta_\mu^\mu (x) = 4V_G (\phi_i) - \sum_i \phi_i \frac{\partial V_G}{\partial \phi_i} \tag{3}$$

$$\equiv 4\varepsilon_{VAC} (\frac{\phi}{\phi_0})^4 \tag{3'}$$

where ϕ_i runs over the scalar fields $\{\sigma, \vec{\pi}, \phi\}$ and ε_{VAC} is the vacuum energy density. Eq. (3) follows simply from Noether's theorem, while the proportionality $\theta_\mu^\mu \propto \phi^4$ in (3') is suggested by the form of the right-hand side of Eq. (1), involving a purely gluonic, color-singlet, scalar, dimension-four operator. The precise form (3') is given in terms of ϕ_0, the vacuum value of the glueball field ϕ, and the proportionality factor is then fixed since at the minimum of the potential $4V_G (\phi_i, min) = 4\varepsilon_{VAC}$.

The physical picture of QCD effects underlying the above choice of L and its split into L_0 and V_G, is as follows: the strong interactions in the gauge sector induce a non-zero gluon condensate in the vacuum, which is presumably related to confinement, and which triggers the spontaneous breaking of chiral symmetry through the

148

formation of a quark condensate. In the effective Lagrangian picture, the glueball potential V_G explicitly breaks the scale invariance with $\langle\phi\rangle = \phi_0$, and this, in turn, allows for spontaneous chiral symmetry breaking to be triggered in L_0. The most general form for $V_G(\phi, \sigma, \vec{\pi})$ involving the chirally invariant combination $\sigma^2 + \vec{\pi}^2$ and satisfying the differential equation in (3) and (3′) is[7]

$$V_G(\phi, \sigma, \vec{\pi}) = A\,\phi^4 \log(\sigma^2 + \vec{\pi}^2) + B\,\phi^4 \log\phi + C\,\phi^4 + D\,\phi^2(\sigma^2 + \vec{\pi}^2) \quad (4)$$

which differs from the glueball potential appropriate to quarkless QCD introduced in Refs. (4) and (5) by the inclusion of the first and last terms in Eq. (4). The complete Lagrangian, written in terms of minimum values $\langle\phi\rangle = \phi_0$ and $\langle\sigma\rangle = f_\pi \equiv \dfrac{\phi_0}{\zeta}$ is of the form

$$L_{eff} = \frac{1}{2}\partial_\mu\sigma\partial^\mu\sigma + \frac{1}{2}\partial_\mu\vec{\pi}\cdot\partial^\mu\vec{\pi} + \frac{1}{2}\partial_\mu\phi\partial^\mu\phi - \frac{1}{4}f_{\mu\nu}f^{\mu\nu}$$

$$+ \bar{N}[\gamma^\mu(i\partial_\mu - g_\omega\omega_\mu) - g(\sigma + i\vec{\pi}\cdot\vec{\tau}\,\gamma_5)]\,N$$

$$+ \frac{1}{2}[G_\phi^2\phi^2 + G_\sigma^2(\sigma^2 + \vec{\pi}^2)]\,\omega_\mu\omega^\mu \qquad (5)$$

$$- \frac{1}{4}\lambda\left(\sigma^2 + \vec{\pi}^2 - \frac{\phi^2}{\zeta^2}\right)^2 - V_G(\phi, \sigma, \vec{\pi})$$

where $f_{\mu\nu} = \partial_\mu\omega_\nu - \partial_\nu\omega_\mu$. This differs from the linear σ-model augmented with ω-exchange considered in Ref. (2) in two important respects: first, the linear sigma model is rendered scale-invariant by the replacement[8] of what is usually f_π in the term proportional to λ by $\dfrac{\phi}{\zeta}$, and by the replacement of an explicit mass term for the ω-meson that would break the scale invariance by the scale invariant quartic interactions[9] involving the couplings G_ϕ^2 and G_σ^2: the mass of the ω is then generated by the breaking of the global chiral and scale symmetries. The parameters G_ϕ^2 and G_σ^2 can then be replaced by m_ω^2

$$= f_\pi^2 (G_\phi^2 \zeta^2 + G_\sigma^2) \text{ and the ratio } r^2 = \frac{G_\sigma^2}{(G_\sigma^2 + G_\phi^2 \zeta^2)}.$$ Note that an interaction of this type was originally introduced on purely phenomenological grounds by Boguta[10].

The specific form of the coefficients A, B, C, D in Eq. (4) is determined by requiring that at the minimum of the combined potential $<\phi> = \phi_0$ and $<\sigma> = f_\pi \equiv \frac{\phi_0}{\zeta}$ and $<\vec{\pi}> = 0$, and that the vacuum energy density in Eq. (3') is given by (introducing the parameter δ)

$$\varepsilon_{VAC} = -\frac{1}{4} B\phi_0^4 (1-\delta) \qquad (6)$$

One then has, referring to Eq. (4), the following relations[7]

$$A = -\frac{1}{2} B\delta$$

$$C = -\frac{B}{4} (1+\delta) - B\ln\phi_0 + B\delta\ln f_\pi \qquad (7)$$

$$D = \frac{1}{2} B\delta\zeta^2$$

Note that stability considerations require $\lambda, B > 0$ while for $\varepsilon_{VAC} < 0$, one needs $\delta < 1$.

Further insight into the meaning of the δ parameter follows upon noticing that the beta-function appearing in the QCD trace anomaly Eq. (1) is proportional to $(1-\frac{2n_f}{3\,3})$ in one-loop order, where n_f is the number of massless quark flavors. This suggests the value $\delta = \frac{4}{3\,3}$ in the present case $n_f = 2$, although one should clearly not put too much faith in this argument. As it turns out, the precise value of δ, as long as it is non vanishing, does not matter much.

3. Nuclear Saturation Properties

The free parameters of the model are therefore δ, B, ζ, r^2, λ and g_ω. The Goldberger-Treiman relation in this model fixes $g = \frac{M_N}{f_\pi}$.

while the vacuum energy is known from independent considerations to be $|\varepsilon_{VAC}|^{\frac{1}{4}}$ = 200-250 MeV; further, while g_ω is rather poorly determined, the convenient parameter $C_\omega^2 = \dfrac{g_\omega^2 M_N^2}{m_\omega^2}$ likely lies in the interval 65-140.

In Ref. 7, the above model was treated in the mean field approximation, with δ fixed at $\dfrac{4}{33}$ and $r^2 = 1$. A satisfactory description of the saturation properties of nuclear matter emerges for choices of parameters such that λ is small: for example, choosing $\lambda = 0.65$, $C_\omega^2 = 50$ and $|\varepsilon_{VAC}|^{\frac{1}{4}}$ = 269 MeV leads[7] to an effective mass at saturation $M_{sat}^* = 0.84 \, M_N$ and to a compression modulus K = 387 MeV, with a binding energy per nucleon of 16.3 MeV and saturation density of 0.153fm^{-3}. The value of K may seem a little high, but in fact the calculated value of the anharmonicity parameter[11] $\dfrac{S}{K}$ = 8.8 is almost consistent with the linear relation postulated by Pearson[12] between K and $\dfrac{S}{K}$ as following from breathing mode data.

The particular value chosen for δ, as long as it is not zero, does not appear to be crucial: factor-of-two changes in δ are absorbed by adjustments in other parameters so that the resulting predictions of saturation properties are essentially unchanged. In an earlier investigation[9], with $\delta = 0$ and $r^2 = 1$ the best possible achievable value of K was of order 700 MeV (the case $\delta = 0$, $r^2 = 0$ was also considered but discarded as it resulted in the even more unphysical result K $\simeq$ 1400 MeV). Thus, the introduction[7] of the modified glueball potential of Eq. (4) with $\delta \neq 0$, leads to a phenomenologically consistent chiral model of nuclear saturation. It is interesting to note that for the preferred (small λ) sets of parameters that lead to the lowest values of K at saturation, the classical glueball field deviates very little from its vacuum value with increasing density: the effect is at most a few percent, and suggests a simplified approximate form[13] to replace the complicated effective Lagrangian just discussed.

The numerical results of the full model described above corresponding to parameter choices with λ small, i.e. $\lambda \lesssim 1$, are recovered to typically better than 10% accuracy in the mean field approximation to a simplified model obtained by simply setting $\lambda = 0$ and $\phi \equiv \phi_0$ in the Lagrangian (5). We have checked[13] that this holds also as r^2 is varied. For example with $r^2 = 0.5$ and $\delta = \frac{4}{3} \frac{1}{3}$, one may compare the predictions of the full model (with $\lambda \lesssim 1$) with those of the approximate model (given in parentheses): $\dfrac{M^*_{sat}}{M_N} = 0.79$ (0.80), $C^2_\omega = 76$ (74), $|\varepsilon_{VAC}|^{\frac{1}{4}} = 259$ MeV (255 MeV), $K = 360$ MeV (332 MeV), $\dfrac{S}{K} = 5.9$ (5.3), corresponding to saturation at $n_0 = 0.16$ fm^{-3}.

In all this, the parameter ζ remains free, and only serves to determine, in the full model, the scalar meson spectrum. For a further discussion of this, see Ref. 7 and 9. It should also be mentioned that terms with explicit chiral symmetry breaking (such as results from non-zero quark masses in QCD) are easily incorporated into the effective Lagrangian: we have checked that in nearly all cases their effects are negligible.

4. Conclusions

We have presented an effective chiral Lagrangian including a scalar glueball that correctly reproduces the low-energy theorems for pion-pion and pion-nucleon scattering and allows at the same time for a phenomenologically successful description of the saturation properties of cold nuclear matter in the mean field approximation. The physics of broken scale invariance as implemented in our model at the hadronic level, is crucial in obtaining this result.

Acknowledgments

It is a pleasure to thank my colleagues Paul Ellis and Erik Heide for a most pleasant and fruitful collaboration in these matters. This work was supported by the U.S. Department of Energy under grant number DE-AC02-83ER40105.

References

1. J. Schwinger, Ann. Phys. **2**, 407 (1957); J. C. Polkinghorne, Nuovo Cimento **8**, 179 (1958); M. Gell-Mann and M. Levy, Nuovo Cimento **16,** 705 (1960).

2. A. K. Kerman and L. D. Miller, "Field Theory Methods for Finite Nuclear Systems and the Possibility of Density Isomerism", in *Proceedings of the Second Relativistic Heavy Ion Summer Study,* (Berkeley, 1974) LBL report LBL-3675 (1974) p. 73.

3. R. J. Crewther, Phys. Rev. Lett. **28**, 1421 (1972); M. S. Chanowitz and J. Ellis, Phys. Lett. **B40**, 397 (1972); J. C. Collins, A. Duncan and S. D. Joglekar, Phys. Rev. **D16**, 438 (1977).

4. J. Schechter, Phys. Rev. **D21**, 3393 (1980).

5. A. A. Migdal and M. A. Shifman, Phys. Lett. **B114**, 445 (1982).

6. E. Huggins, Thesis, Caltech 1962 (unpublished); F. Gursey, Ann. Phys. **24**, 211 (1963); C. Callan, S. Coleman and R. Jackiw, Ann. Phys. **59**, 42 (1970).

7. E. K. Heide, S. Rudaz and P. J. Ellis, "Implications of a Modified Glueball Potential for Nuclear Matter", Minnesota preprint UMN-TH-1101/92, July 1992, to appear in Phys. Lett. B.

8. H. Gomm and J. Schechter, Phys. Lett. **158B**, 499 (1985).

9. P. J. Ellis, E. K. Heide and S. Rudaz, Phys. Lett. **B282**, 271 (1992); Phys. Lett **B287**, 413 (1992) (E).

10. J. Boguta, Phys. Lett. **B120**, 34 (1983).

11. S. Rudaz, P. J. Ellis, E. K. Heide and M. Prakash, Phys. Lett. **B285**, 183 (1992).

12. J. M. Pearson, Phys. Lett. **B271**, 12 (1992).

13. P. J . Ellis, E. K. Heide and S. Rudaz, preprint in preparation.

THE GLUON CONDENSATE IN THE CHIRAL PHASE TRANSITION

GEORGES RIPKA
Service de Physique Théorique
Centre d'Etudes de Saclay
F-91191 Gif-sur-Yvette, France

and

MARTINE JAMINON
Institut de Physique B5
Université de Liège
B-4000 Liège, Belgium

ABSTRACT

Two chiral models which implement the QCD scale anomaly are presented. Both models predict that the gluon consensate remains constant across the chiral phase transition. However the two models differ in the nature of the chiral phase transition, the predicted mass of the σ-meson as well as the behaviour of the pion mass with density.

1. Introduction

Several years ago the Syracuse group [1,2] proposed to implement the QCD scale anomaly in low energy effective lagrangians by introducing an extra scalar field χ, proportional to the gluon condensate, often called the dilaton field:

$$\chi \approx \left\langle G_{\mu\nu}^2 \right\rangle^{1/4} \tag{1}$$

The power of the gluon condensate is chosen such as to give the dilaton field a scale dimension 1: $\chi(x) \Rightarrow \lambda\chi(\lambda x)$. The dynamics of the dilaton field is represented by the lagrangian:

$$\mathcal{L}_\chi = -\frac{1}{2}(\partial_\mu\chi)^2 + V(\chi) \qquad V(\chi) = \frac{b^2}{16}\left[\chi^4 \ln\frac{\chi^4}{\chi_G^2} - \left(\chi^4 - \chi_G^4\right)\right] \tag{2}$$

Throughout we use a euclidean metric with $x_\mu = x^\mu = (\tau, \vec{r})$ and $\gamma_\mu = \gamma^\mu = (\beta, \vec{\gamma})$. The potential $V(\chi)$ has a shape reminiscent of the potential in the linear σ-model with a minimum at $\chi = \chi_G$. Its log term breaks scale

154

invariance and the shape of the potential is chosen such that the divergence of the current s_μ, associated to the scale transformation, should obey the continuity equation equation $\partial_\mu s^\mu \approx \chi^4$. Thus, if we interpret χ as the gluon condensate as in (1), then the continuity equation becomes the QCD scale anomaly equation $\partial_\mu s^\mu \approx \left\langle G^2_{\mu\nu} \right\rangle$ and this is what we mean when we say that the QCD scale anomaly is implemented by the effective action. The kinetic term in (2) is arbitrary and serves only to normalize the χ field. It is chosen such that the lagrangian (2) yields a (glueball) meson with mass $b\chi_G$.

The dilaton field is introduced into effective low energy lagrangians in such a way as to make them scale invariant. For example, the scaled Weinberg lagrangian would be modified as follows:

$$\mathcal{L} = \frac{\chi^2}{\chi^2_G} \frac{f^2_\pi}{4} \, \mathrm{tr}(\partial_\mu U)\left(\partial_\mu U^\dagger\right) - \frac{\chi^3}{\chi^3_G} \frac{f^2_\pi m^2_\pi}{4} \, \mathrm{tr}(U + U^\dagger - 2) + \mathcal{L}_\chi \qquad (3)$$

where $U = \exp(i\vec\theta.\vec\tau)$ is a chiral field and where $\vec\theta(x)$ has scale dimension zero. If χ is set to its vacuum value χ_G, the lagrangian (3) reduces to the Weinberg lagrangian, describing pions with mass m_π in the *non-linear* σ-model. The first term is multiplied by χ^2 so as to make it scale invariant. However the second term, which gives the pion its mass m_π, is multiplied by χ^3 so as to give it the same scale dimension (d=3) as the corresponding term $m\bar{q}q$ of the QCD lagrangian. The effect of the χ field is to modify f_π and the pion mass. They become:

$$f^*_\pi = (\chi/\chi_G) f_\pi \qquad\qquad m^{*2}_\pi = m^2_\pi (\chi/\chi_G) \qquad\qquad \frac{m^{*2}_\pi}{m^2_\pi} = \frac{f^*_\pi}{f_\pi} \qquad (4)$$

However the modified Weinberg lagrangian (3) is nothing but the *linear* σ-model with a potetial $V(\chi)$, given in (1) and which, in the vicinity of χ_G, has the familiar mexican hat shape $V(\chi) \approx (b^2/8)\left(\chi^2 - \chi^2_G\right)^2$. In this sense the scaling of the Weinberg lagrangian is nothing but a substitution of the non-linear σ-model by its linear version in which the dilaton field plays the role of the chiral partner of the pion. This is essentially what is done in some of the more recent implementations of the QCD scale anomaly recently suggested [3,4]. It is in sharp contrast to the Nambu mechanism of spontaneously broken chiral symmetry, in which the scalar field, which is the chiral partner of the pion, represents the chiral condensate $\langle \bar{q}q \rangle$ rather than the gluon condensate. The resolution of this contradiction is the main motivation of the following study.

2. Two scale invariant effective actions

We consider two scaled effective lagrangians. The first, which we refer to as model A, is:

$$\mathcal{L}_A = \bar{q}(-i\partial_\mu\gamma_\mu + m + \varphi U)q + \frac{a^2\chi^2\varphi^2}{2} + \mathcal{L}_\chi \tag{5}$$

where $\mathcal{L}_\chi$ is the dilaton lagrangian (2). The lagrangian (5) may be viewed as a scaled Nambu Jona-Lasinio lagrangian. The first term is only apparently scale invariant because the regularization of the quark loop breaks scale invariance, which may however be recovered by making the cut-off proportional to the dilaton field χ. Let us integrate out the quark fields, redefine the chiral field as $\varphi'U' \equiv \varphi U + m$, and drop the primes. The action of model A becomes then:

$$I_A(\varphi U,\chi) = - \mathrm{Tr}_{\Lambda\chi}\ln(-i\partial_\mu\gamma_\mu + \varphi U) + \int d_4 x \, \frac{a^2\chi^2}{2}((\varphi-m)^2 - m\varphi(U+U^\dagger-2)) + \mathcal{L}_\chi \tag{6}$$

where $\mathrm{Tr}_{\Lambda\chi}$, in the quark loop term, means a trace regularized with a cut-off $\Lambda\chi$. The only scale breaking term of the action (6) is the log term of the dilaton potential $V(\chi)$. By straightforward differentiation we see that $\varphi = \langle \bar{q}q \rangle/a^2$. The action introduces non trivial couplings between the dilaton field χ, which represents the gluon condensate as in (1), and the φ field, proportional to the quark condensate $\langle \bar{q}q \rangle$. The model may therefore be used to study the coupling between the quark and gluon condensates [6]. The field φ represents the constituent quark mass. Hadron masses are expected to vary in proportion to the constituent quark mass so that we expect model A to produce hadrons with masses proportional to the quark condensate $\langle \bar{q}q \rangle$.

There have been suggestions [4,5] that hadron masses should scale as $\langle \bar{q}q \rangle^{1/3}$ rather than as $\langle \bar{q}q \rangle$ as in model A. Such behaviour can be described by another model, which we refer to as model B, and which is defined by the action:

$$I_B(\varphi U,\chi) = - \mathrm{Tr}_{\Lambda\chi}\ln(-i\partial_\mu\gamma_\mu + \varphi U) + \int d_4 x \, \frac{a^2}{2}((\varphi-m)^2 - m\varphi(U+U^\dagger-2))^2 + \mathcal{L}_\chi \tag{7}$$

There is not a very big difference between the actions of the models A and B. We shall see however that the systems they describe have quite different properties. Model B also involves coupled quark and gluon condensates represented by the scalar fields φ and χ. However, in model B, the φ filed is proportional to $\langle \bar{q}q \rangle^{1/3}$ as may be checked by straightforward differentiation. We therefore expect that model B will yield hadron masses peroportional to $\langle \bar{q}q \rangle^{1/3}$ in agreement with the speculated scaling law of hadron masses.

3. The equivalent sigma model

Several properties of systems described by actions such as (6) and (7) can be calculated by deriving an equivalent sigma model. Let $\varphi=\varphi_0$ and $\chi=\chi_0$ be a stationary point. We can expand the action in powers of $\left(\varphi^2-\varphi_0^2\right)$ and $\left(\chi^2-\chi_0^2\right)$. If we expand up to second order in these quantities as well as to second order in $(\partial_\mu\varphi)$, we can derive the following equivalent sigma model [6]:

$$\mathcal{L}_{\sigma\text{-model}}(\sigma,\pi,\chi) = \left[\frac{1}{2}(\partial_\mu\pi)^2 + \frac{m_\pi^2}{2}\,\pi^2 + \frac{1}{2}(\partial_\mu\sigma)^2 + \frac{m_\sigma^2}{8\sigma_0^2}\left(\sigma^2 - \frac{\sigma_0^2}{\chi_0^2}\,\chi^2\right)^2\right]$$

$$+ \frac{1}{2}(\partial_\mu\chi)^2 + \frac{m_{GL}^2}{8\chi_0^2}\left(\chi^2-\chi_0^2\right)^2 \tag{8}$$

It has precisely the form of the scaled σ-model considered by the Syracuse group [1,2]. The fields σ and π are related to the chiral field $\varphi U = \varphi \exp\left(i\gamma_5\vec{\theta}.\vec{\tau}\right)$ by a normalization factor Z:

$$\sigma = \sqrt{Z}\,\varphi \qquad \sigma_0 = \sqrt{Z}\,\varphi_0 = f_\pi \qquad \pi = \sqrt{Z}\,\varphi\theta$$

$$Z = \frac{2\upsilon}{16\pi^2}\left(\ln\frac{\Lambda^2\chi_0^2 + \varphi_0^2}{\varphi_0^2} - \frac{\Lambda^2\chi_0^2}{\Lambda^2\chi_0^2 + \varphi_0^2}\right) \tag{9}$$

Because of the coupling between the φ and χ fields, the glueball mass m_{GL}, appearing in (8), is not equal to $b\chi_G$. Instead, it is given by:

$$m_{GL}^2 = \left(b^2 - \frac{\upsilon\Lambda^4}{4\pi^2}\right)\chi_0^2 \tag{10}$$

(The evaluation of the *physical* glueball mass requires a trivial diagonalization of a 2x2 mass matrix.) The pion mass obeys the Gell Mann-Oakes Renner formula $m_\pi^2 f_\pi^2 = -\,m\langle\overline{q}q\rangle$.

The form (8) of the equivalent sigma model, the normalization (9) of the chiral field and the glueball mass m_{GL} are the same for models A and B. However the models predict different sigma and pion masses:

$$m_\sigma^2 = 4\varphi_0^2 \quad \text{(for model A)} \qquad m_\sigma^2 = 4\varphi_0^2\left(1 + \frac{a^2\varphi_0^2}{f_\pi^2}\right) \quad \text{(for model B)}$$

$$m_\pi^2 = \frac{a^2 m\varphi_0 \chi_0^2}{f_\pi^2} \quad \text{(for model A)} \qquad m_\pi^2 = \frac{a^2 m\varphi_0^3}{f_\pi^2} \quad \text{(for model B)} \quad (11)$$

Thus model A predicts that the scalar meson, which is the chiral partner of the pion, has a mass equal to $2\varphi_0$ as in the Nambu Jona-Lasinio model. However, in model B, the scalar meson has a higher mass, which turns out to be about twice higher, in the 1.5 GeV region. We shall see below that model B predicts a pion mass which decreases with density, whereas, in model A, it increases as it does in the Nambu Jona-Lasinio model [6,7].

There are five model parameters, namely a^2, Λ, m, χ_G and b^2. They are determined by fitting f_π = 93 MeV, m_π = 138 MeV, and by choosing χ_0 = 350 MeV and m_{GL} = 1.3 GeV. The choice of χ_0 corresponds to the QCD sum rule estimate [8] and m_{GL} is chosen close to current glueball mass estimates. This leaves one undetermined parameter which we chose to be the constituent quark mass φ_0 = 450 MeV, close to the value required to fit baryon and meson masses. Table 1 gives the model parameters and some vacuum properties obtained this way:

	a^2	Λ	b^2	χ_G (GeV)	m (MeV)	$\langle\overline{q}q\rangle^{1/3}$ MeV	m_σ (GeV)
model A	0.34	2.08	19.5	0.38	9	- 264	0.9
model B	0.11	2.08	19.5	0.37	9	- 264	1.7

Table 1: Parameters and some vacuum properties of models A and B.

4. Finite baryonic density

By adding a Fermi sea of quarks, we can seek stationary points of the action at finite baryonic density [6]. This allows us to study the quark and gluon condensates as we approach the chiral phase transition. Figs. 1 and 2 show the behaviour of the quark and gluon condensates as a function of density. We see that the gluon condensate remains almost constant so that it does not participate to the dynamics of the chiral phase transition. Model A shows a gradual restoration of chiral symmetry whereas model B shows a sharp phase transition. Fig.3 shows the bahaviour of the pion masses. Model B predicts an increase in the pion mass whereas model B a decrease. Since χ remains almost constant, model A is quite similar to the Nambu Jona-Lasinio model.

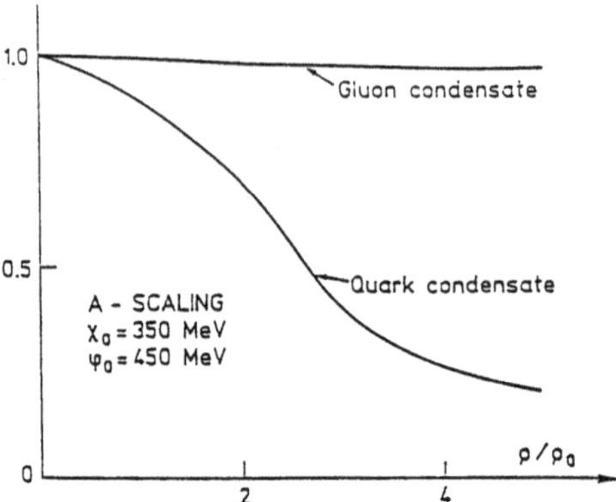

Figure 1: Quark and gluon condesates plotted as a function of density in model A.

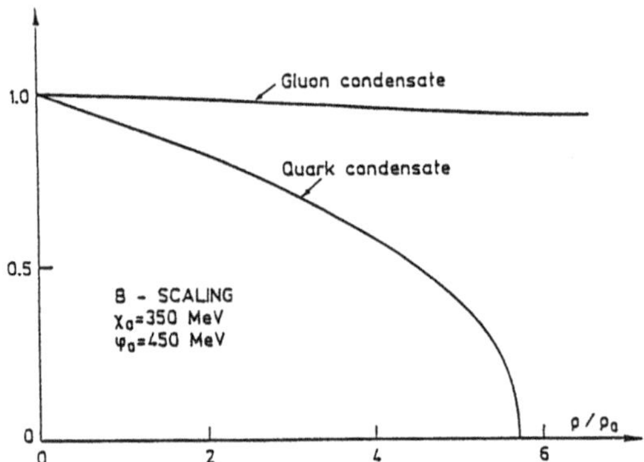

Figure 2: Quark and gluon condesates plotted as a function of density in model B.

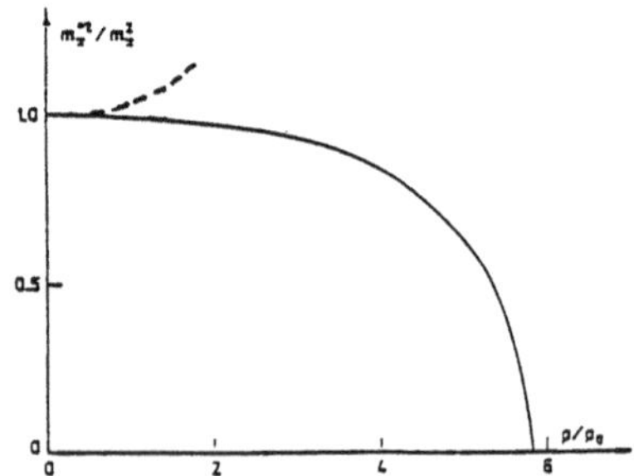

Figure 3: The squared pion mass plotted as function of density. The dashed line is for model A and the full line for model B.

Model B however is different and in fact very close to the one proposed
by Adami and Brown [5] who suggest to make the coupling constant of the
Nambu Jona-Lasinio model inversely proportional to m_σ^2 while keeping the
cut-off constant.

5. References

[1] H.Gomm, P.Jain, R.Johnson and J.Schechter, Phys.Rev. D33 (1986)
    801.
[2] P.Jain, R.Johnson and J.Schechter, Phys.Rev. D38 (1988) 1571.
[3] B.A.Campbell, J.Ellis and K.A.Olive, Nucl.Phys. B345 (1990) 57.
[4] G.E.Brown and M.Rho, Phys.Rev.Lett. 66 (1991) 93.
[5] C.Adami and G.E.Brown, Zeit.Phys. A340 (1991) 93
[6] G.Ripka and M.Jaminon, Saclay preprint SPhT/92-018, to be published
    in Annals of Physics, 1992.
[7] K.Kusaka and W.Weise, University of Regensburg preprint TPR-91-46.
[8] M.A.Schifman, A.I.Vainstein and V.I.Zakharov, Nucl.Phys. B147
    (1979) 385 and 448.

MESON MASSES IN A CHIRALLY SYMMETRIC,

COVARIANT EFFECTIVE QUARK MODEL
WITHOUT FREE QUARKS

S.KREWALD and M.BUBALLA
Institut für Kernphysik, Forschungszentrum
517 Jülich, Germany

ABSTRACT

A new chirally symmetric, covariant effective quark model is presented which suppresses the appearance of free quarks by the same many-body mechanism which spontaneously breaks chiral invariance. Meson masses are evaluated.

In order to improve the knowledge about the structure of mesons, theoretical models based on the underlying quark dynamics are needed which produce both the rest masses and widths of mesons. Since pions play an important role at intermediate energies, chiral symmetry should be taken care of. Likewise, a relativistic treatment appears necessary at intermediate energies, especially when considering decay processes. Furthermore, in effective quark models addressing the widths of mesons, quarks have to be confined in order to suppress unphysical contributions to the widths.

Effective quark models, which are chirally symmetric, covariant and confine, have emerged only very recently [1,2]. In this letter, we want to present a new effective covariant quark model which links the confinement of quarks to the spontaneous breaking of chiral symmetry. As a first step, we derive meson masses and show that mesons are stable, if a quark-antiquark structure is assumed.

Our model is defined by the following effective Lagrangian in Euclidean space:

$$
\begin{aligned}
\mathcal{L}_E(x) \;=\; & \bar{q}(x)\,(i\gamma_E^\mu \frac{\partial}{\partial x^\mu} - m_o)\,q(x) \\
& + \int d^4x_1\, d^4x_2\, d^4x_3\, d^4x_4\; \alpha(x; x_3, x_4; x_1, x_2) \\
& \times \quad (\bar{q}(x_3)\,\gamma_E^\mu\, 1_f \lambda_c^k\, q(x_1))\,(\bar{q}(x_4)\,\gamma_E^\mu\, 1_f \lambda_c^k\, q(x_2))
\end{aligned}
\tag{1}
$$

Here the in- and outgoing quark fields are denoted by q and $\bar{q}$. The first term is the usual kinetic part of the Lagrangian with a small current quark mass m_o, which allows to manifestly break chiral symmetry. Euclidean γ-matrices γ_E^μ are defined by $\gamma_E^o = i\gamma^o$ and $\gamma_E^k = \gamma^k$.

The second integral defines an effective non-local interaction between the quarks which is motivated as follows. When the space-like momenta exchanged between two quarks approach the QCD scale parameter Λ_{QCD}, the running coupling constant becomes very large. Brueckner's theory of nuclear matter has shown that even infinitely strong bare interactions give finite two-fermion scattering amplitudes in a medium, provided one sums an infinite class of diagrams. The resulting two-fermion scattering amplitude is non-local. By analogy, we expect that for low-energy phenomena the underlying QCD Lagrangian leads to an effective Lagrangian which is non-local.

The non-locality has no influence on the internal global symmetries of the Lagrangian. In particular the interaction is invariant under global chiral symmetry transformations $SU(n_f)_L \otimes SU(n_f)_R$, with n_f being the number of quark flavors. In this communication we restrict ourselves to $n_f = 2$. In addition to chiral symmetry our model is symmetric under global $U(1)_V$ and color $SU(3)$ transformations.

Expanding the quark fields in the interaction Lagrangian into a Taylor series around the point x one obtains derivatives of all orders. As a consequence the conserved Noether currents of non-local theories have extra terms in addition to the familiar currents of local theories. A practical method to calculate the corresponding conserved charges can be found in ref.[3].

The non-locality function α can be transformed into Euclidean momentum space:

$$\alpha(x; x_3, x_4; x_1, x_2) = \int (\prod_{j=1}^{4} \frac{d^4 k_j}{(2\pi)^4} \, e^{ik_j \cdot (x - x_j)}) \, \tilde{\alpha}(k_3, k_4; k_1, k_2) \qquad (2)$$

Integrating α over x leads to a four-momentum conserving δ-function. This guarantees the invariance of the action under translations $x \to x + \Delta x$. Since the interaction should be symmetric under the simultaneous exchange of the two incoming and the two outgoing particles, we can assume $\tilde{\alpha}(k_3, k_4; k_1, k_2)$ to be equal to $\tilde{\alpha}(k_4, k_3; k_2, k_1)$.

In the model chiral symmetry is spontaneously broken. This gives rise to a quark selfenergy Σ which is in general a momentum dependent operator. In the Hartree-Fock approximation Σ is given by the following selfconsistency relation:

$$\begin{aligned}
\Sigma(p) &= \mathcal{F}(p^2) + \mathcal{G}(p^2)\, \slashed{p} \\
&= (4n_f n_c) \int \frac{d^4 k}{(2\pi)^4} \, \frac{16}{9} \, \tilde{\alpha}(-k, -p; p, k) \, \frac{m_o + \mathcal{F}(k^2) + \frac{1}{2}(1 + \mathcal{G}(k^2))\, \slashed{k}}{[1 + \mathcal{G}(k^2)]^2 \, k^2 + [m_o + \mathcal{F}(k^2)]^2}
\end{aligned} \qquad (3)$$

In order to solve this equation we have to specify the function $\tilde{\alpha}$. In the local limit $\tilde{\alpha} = const.$, the selfenergy is a scalar constant, which is usually identified with the mass gap between constituent and current quark mass. Other models are motivated by the one-gluon exchange [1], which is a reasonable approximation for large Euclidean transferred momenta. On the other hand there is no theoretical prescription, how to extrapolate the gluon propagator and the strong coupling constant into the time-like regime. As already pointed out it is even very unlikely that the interaction in this region can be expressed in terms of a single effective gluon.

In the present work we want to develop a model which in contrast to models of the Nambu Jona-Lasinio type [4] prevents the emission of quarks into the continuum, and which is on the other hand not too complicated for applications to more complex processes, like hadronic form factors. In particular we would like to have both, the quark selfenergy and the four-quark vertex function, in an analytical form, since both functions enter into the integrands of quark-antiquark loops, triangles, rectangles and so on.

A quite simple structure can be achieved by the assumption of a separable interaction:

$$\tilde{\alpha}(k_3, k_4; k_1, k_2) = \prod_{j=1}^{4} f^{\frac{1}{2}}(k_j^2) \tag{4}$$

Inserting this into Eq.(3) the whole momentum dependence of Σ is contained in a factor $f(p^2)$ which can be written in front of the integral. Thus the selfenergy becomes a function, exactly proportional to $f(p^2)$. Furthermore $\mathcal{G}(p^2)$ vanishes identically:

$$\Sigma(p) \equiv \mathcal{F}(p^2) = \frac{16}{9} f(p^2) (4 n_f n_c) \int \frac{d^4 k}{(2\pi)^4} f(k^2) \frac{m_o + \mathcal{F}(k^2)}{k^2 + \lfloor m_o + \mathcal{F}(k^2) \rfloor^2} \tag{5}$$

The mass M of a physical particle is given by the pole of its dressed propagator. For the quarks of our model this leads to the condition

$$M^2 = (m_o + \mathcal{F}(-M^2))^2 \qquad . \tag{6}$$

The idea to model the absence of free quarks by pole free quark propagators has been suggested almost twenty years ago [5,6] . In the present model, the chiral symmetry breaking many-body mechanism may generate pole free quark propagators, if a function $\mathcal{F}$ is chosen such that Eq.(6) does *not* have a solution. Since poles of a fermion propagator in the complex plane correspond to quasiparticles with a finite lifetime, as is well known in many-body theory, even all complex solutions for M^2 have to be excluded. By demanding that Eq.(6) has no solution in the whole complex plane, large classes of functions, such as polynomials are excluded when specifying $\mathcal{F}$.

A function which fulfills our requirements is given by

$$\mathcal{F}(p^2) = \lfloor \sqrt{(\frac{-p^2 + a^2}{2})^2 + b^4} + \frac{-p^2 + a^2}{2} \rfloor^{\frac{1}{2}} \qquad , \tag{7}$$

with a and b being free parameters. For complex momenta the positive squareroot is continued analytically by making the cut at the negative real axis.

For large Euclidean momenta $\mathcal{F}(p^2)$ behaves like $b^2/\sqrt{p^2}$. This is not sufficient to make the integral in Eq.(5) convergent. We therefore have to introduce a regularization scheme. In this work we are using a four dimensional sharp cutoff Λ.

As a first application we compute meson masses within our model. In this letter we assume the mesons to be pure quark-antiquark states, i.e. we neglect final state interactions between their hadronic decay products [7].

Our approximations are consistent with chiral symmetry. In the chiral limit $m_o = 0$ we find the pion to be massless, as it is demanded by the Goldstone theorem.

Our numerical calculations are performed with the parameters a = 300 MeV and b = 415 MeV. The cut-off employed, Λ = 1000 MeV, as well as the current quark mass m_o = 7.5 MeV are of the same magnitude as in calculations based on the Nambu Jona-Lasinio model. Using these parameters we obtain m_π = 138.3 MeV, m_σ = 666MeV, $m_\rho = m_\omega$ = 781 MeV, and m_{a_1} = 1265 MeV. In contrast to the results obtained in ref. [8], the a_1 meson is stable against quark-antiquark decay. In order to describe the finite widths of vector and axial mesons one has to go beyond the $q\bar{q}$-approximation and to take into account final state interactions in the corresponding strong decay channels [7]. The effect of πa_1-mixing is an enhancement of the pion mass of about 10 %. Without mixing we get m_π = 127.3 MeV.

The main goal of the present work is an effective model for the strong interaction between quarks. The coupling of the electromagnetic or weak interaction to the effective Lagrangian Eq.(1) requires more effort than in local models because of the extra terms in the conserved currents, as pointed out in ref. [3]. As an example we briefly sketch how to couple the electromagnetic interaction to our model.

A Lagrangian which is invariant under local $U(1)$ gauge transformations can be constructed by multiplying the quark fields $q(x_i)$ in Eq.(1) with a Bloch function:

$$q(x_i) \quad \rightarrow \quad exp(ie \int_x^{x_i} d^\mu z\, A_\mu(z))\quad q(x_i) \tag{8}$$

In local models these phases contribute to the kinetic part of the Lagrangian only and are equivalent to minimal substitution. In non-local models the gauge field A_μ also enters into the interaction part. The physical interpretation of this fact is obvious: Because of charge conservation there must be a current which corresponds to the charge transport between the different corners x_i of the non-local vertex. Hence the photon field couples not only to the quark currents, but also to the non-local strong vertices. The explicit form of the resulting electromagnetic vertex function can be derived from the gauge invariant Lagrangian. Details will be discussed elsewhere.

To summarize, we have developed a simple covariant model to illustrate that the same many-body mechanism which spontaneously breaks chiral symmetry may suppress the presence of free quarks.

1. M.R. Frank, P.C. Tandy and G. Fai, *Phys. Rev.* **C43** (1991) 2808.

2. F. Gross and J. Milana, *Phys. Rev.* **D45** (1992) 969.

3. J.W. Bos, J.H. Koch and H.W.L. Naus, *Phys. Rev.* **C44** (1991) 485.

4. Y. Nambu and G. Jona-Lasinio, *Phys. Rev.* **122** (1961) 345.

5. R. Fukada and T. Kugo, *Nucl. Phys.* **B117** (1976) 250.

6. G. Preparata, *Nucl. Phys.* **B80** (1974) 299.

7. S. Krewald, K. Nakayama and J. Speth, *Phys. Lett.* **B272** (1991) 190.

8. V. Bernard and U.-G. Meissner, **Nucl. Phys. A489** (1988) 647.

NONLOCAL CONDENSATES IN A SIMPLE NUCLEON MODEL

P. C. Tandy

Center for Nuclear Research, Department of Physics
Kent State University, Kent, Ohio 44242, USA

ABSTRACT

We review the features and performance of a simple mean field model of
the nucleon in which valence quarks are absolutely confined and couple
to extended $\bar{q}q$ Nambu-Goldstone meson modes in a sigma model format.
The nonlocal features are interpreted in terms of the quark condensates of
the model and their influence upon nucleon properties, including the pion-
nucleon form factor, is discussed.

1. Introduction

Chiral quark-meson models, usually in the linear sigma format, have formed the
basis for many studies of the nucleon, usually at the mean field level.[1] Instead of
treating the mesons as effective elementary fields, it is possible to bridge with some
elements of QCD by generation of the meson sector as the $\bar{q}q$ modes from bosonization[2] of the Nambu–Jona-Lasinio[3] (NJL) mechanism. The chiral symmetry breaking
fermion mass then has its origin in the scalar quark condensate $< \bar{q}(0)q(0) >$ which
also provides the quark-meson coupling constant. The mesons then are point-like.
This path integral bosonization can be extended to endow the mesons with size if the
current-current interaction is mediated by a gluon propagator with finite range.[4,5]
The condensates arising are then nonlocal. We give an overview of a simple one-
parameter confining nucleon model of this type[6] and discuss the condensates and
their influence upon the solution. Our original purpose in investigating this model
was not the study of condensates, but a test of whether a reasonable nucleon solution survives the dynamical nonlocalities associated with extended structure for the
meson sector obtained from the vacuum. The fact that it does suggests an important
physical role for the nonlocality of condensates in baryons.

2. Quark Condensate Nucleon Model

We begin with the Global Color-symmetry Model(GCM)[5] as our link to QCD
degrees of freedom. Unlike the NJL model, the GCM can accommodate absolute
confinement. A chiral quark-meson soliton model arising from the GCM was put
forward some time ago,[4] however no attempt was made at solutions that retain the

intrinsic nonlocalities. The GCM action is defined in Euclidean space as[5]

$$S[\bar{q}, q] = \int d^4x d^4y \left[\bar{q}(x)\gamma \cdot \partial\delta(x - y)q(y) + \frac{g^2}{2} j_\mu^a(x)D_{\mu\nu}(x - y)j_\nu^a(y) \right], \quad (1)$$

where the quark current is $j_\mu^a(x) = \bar{q}(x)\frac{\lambda^a}{2}\gamma_\mu q(x)$. The GCM has global chiral and color symmetries. With phenomenology entering through the effective gluon two point function $D_{\mu\nu}$, this action has provided a successful modeling of meson properties and dynamics.[4,5] The bosonisation procedure[4,7] exactly reformulates the quartic quark term as a functional integration over auxiliary bilocal Bose fields $B^\theta(x, y)$ having the transformation properties of $\bar{q}(y)\Lambda^\theta q(x)$. Here Λ^θ are direct product combinations of color, flavor and spin matrices from Fierz reordering. Vacuum fluctuations in these fields generate the meson sector. To obtain a mean field model for a baryon one can[8] use a canonical transformation to introduce chemical potentials μ to fix the baryon number and flavor. Then integration over the fermions yields the bosonized action[6]

$$S = -TrLn\left[G^{-1}(\mu)G(0)\right] - TrLnG^{-1}(0) + \frac{1}{2}\int d^4(xy)\frac{B^\theta(x, y)B^\theta(y, x)}{g^2 D(x - y)}, \quad (2)$$

with the inverse propagator $G^{-1}(\mu) = e^{\mu x_4}[\gamma \cdot \partial\delta(x - y) + \Lambda^\theta B^\theta(x, y)]e^{-\mu y_4}$. The first term of $S[\mathcal{B}]$ represents valence quarks. The remainder is the action for the vacuum sector which, upon minimization, produces the classical vacuum configurations $B_0^\theta(r) = g^2 D(r)tr[\Lambda^\theta G(r)]$ with $r = x - y$. This is equivalent to the rainbow or ladder approximation to the Schwinger-Dyson equation for the quark self-energy $\Sigma(r) = \Lambda^\theta B_0^\theta(r)$. Only the modes for which Λ^θ is color singlet, flavor singlet, spin-scalar and spin-vector produce non-zero vacuum values and one obtains $\Sigma(p) = i\gamma \cdot p[A(p^2) - 1] + B(p^2)$.

Propagating boson modes are exposed by expansion of the vacuum sector action about the classical configurations. If just the color singlet $(\sigma, \vec{\pi})$ modes are retained a composite extension of the linear sigma model is obtained.[4,6] The quadratic fluctuation term in the vacuum sector produces solutions for the $\bar{q}q$ vertex for free mesons equivalent to the ladder Bethe-Salpeter equation. At zero momentum, chiral symmetry forces an equivalence to the ladder Schwinger-Dyson equation and the scalar self-energy amplitude $B(r)$ describes the internal structure of the lowest mass chiral mesons.[9] The fluctuation chiral modes are represented as[4,9] $B(r)f_\pi^{-1}V(R)$ where $V(R) = \chi(R)e^{i\gamma_5\vec{\tau}\cdot\vec{\phi}(R)/f_\pi}$ and $R = (x + y)/2$. Only the lowest order derivatives of the propagation fields χ and ϕ from the fermion loop $TrLnG^{-1}$ are retained. Expressions for f_π and the χ mass are obtained as loop integrals[6,9] that are automatically regulated by the vertex amplitude $B(p^2)$.

Consider the simplest case without a pion. The scalar model action is

$$S[\chi] = -TrLn\left[G^{-1}(\mu)G(0)\right] + \int d^4R \left[\frac{1}{2}(\partial_\mu\chi)^2 + U(\chi^2)\right], \quad (3)$$

166

where $G^{-1}(0) = \gamma \cdot \partial A(r) + f_\pi^{-1} B(r)\chi(R)$. The one-loop vacuum effective potential $U(\chi^2)$ is easily evaluated and has a mexican hat shape due to chiral symmetry.[6] Standard mean field methods produce the energy functional of the baryon as

$$E[\chi] = 3\epsilon_0[\chi] + \int d^3x \left[\frac{1}{2}\left(\vec{\nabla}\chi\right)^2 + U\left(\chi^2\right)\right]. \tag{4}$$

where ϵ_0 is the energy of the degenerate lowest S state. The field equation for χ is $\delta E/\delta\chi = 0$. The quark eigenenergy is obtained from

$$\left[i\gamma \cdot pA(p^2) + B(p^2)\right] u(\vec{p}) + f_\pi^{-1} \int \frac{d^3k}{(2\pi)^{3/2}} B\left(\frac{p+k}{2}\right) \hat{\chi}(\vec{p} - \vec{k})u(\vec{k}) = 0, \tag{5}$$

where the eigenvalue $p_4 = k_4 = i\epsilon_0$ enters in a nonlinear way. The spatial dependent part of the χ field has been separated out, that is $\chi = f_\pi + \hat{\chi}$. Equation(4) is the standard result except that here, the potential U and the Dirac equation relating ϵ_0 and χ, contain dynamical nonlocalities that express both vacuum quark dressing and meson structure. We enforce confinement through the sufficient condition that there is no mass-shell pole in the vacuum propagator.[12] That is, there should be no real solution to $p^2 + M^2(p^2) = 0$, where $M = B/A$ is the mass function. We use a simple realization[10] of this via use of $g^2 D(p) = (2\pi)^4 \frac{3}{16} \alpha^2 \delta^{(4)}(p)$ to produce confining self-energy amplitudes in convenient closed form.[11] Several studies have obtained useful results with this simple ansatz.[4,13,14] The important result for the present application is that the Dirac equation (5), with a finite range $\hat{\chi}$ field, has only a discrete spectrum.[13] The self-consistent baryon environment induces $\chi \neq f_\pi$ over a finite region of space and creates a constituent mass-shell.[6]

3. Condensates

In the nucleon model defined above, the saddle-point configurations of the bilocal Bose fields $B_0^\theta(r) = g^2 D(r) tr[\Lambda^\theta G(r)]$ are proportional to the nonlocal condensates from a classical treatment of the vacuum. The only condensates that enter are the scalar $< \bar{q}(r)q(0) >= trG(r)$ and the vector $< \bar{q}(r)\gamma_\mu q(0) >= tr\gamma_\mu G(r)$. They generate the self-energy amplitude $\Sigma(r)$. The scalar condensate may be calculated from

$$< \bar{u}(r)u(0) > = \frac{1}{8\pi^2} \int_0^\infty dp p^3 \frac{J_1(pr)}{pr} tr[G(p)], \tag{6}$$

where $J_1(z)$ is the Bessel function and now $r = \sqrt{r_\mu r_\mu}$. This simplifies considerably with the delta function ansatz: the shape is given by $B(r)$ and $< \bar{u}(r)u(0) >= (3\alpha/4\pi^2 r^2)j_2(\alpha r/2)$ is the result with j_2 being the spherical Bessel function. The shape is displayed in Fig. 1 with $\alpha = 1.04 GeV$ from the nucleon solution. The oscillatory behavior arises from the unrealistic feature that here $B(p^2)$ is zero for $p^2 > \alpha^2/4$.[†] The condensate at $r = 0$ is $(113 MeV)^{\frac{1}{3}}$. This is about one half of

---

[†]It is interesting to note that with the Matsubara finite temperature formalism, this implies chiral resoration with $f_\pi = 0$ at $T = \frac{\alpha}{2\pi} = 166 MeV$.

the accepted empirical value and would rise significantly with addition of a realistic tail at high Euclidean momenta for $B(p^2)$. The range of the nonlocality from this simple model can perhaps be better trusted. The 4-space rms radius is $0.76fm$ corresponding to a 3-space size of $0.5fm$ for the Goldstone pion. A recent range estimate[15] for the NJL scalar condensate is $1.7fm$. It is the time-like behavior of $B(p^2)$ that implements confinement in the present model and that information is hidden in a Laplace transform of the shape in Fig. 1. In contrast to color dielectric representations of gluonic effects,[16] confinement here resides in the time-like structure of

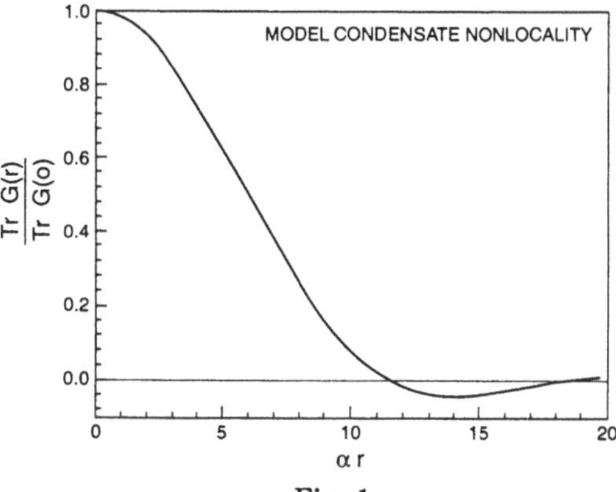

Fig. 1.

the quark condensate $< \bar{q}(x)q(y) >$ and the associated fluctuation field is the chiral partner of the $\bar{q}q$ pion.

4. Nucleon Results

The results for the mean field nucleon calculations[6] are summarized in the Table. The strength α for the gluon propagator is the only free parameter and it is used to fit f_π. The first column retains the confining quark self-energy and the related nonlocal quark-meson coupling. Inclusion of the pion would lower the soliton mass by about 200 MeV. The second column shows the limit of local coupling and no confinement where the amplitudes A and B in the Dirac equation are held constant at their $p^2 = 0$ values with the meson sector left unchanged. The result is a local soliton model with a constant constituent quark mass $M_q = \alpha/2 = 520MeV$, and point coupling via $g = M_q/f_\pi$. The main

	Full Model	Local Limit	Expt.
α (GeV)	1.04	1.04	
ϵ (MeV)	390	434	
χ_{PE} (MeV)	119	119	
χ_{KE} (MeV)	256	82	
m_χ (MeV)	465	465	
M_s (MeV)	1359	1406	1086
R_{rms} (fm)	0.79	1.0	0.83
g_A	1.23	1.42	1.24
f_π (MeV)	93	93	93
Z	1.8	2.0	

effect of the removal of the confinement mechanism and nonlocal coupling is a 25% increase in the rms radius of the baryon density. This is the result of two mechanisms. The confining self-energy causes the quark states to fall off faster with distance than the exponential behavior typical of bound states. The dynamical nature of the coupling vertex provided by $B(\frac{p+k}{2})$ in (5) introduces an energy-dependent and finite range spatially-dependent coupling strength. For low three-space quark momenta, the nonlocal model has much stronger coupling and the reverse is true for high momentum components. Both mechanisms induce a more compact behavior for the baryon density.

For the axial coupling constant, we use a simple minimal substitution approximation[6] with produces

$$g_A = \sum_j{}' Z_j^{-1} \int d^3p\, \bar{u}_j(\vec{p}) \left\{ \frac{\partial}{\partial p_3} i\gamma \cdot p A(p^2) \right\} \gamma_5 \tau_3 u_j(\vec{p}), \tag{7}$$

where $p = (i\epsilon_j, \vec{p})$, and the spin-flavor summation over j is weighted by the occupation probabilities of the standard $SU(4)$ valence quark model. The dependence upon quark momentum is a consequence of the quark self-energy dressing. Since g_A is usually overestimated in local models, the reduction produced by the confining vacuum condensate $< \bar{q}(x)q(y) >$, and evident in the Table, is an interesting phenomenon. The pion-nucleon vertex may be calculated and is given by

$$\vec{V}(\vec{q}) = \frac{-i}{f_\pi \sqrt{2\omega_q}} \sum_j{}' \frac{1}{Z_j} \int \frac{d^3p, k}{(2\pi)^3} \bar{u}_j(\vec{p}) B\left(\frac{p+k}{2}\right) \frac{\chi(\vec{p}-\vec{k}-\vec{q})}{f_\pi} \gamma_5 \vec{\tau} u_j(\vec{k}). \tag{8}$$

The extracted form factor of the confining nonlocal model is plotted in Fig. 2 compared to the local non-confining limit and the simple MIT bag result. The hardening of the form factor due to the nonlocalities is the opposite of what one expects for extended mesons. The dominant effect in (8) comes from the compact confined quark states rather than the amplitude B. The standard coupling constant $g_{\pi NN}$ from the $q = 0$ limit may be related to g_A by using the Dirac equation and we find that the full model contains the Goldberger-Trieman relation $f_\pi g_{\pi NN} = M_N g_A$.

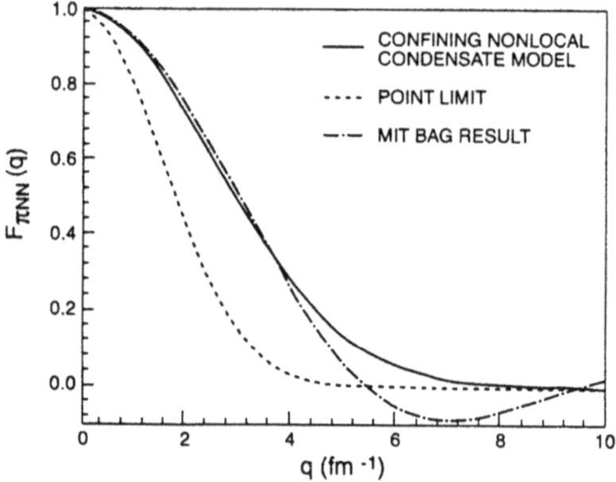

Fig. 2.

5. Summary

The field content of this mean field model is quite primitive, but the dynamics is rather novel. The extended Goldstone $\bar{q}q$ modes are dynamicaly related to nonlocal condensates which introduce a 3-space dependence and a self-consistent energy-dependence for the quark-meson vertex. A valence quark environment induces a change in the vacuum scalar condensate which can overcome the absence of a vacuum mass-shell and produce a constituent mass-shell. The nonlocalities of the model tend to improve the results for a number of nucleon properties. It would be of interest to see whether these mechanisms survive with a more realistic representation of the quark propagator particularly in the ellusive time-like region. A recent confining model[11] that includes the important dressing of the quark-gluon vertex is of interest in this regard.

Acknowledgements

This work was supported in part by the National Science Foundation under Grant Nos. PHY88-05633 and PHY91-13117. We thank V. K. Mishra and D. Kahana for helpful discussions and S. Banerjee for assistance with calculations.

References

1. For a review see M. C. Birse, *Prog. Part. Nucl. Phys.* **25** (1990) 1.

2. T. Eguchi, *Phys. Rev.* **D14** (1976) 2755.

3. Y. Nambu and G. Jona-Lasinio, *Phys. Rev.* **122** (1961) 345; **124** (1961) 246.

4. R. T. Cahill and C. D. Roberts, *Phys. Rev.* **D32** (1985) 2419.

5. J. Praschifka, C. D. Roberts, and R. T. Cahill, *Phys. Rev.* **D36** (1987) 209; J. Praschifka, R. T. Cahill and C. D. Roberts, *Int. J. Mod. Phys.* **A4** (1989) 4929.

6. M. R. Frank and P. C. Tandy, *Phys. Rev.* **C46** (1992) 338.

7. E. Shrauner, *Phys. Rev.* **D16** (1977) 1887; H. J. Munczek, *Phys. Rev.* **D25** (1982) 1579.

8. M. R. Frank, P. C. Tandy, and G. Fai, *Phys. Rev.* **C43** (1991) 2808.

9. C. D. Roberts, R. T. Cahill and J. Praschifka, *Ann. Phys. (N.Y.)* **188** (1988) 20.

10. H. J. Munczek and A. M. Nemirovsky, *Phys. Rev.* **D28** (1983) 181.

11. C. J. Burden, C. D. Roberts and A. G. Williams, *Phys. Lett.* **B285** (1992) 347; see also C. D. Roberts, these proceedings.

12. J. C. Polkinghorne, *Nucl. Phys.* **B93** (1975) 515; C. D. Roberts, A. G. Williams and G. Krein, *Int. J. Mod. Phys.* **A7** (1992) 5607.

13. P. C. Tandy and M. R. Frank, *Aust. J. Phys.* **44** (1991) 181.

14. C. M. Shakin, *Ann. Phys.* **192** (1989) 254.

15. D. Kahana and M. Lavelle, *Phys. Lett.* **B275** (1992) 129.

16. G. Fai, R. J. Perry and L. Wilets, *Phys. Lett.* **B208** (1988) 1; G. Krein, P. Tang, L. Wilets and A. G. Williams, *Phys. Lett.* **B212** (1988) 362.

Nambu – Jona-Lasinio – Models
of chiral symmetry breaking

Detlev Bückers *and* Herbert Müther
Institut für Theoretische Physik, Universität Tübingen
Auf der Morgenstelle 14, D–7400 Tübingen, Germany

ABSTRACT

A chiral invariant Lagragian of Nambu – Jona-Lasinio (NJL) type with a momentum dependent quark–quark interaction is used to calculate effective quark masses within a Hartree–Fock mean–field approximation. This approach is compared with models using a rigid cutoff in momentum space for regularization.

1. Introduction

Due to the strong interaction and the complex structure of the QCD vacuum it is not possible to calculate quark self–energy contributions and bound–state problems in perturbation theory. Thus, we have to consider effective theories which allow an approach to non-perturbative aspects of QCD.

One of these effective theories which has received increasing attention in the last decade is a fermionic model proposed by Y. Nambu and G. Jona-Lasinio (NJL model) [1,2] in the early sixties. It has been applied to calculations in the non-perturbative sector of QCD [3,4,5,6]. One of the basic properties of this model is a chiral invariant four–fermion point like interaction of simplest type. Like other effective theories it is not renormalizable. Usually, a rigid cutoff in momentum space is used for regularization. Unfortunately, this cutoff turns out to be rather small compared to the scale given by the meson masses – even in flavor $SU(2)$ – and all results depend strongly on the chosen cutoff. Guided by the "running" coupling constant of perturbative QCD we use instead of a rigid cutoff a momentum dependent quark–quark interaction which acts like a form–factor in the gluonic sector. The self energy of the quarks is evaluated within a Hartree–Fock mean field approximation.

2. An effective Lagrangian

Let us consider an effective Lagrangian $\mathcal{L}$ of NJL type [1,2]

$$\mathcal{L} = \bar{\psi}\left[i\not{\nabla} - m\right]\psi + \mathcal{L}_{int} \tag{1}$$

with a chiral invariant four-fermion-point-interaction $\mathcal{L}_{int}$

$$\mathcal{L}_{int} = \mathcal{L}_{int}^{dir} - \mathcal{L}_{int}^{exch} \tag{2}$$

where the direct part $\mathcal{L}_{int}^{dir}$ of the interaction $\mathcal{L}_{int}$ is given by

$$\mathcal{L}_{int}^{dir} = G_{\sigma,\pi}(q^2)\left[(\bar{\psi}\mathbf{1}^J\lambda\psi)^2 + (\bar{\psi}i\gamma_5\tau\lambda\psi)^2\right] - G_\omega(q^2)\left[(\bar{\psi}\gamma_\mu\mathbf{1}^J\lambda\psi)^2\right] \tag{3}$$

while the exchange term $\mathcal{L}_{int}^{exch}$ is determined by Fierz transforming $\mathcal{L}_{int}^{dir}$. Here we use a quark–quark coupling $G_\eta(q^2)$ with

$$\eta = \begin{cases} \sigma,\pi & \text{scalar and pseudo scalar channel} \\ \omega & \text{vector channel} \end{cases} \tag{4}$$

which depends on the momentum transfer q between the interacting fermions [6]. λ denotes the Gell–Mann $SU(3)$ generators of color space and τ the corresponding generators of $SU(n_f)$ flavor space. For $n_f = 2$ they are identical to the well–known isospin matrices. $\mathbf{1}^J$ is the unit operator in flavor space. ψ represents the quark spinor field. In the limit of vanishing quark current masses $m \to 0$ this Lagrangian is invariant under a chiral transformation if we require $G_\sigma(q^2) \equiv G_\pi(q^2) \equiv G_{\sigma,\pi}(q^2)$.

The usage of an momentum dependent coupling constant $G_\eta(q^2)$

$$G_\eta(q^2) = g_\eta \times \Gamma_\eta(q^2)) \tag{5}$$

is guided by the QCD "running" coupling constant. It is well accepted that the quark–quark interactions becomes weak for large momentum transfers while it gets strong for small momentum transfers.

While it is not possible to derive the functional dependence of $G_\eta(q^2)$ from a more fundamental theory, we make a phenomenological ansatz for $\Gamma_\eta(q^2)$ which reflects the behavior discussed above in dependence of the exchange momentum q.

$$\Gamma_\eta(q^2) = \exp(-\frac{q^2}{\Delta_\eta^2}) \tag{6}$$

Like g_η – giving the overall coupling strength – Δ_η are just model parameters determining the momentum dependence of the qq-interaction. In principle, the functional $\Gamma_\eta(q^2)$ itself is a model parameter. The chosen one turns to be well suited in this model framework. The scalar density $< \bar{u}u > \simeq < \bar{d}d > \simeq -(250GeV)^3$ for instance cannot be reproduced with a logarithmic functional suggested by perturbative QCD. This is not surprising because the point like effective interaction in this framework is quite different from a gluon exchange potential.

3. Effective Quark Masses

In isospin–flavor–space $SU(2)$ the current quark masses $m \simeq 5-7 MeV$ of up and down quarks are small compared with effective quark masses $M^* \simeq 300-350 MeV$ used in constituent quark models. These constituent quark masses contain self–energy contributions $\Sigma(p)$ which are of non-perturbative origin.

$$M(p) = m + \Sigma(p) \qquad (7)$$

Although the interaction terms of our Lagragian are chiral invariant, chiral symmetry is dynamically broken by the self–energy contributions $\Sigma(p)$.

In order to calculate the quark self–energy contributions in a Hartree–Fock mean–field approximation we include the Fock exchange terms which are calculated by Fierz–transforming the interaction terms $\mathcal{L}_{int}^{dir}$ of our Lagrangian $\mathcal{L}_{int}$

$$(\bar{\psi}\mathbf{1}^f\lambda\psi)^2 \xrightarrow{Fierz} \frac{4}{9n_f}(\bar{\psi}\mathbf{1}^f\mathbf{1}^c\psi)^2 + \cdots \qquad (8.a)$$

$$(\bar{\psi}i\gamma_5\tau\lambda\psi)^2 \xrightarrow{Fierz} -\frac{8}{9}\frac{(n_f^2-1)}{n_f^2}(\bar{\psi}\mathbf{1}^f\mathbf{1}^c\psi)^2 + \cdots \qquad (8.b)$$

$$(\bar{\psi}\gamma_\mu\mathbf{1}^f\lambda\psi)^2 \xrightarrow{Fierz} \frac{16}{9n_f}(\bar{\psi}\mathbf{1}^f\mathbf{1}^c\psi)^2 + \cdots \qquad (8.c)$$

Non–scalar terms in flavor or color space are not denoted here because their vacuum expectation values vanish in the HF approximation, i.e.

$$< (\bar{\psi}\mathbf{1}^f\lambda\psi) >_{HF} =< (\bar{\psi}i\gamma_5\tau\lambda\psi) >_{HF} = \cdots =< (\bar{\psi}\gamma_\mu\mathbf{1}^f\lambda\psi) >_{HF} = 0 \qquad (9)$$

Thus, only Fock exchange terms contribute to the self-energy contributions $\Sigma(p)$. $\mathcal{L}_{int}^{dir}$ contains only non-scalar terms.

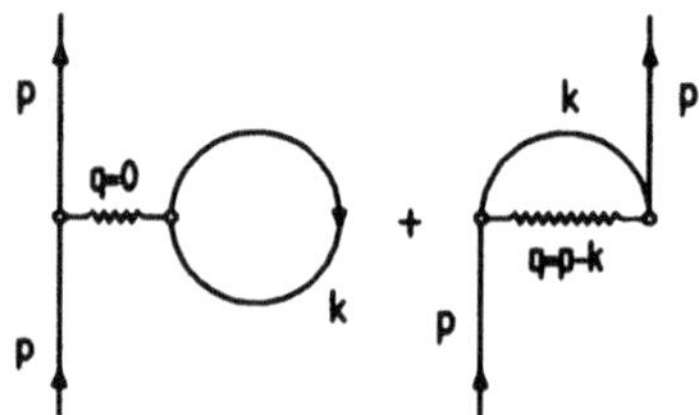

FIG. 1 Hartree–Fock diagrams taken into account for the calculations of the self-energy contributions $\Sigma(p)$.

With the self energy $\Sigma(p)$ given by

$$\Sigma(p) = i \, \mathrm{tr} \int \frac{d^4k}{(2\pi)^4}\left[g_{\sigma,\pi}\frac{8n_f^2-4n_f-8}{9n_f^2} \times \frac{\Gamma_{\sigma,\pi}(p-k)}{\not{k}-m-\Sigma(k)} + g_\omega\frac{16}{9n_f} \times \frac{\Gamma_\omega(p-k)}{\not{k}-m-\Sigma(k)}\right] \qquad (10)$$

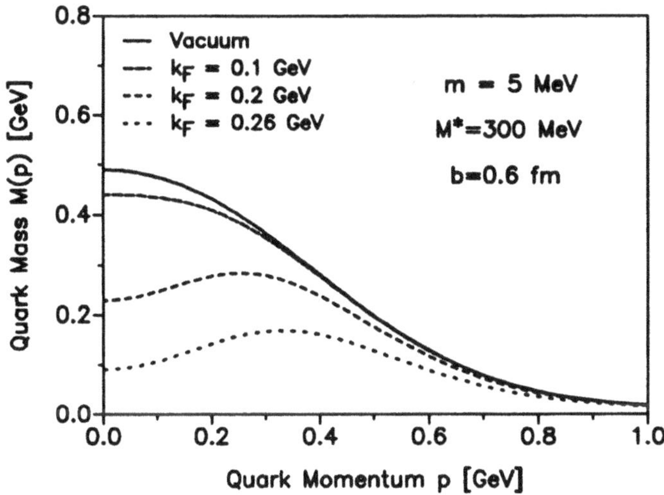

FIG. 2 Self-consistent solutions of the non-covariant gap equation (14) in isospin flavor space SU(2) for the effective quark mass $M(p) = m + \Sigma(p)$ in systems with various Fermi momenta k_F. The mean constituent quark mass M^* has been calculated by weighting the momentum dependent vacuum quark mass $M(p)$ with an $0s$–oscillator wave function. b denotes the oscillator width.

the effective quark mass is obtained for flavor $SU(n_f = 2)$

$$M(p) = m + \int \frac{d^4k}{(2\pi)^4} \left[\frac{32}{3} G_{\sigma,\pi}((p-k)^2) + \frac{64}{3} G_\omega((p-k)^2) \right] \frac{M(k)}{k^2 + M^2(k)} \tag{11}$$

This self consistent gap equation yields finite non-trivial solutions for the quark self–energy contributions $\Sigma(p)$ w i t h o u t the need of a rigid cutoff in the single particle momenta.

4. Comparison with the "cutoff" model

In order to compare this model with the "cutoff"–model we consider a system of quark matter. Here we have chosen a non–covariant formulation because in an infinite quark–medium where positive energy eigenstates up to a Fermi momentum k_F are occupied, covariance is already destroyed by the non–covariant Fermi momentum k_F which is a vector in three–dimensional euclidean space.

$$G(|\vec{q}|^2) = g \times \exp\left(-\frac{|\vec{q}|^2}{\Delta^2}\right) \tag{12}$$

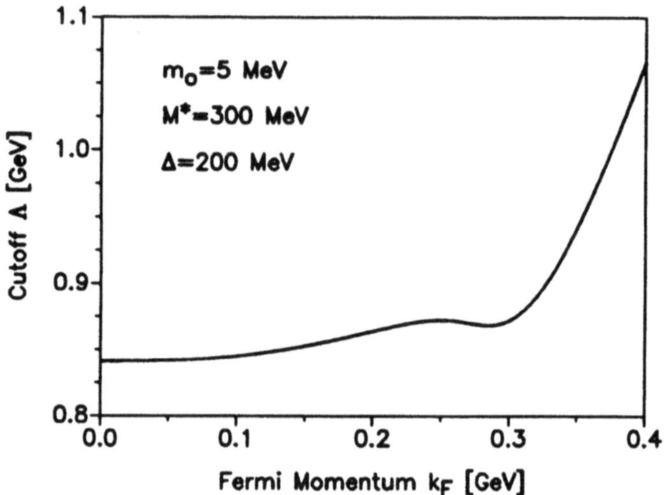

FIG. 3 The self-consistent solutions for the effective quark in an infinite quark matter yield by the present approach are compared with results obtained in the conventional NJL model with a cutoff Λ. Therefore Λ is fitted under the condition that the effective masses predicted by the model proposed here $M_\Lambda(k_F) =$.

The effective quark mass in a system with Fermi momentum k_F is then given self–consistently by

$$M(p, k_F) = m + \frac{16}{3}(n_f^2 - 2) \int_{k_F}^{\infty} \frac{d^3p'}{(2\pi)^3} \frac{M(p', k_F)}{\sqrt{p'^2 + M^2(p', k_F)}} G(|\vec{p} - \vec{p}|^2) \tag{13}$$

Numerical solutions of this self–consistent gap equation are plotted in Figure 2 for several Fermi momenta k_F and in the vacuum limit $k_F \to 0$.

In a medium with a non-vanishing quark–density the effective quark mass is not only reduced with increasing density but the momentum dependence of the quark mass is effected, too. If we compare the results of our approach to those obtained by the conventional method with a sharp cutoff Λ in momentum space – where the gap equation is given by

$$M_\Lambda(k_F, n_F = 2) = m + 24 g_\Lambda \int_{k_F}^{\Lambda} \frac{d^3p'}{(2\pi)^3} \frac{M_\Lambda(k_F)}{\sqrt{p'^2 + M_\Lambda^2(k_F)}} \tag{14}$$

it turns out that the medium dependence of the effective quark masses predicted by both models are nearly equivalent up to a as long as the Fermi momentum k_F

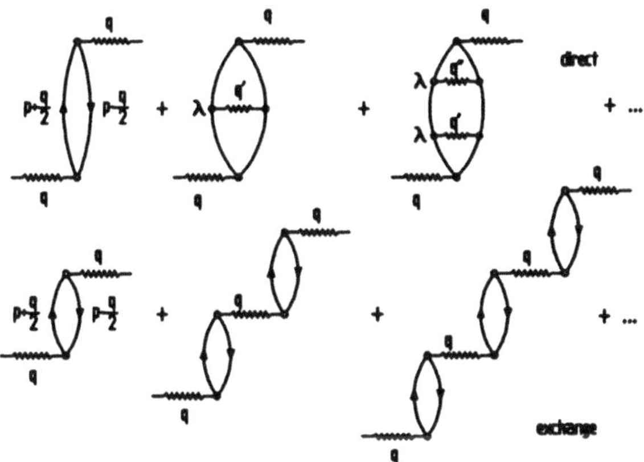

FIG. 4 Ladder approximation

is smaller then $\Lambda/3$. In systems with densities above the conventional NJL model seems to predicts a too strong decrease of the effective quark masses which can be corrected by the usage of a medium dependent cutoff $\Lambda(k_F)$ as shown in Figure 3. The influence of the cutoff seems to be visible at rather small energies.

5. Mesons

For the solutions of the Bethe–Salpeter equation in the ladder approximation we want to give just a brief sketch of the calculation of the bound state problem in the framework of this momentum dependent model.

In the case that $g_{\sigma,\pi}$ is set to zero, the repulsive vector part of the direct channel would not lead to bound states. Thus, we only have to consider exchange terms obtained by Fierz transformation. Non-scalar parts in color space are projected out by the requirement that we want to obtain physical, i.e. colorless mesons. Therefore the polarization in lowest order is given by

$$\Pi_\eta^{(0)}(q) = 2ig(q) \ \mathrm{tr} \ \int \frac{d^4p}{(2\pi)^4} \ O_\eta S_F'(p+\tfrac{q}{2}) O_\eta S_F'(p-\tfrac{q}{2}) \tag{15}$$

where $S_F'(p)$ is the propagator of the dressed quark

$$iS_F'(k) = \frac{i}{\not{k} - m - \Sigma(k)} \tag{16}$$

and O_η are operators concerning the different mesonic modes.

If one includes the scalar and pseudo-scalar part by using a $g_{\sigma,\pi}$ different from zero one has to consider contributions from the direct part with a quite complicated momentum dependence. While the direct channel becomes attractive second, forth order diagrams can produce colorless bound states. This scenario has not been studied by us up to now.

6. Conclusion

Effective theories like that we presented here are not renormalizable. Thus, they must be regularized either by a sharp cutoff or a form-factor in momentum space. The usage of a momentum dependent coupling constant as a regularizing form-factor seems to arise quite naturally from QCD.

The NJL model has been quite useful because it allows the calculation of non–perturbative aspects of QCD. Some of it's limitations can be circumvented by using a momentum–dependent interaction instead of a rigid cutoff. This concerns calculations in a larger flavor space than isospin SU(2) with respect to effective quark masses even larger than the momentum cutoff and those in hot and dense matter where the fermi momentum k_F is not small compared to the cutoff Λ.

7. Acknowledgements

These investigations have been partly supported by the Deutsche Forschungs-Gemeinschaft (DFG). This support is gratefully acknowledged.

8. References

1. Nambu, Y., Jona-Lasinio, G., *Phys. Rev.* **122** (1961) 345.
2. Nambu, Y., Jona-Lasinio, G., *Phys. Rev.* **124** (1961) 246.
3. Bernard, V., Meissner, U.G., Zahed, I., *Phys. Rev. Lett.* **59** (1987) 966.
4. Henley, E.M, Müther, H., *Nucl. Phys.* **A513** (1990) 667.
5. Buck, A., Alkofer, R., Reinhard, H., *Phys. Lett.* **B** to be published.
6. Bückers, D., Müther, H., *Nucl. Phys.* **A523** (1991) 629.

PHOTO–NUCLEON PROCESSES IN CHIRAL PERTURBATION THEORY

V. BERNARD

Centre de Recherches Nucléaires et Université Louis Pasteur de Strasbourg
Physique Théorique
BP 20 Cr, 67037 Strasbourg Cedex 2, France

ABSTRACT

Chiral perturbation theory allows to investigate the constraints imposed by the spontaneously broken chiral symmetry on processes involving low momenta. I consider here two particular reactions which have attracted considerable experimental attention over the last years. These are the production of neutral pions from protons by real and virtual photons in the threshold region and the electromagnetic polarizabilities of the nucleon. The latter are a pure loop effect in the one–loop approximation and serve as a good probe of our understanding of the chiral sector of QCD.

1. Introduction

Over the last few years, experimental techniques in photo–nuclear physics have reached an unprecedented level of accuracy. This has allowed to study in detail two hitherto not too well determined quantities, namely the threshold production of neutral pions as well as the electric and magnetic polarizabilities of the neutron and the proton. There exist many models which have some links to QCD and are used to make theoretical predictions of these processes, but they are not sufficiently accurate to deepen our understanding of the structure of the nucleon as revealed by these photo–nuclear experiments. However, almost parallel to the experimental improvements, a high precision tool has been developed also on the theoretical side–chiral perturbation theory (CHPT). It allows for a systematic expansion of the QCD Green functions in power of the small quark masses and external momenta (small compared to the typical chiral symmetry breaking scale $\Lambda_\chi \approx 1$ GeV). While CHPT enjoys considerable success in the meson sector[1], a systematic formulation involving baryons has only become available since the work of Gasser *et al.*[2] The starting point is an effective Lagrangian of interacting nucleons, pions and external sources like *e.g.* the photon. To leading order one constructs tree diagrams which leads to current algebra results. However, to restore unitarity and for other reasons one has to include pion loops. In general these are suppressed by powers of the external momenta and/or quark masses. In the one–loop approximation, the effective Lagrangian has the generic form

$$\mathcal{L}_{eff} = \mathcal{L}_{\pi N}^{(1)} + \mathcal{L}_{\pi N}^{(2)} + \mathcal{L}_{\pi N}^{(3)} + \Delta\mathcal{L}_{\pi N}^{(0)} + \Delta\mathcal{L}_{\pi N}^{(1)} + \mathcal{L}_{\pi\pi}^{(2)} + \mathcal{L}_{\pi\pi}^{(4)} \tag{1}$$

The superscript (i) denotes the low energy dimension (number of derivatives and quark mass insertions). The first term is the non–linear σ–model coupled to nucleons, the next two terms embody the counterterms which one needs to perform the renormalization of the pion loops. $\Delta\mathcal{L}^{(0,1)}_{\pi N}$ are necessary to give a finite nucleon mass ($\mathring{m}$) and axial–vector coupling constant ($\mathring{g}_A$) in the chiral limit. The last two terms concern only the meson sector, $i.e.$ the interacting Goldstone bosons. To lowest order in momenta and quark masses, the $\pi N\gamma$ Lagrangian is specified by four parameters, $\mathring{m}$, $\mathring{g}_A$, F (the pion decay constant in the chiral limit) and M^2, the leading term in the quark mass expansion of the pion mass squared. Let me now apply this formalism to two particular processes.

2. Threshold production of neutral pions by photons

20 years ago, Vainhstein and Zakharov and de Baenst[3] derived a so–called low energy theorem for the electric dipole amplitude $E_{0+}(\gamma p \to \pi^0 p)$ at threshold, which states that at next–to–leading order in powers of $\mu = M_\pi/m \approx 1/7$ (M_π is the pion mass, m the nucleon mass) all coefficients are given in terms of a few measurable quantities. The numerical prediction for this S–wave threshold multipole (remember that as the pion momentum goes to zero, $d\sigma/d\Omega \sim (E_{0+})^2$) is $-2.3 \cdot 10^{-3}/M_{\pi+}$. This result was challenged by the Saclay and Mainz[4] data, which gave numbers by factors $4\cdots10$ below the LET prediction (with relatively small error bars). This spurred hectic theoretical activity, however, later on it was noted that both experimental analyses were flawed. The Saclay group had subtracted some large rescattering correction and the Mainz group ignored the constraints from the total cross section. The reanalyzed data seem to converge around $E_{0+} = -(2.0 \pm 0.2) \cdot 10^{-3}/M_{\pi+}$, close to the old LET value.

Clearly, having a prediction not only from the leading order term based solely on PCAC, gauge invariance, crossing and some other seemingly harmless assumptions is quite amazing. Therefore, it appears natural to check this expansion within CHPT. For doing that, one has to consider approximatively 60 Feynman diagrams. Two of these lead to a modification of the LET at next–to–leading order. Thus, the correct form of the LET for $\gamma p \to \pi^0 p$ at threshold in QCD reads[5]

$$E_{0+}^{\pi^0 p}(s_{\text{th}}) = -\frac{eg_{\pi N}}{8\pi m}\mu\left\{1 - \left[\frac{1}{2}(3 + \kappa_p) + (\frac{m}{4F_\pi})^2\right]\mu + \mathcal{O}(\mu^2)\right\} \tag{2}$$

where the new term $\sim m^2/F_\pi^2$ stems entirely from the so–called triangle diagram and its crossed partner. The effect is due to the pions in the loop – in the chiral limit this diagram diverges inducing a discontinuity under $s \leftrightarrow u$ crossing. Previous derivations tentatively (and wrongly) assumed continuity. Clearly, the LET (2) is not sufficient – it is a slowly converging series in μ and can not be stopped at order μ^2. In Fig.1 is shown the result for the total cross section in the threshold region for the full one–loop calculation (no expansion in μ). It agrees rather well with the data, only at very low photon energies some S–wave strength is lacking (this can be traced back to a particular two–loop effect as pointed out in Ref.6). The detailed

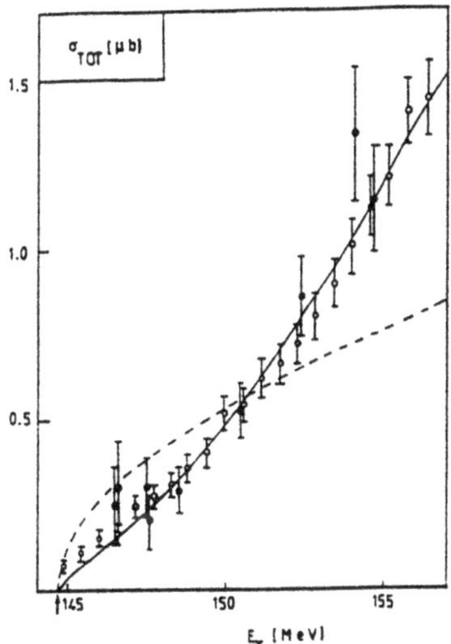

Fig. 1:   Total cross section for $\gamma p \to \pi^0 p$ in comparison to the Saclay and Mainz data.

analysis of the one–loop results for neutral and charged pion photoproduction is given in Ref.7.

A related issue is the electroproduction of pions in the threshold region. In particular, knowledge of the isospin odd electric dipole amplitude $E_{0+}^{(-)}$ allows one to extract the axial root mean square radius of the nucleon. The latter can also be determined by (anti)neutrino proton scattering. If one parametrizes the axial form factor by a dipole form, electroproduction experiments systematically give a ten percent larger cut–off mass (smaller radius) than the neutrino experiments. In Ref.8 it was shown that this effect is real. Pion loops generate a term which generalizes the ancient low energy theorem of Nambu, Lurié and Shrauner to the case of non–vanishing pion mass so that in fact, previously one extracted the modified axial radius $\tilde{r}_A^2$:

$$\tilde{r}_A^2 = r_A^2 + \frac{3}{64 F_\pi^2}\left(1 - \frac{12}{\pi^2}\right) \tag{3}$$

The second term is a model–independent correction which is due to the fact that one can not interchange the order of taking the chiral limit $(M_\pi \to 0)$ and the derivative at zero momentum transfer. This correction amounts exactly to the ten percent reduction of the axial radius previously found from the electroproduction data. Taking it into account as one should, one finds perfect consistency with the neutrino scattering data. This neat little modification nicely demonstrates the strength of CHPT, *i.e.* of a consistent calculation at next–to–leading order.

3. Electromagnetic Polarizabilities

The electric charge radii of the neutron and the proton as well as their magnetic moments are known to a high accuracy. Low–energy Compton scattering reveals further important information about the internal structure of the nucleon, parametrized by the so–called electric and magnetic polarizabilities. These are the first non–trivial structure constants in two–photon observables. They characterize the ease with which a dipole moment can be induced in a composite system.

Before discussing the CHPT calculation of the polarizabilities, let me give a brief overview about the experimental situation:

- The sum of the electric and the magnetic polarizabilities of the proton and the neutron can be determined rather precisely by use of a forward dispersion relation sum rule,

$$\bar{\alpha} + \bar{\beta} = \frac{1}{2\pi^2} \int_{E_{\gamma,\text{thr}}}^{\infty} d\omega \frac{\sigma_\gamma(\omega)}{\omega^2} \tag{4}$$

  with $\sigma_\gamma(\omega)$ the total photoabsorption cross section of the proton or neutron, $\omega = E_\gamma$ the photon energy in the lab frame and $E_{\gamma,\text{thr}}$ its threshold value. The most commonly quoted values are $\bar{\alpha}_p + \bar{\beta}_p = (14.3 \pm 0.3) \cdot 10^{-4}\,\text{fm}^3$ and $\bar{\alpha}_n + \bar{\beta}_n = (15.8 \pm 0.5) \cdot 10^{-4}\,\text{fm}^3$.

- The separation into the individual contributions $(\bar{\alpha}_{p,n},\ \bar{\beta}_{p,n})$ is afflicted with much larger uncertainties. However, there exists clear evidence that the proton and the neutron behave as *electric* dipoles, $\bar{\alpha}_{p,n} >> \bar{\beta}_{p,n}$. Typical values quoted in the literature span the ranges $\alpha_n \approx \bar{\alpha}_p \approx 8 \ldots 12 \cdot 10^{-4}\,fm^3$ and $\bar{\beta}_n \approx \bar{\beta}_p \approx -1 \ldots 4 \cdot 10^{-4}\,fm^3$. There is also some evidence that the intrinsic electric polarizability is larger for the neutron than for the proton.

What is most amazing about the CHPT calculation of $\bar{\alpha}_{p,n}$ and $\bar{\beta}_{p,n}$ to one–loop order is that there are _no_ contributions from polynomial counterterms. This is essentially due to the fact that the kinematics is rather special and that there are no external pions. The complete discussion of the vanishing of possible counterterms from $\mathcal{L}_{\pi N}^{(0,1,2,3)}$ is given in Ref.9. To my knowledge, the electric and magnetic polarizabilities are the first observables including baryons in which this intriguing phenomenon appears.

For the proton/neutron one has to evaluate 52/22 one-loop diagrams. This task is, however, tremendously simplified by the special kinematics. In fact, one can perform all Feynman parameter integrations. The full expressions are given in Ref.9, here let me only discuss the chiral expansion of the polarizabilities. Since we are dealing with a forward amplitude, it does not come as a surprise that they diverge in the chiral limit. The chiral expansion reads

$$\bar{\alpha}_p = \frac{e^2 g_{\pi N}^2}{192\pi^3 m^3}\left\{\frac{5\pi}{2\mu} + 18\,ln\mu + \frac{33}{2} + \mathcal{O}(\mu)\right\}$$

$$\bar{\alpha}_n = \frac{e^2 g_{\pi N}^2}{192\pi^3 m^3}\left\{\frac{5\pi}{2\mu} + 6\,ln\mu - \frac{3}{2} + \mathcal{O}(\mu)\right\}$$

$$\bar{\beta}_p = \frac{e^2 g_{\pi N}^2}{192\pi^3 m^3}\left\{\frac{\pi}{4\mu} + 18\,ln\mu + \frac{63}{2} + \mathcal{O}(\mu)\right\}$$

$$\bar{\beta}_n = \frac{e^2 g_{\pi N}^2}{192\pi^3 m^3}\left\{\frac{\pi}{4\mu} + 6\,ln\mu + \frac{5}{2} + \mathcal{O}(\mu)\right\}$$

$$(5)$$

The first interesting observation from (5) is that the empirical result $\bar{\alpha}_p + \bar{\beta}_p \approx \bar{\alpha}_n + \bar{\beta}_n$ follows naturally from the leading term, in fact, we find $\bar{\alpha}_p + \bar{\beta}_p = \bar{\alpha}_n + \bar{\beta}_n = 15 \cdot 10^{-4} fm^3$, almost too close to the data. Furthermore, the coefficient of the term proportional to M_π^{-1} is ten times bigger for $\bar{\alpha}_{p,n}$ than for $\bar{\beta}_{p,n}$, in agreement with the empirical trends. This can be traced back to the fact that there are 8 diagrams which contribute to the leading singularity – and they have the same isospin factors for the proton and the neutron.

The full one–loop prediction has also been worked out with the following results

$$\bar{\alpha}_p = 7.4 \cdot 10^{-4}\ fm^3 \qquad \bar{\beta}_p = -2.0 \cdot 10^{-4}\ fm^3$$
$$\bar{\alpha}_n = 10.1 \cdot 10^{-4}\ fm^3 \qquad \bar{\beta}_n = -1.2 \cdot 10^{-4}\ fm^3$$

$$(6)$$

The values for the proton and neutron electric polarizabilities compare favorably with the data. The empirical trend $\bar{\alpha}_n > \bar{\alpha}_p$ is supported by this result. The values for $\bar{\beta}_p$ and $\bar{\beta}_n$ are considerably smaller than the corresponding electric ones. The negative sign for $\bar{\beta}_{p,n}$ leads, of course, to too small values of the sums $(\bar{\alpha} + \bar{\beta})_{p,n}$. There are, however, important higher loop effects which eventually have to be considered like e.g. the contribution from the $\Delta(1232)$ resonance which in many hadron models gives a large positive magnetic polarizability. While a two–loop calculation by itself is a formidable task, there is a "cheap way" of getting a handle on the two-loop effects for the sum $\bar{\alpha} + \bar{\beta}$ based on the dispersion sum rule (4). Once one has performed the calculation of the pion photopoduction amplitudes (neutral and charged pions) to one–loop order and of tree level processes like $\gamma N \to \pi\pi N$, the dispersive integral can be evaluated to give $\bar{\alpha} + \bar{\beta}$ to two-loop order (in the isospin limit). This can be most easily understood if one considers a typical two-loop Feynman diagram for Compton scattering and cuts it in the interaction region (after/before the absorption/emission of the incoming/outgoing photon). Since these questions are presently under investigation, I will finish the discussion here. I only want to stress one more time that for the baryon sector the polarizabilities are a clean probe of loop effects.

4. Summary and Outlook

Chiral symmetry poses important constraints on the QCD Green functions at low energies. As one particular example, I have shown that the nucleon polarizabilities fall in a very special category of processes – to one–loop order they are pure loop effects and thus one directly tests the chiral sector of QCD without *any* unknown parameter. Let me stress here again that the chiral expansion is not a conventional perturbation theory due to the cuts in the spectrum induced by the massless Goldstone excitations. Furthermore, CHPT can be used to unravel the dynamics in the threshold production of pions from nucleons. I have discussed the first systematic investigation of chiral symmetry constraints on these processes here. However, more work is needed to deepen our understanding of the photo–nuclear dynamics.

Finally, I wish to draw your attention to some recent developments in baryon CHPT. In the relativistic formulation of Ref.2 baryon four momenta can never get small since the nucleon mass is of the order of the chiral symmetry breaking scale. This can be circumvented in a $1/m$ expansion which makes use of effective field theory methods for systems containing heavy quarks[10]. In this framework, it can be proven that the form of the LET (2) is in fact not modified by higher loop contributions. Also, in the heavy mass formulation the nucleon polarizabilities are simply given by the leading singular term (4). A systematic study of this approach has only begun[11] and is pursued actively.

5. Acknowledgements

It is my pleasure to thank the organizers for their invitation and hospitality. The work reported here has been done in collaborations with Jürg Gasser, Norbert Kaiser, Joachim Kambor and Ulf-G. Meißner, to whom I express my gratitude.

6. References

1. S. Weinberg, *Physica* **96A** (1979) 327;
   J. Gasser and H. Leutwyler, *Ann. Phys. (N.Y.)* **158** (1984) 142.
2. J. Gasser, M.E. Sainio and A. Švarc, *Nucl. Phys.* **B307** (1988) 779.
3. A. I. Vainshtein and V. I. Zakharov, *Nucl. Phys.* **B36** (1972) 589;
   P. de Baenst, *Nucl. Phys.* **B24** (1970) 633.
4. E. Mazzucato *et al.*, *Phys. Rev. Lett.* **57B** (1986) 3144;
   R. Beck *et al.*, *Phys. Rev. Lett.* **65B** (1990) 1841.
5. V. Bernard, J. Gasser, N. Kaiser and Ulf-G. Meißner, *Phys. Lett.* **B268** (1991) 291;
6. V. Bernard, N. Kaiser and Ulf-G. Meißner, "Testing Nuclear QCD: $\gamma p \to \pi^0 p$", Bern University preprint BUTP-92/14, 1992, to app. in πN Newsletter 7.
7. V. Bernard, N. Kaiser and Ulf-G. Meißner, "Threshold Pion Photoproduction in Chiral Perturbation Theory", Bern University preprint BUTP-92/01, 1992, to app. in Nucl. Phys. B.
8. V. Bernard, N. Kaiser and Ulf-G. Meißner, "Measuring the axial radius of the nucleon in pion electroproduction", Bern University preprint BUTP-92/30, 1992.

9. V. Bernard, N. Kaiser and Ulf-G. Meißner, *Phys. Rev. Lett.* **67** (1991) 1515; *Nucl. Phys.* **B373** (1992) 346.

10. E. Jenkins and A.V. Manohar, *Phys. Lett.* **B255** (1991) 558; *Phys. Lett.* **B259** (1991) 353.

11. V. Bernard, N. Kaiser, J. Kambor and Ulf-G. Meißner, "Chiral Structure of the Nucleon", Bern University preprint BUTP-92/15, 1992; "Hyperon Polarizabilities", Bern University preprint BUTP-92/31, 1992.

VACUUM STRUCTURE OF PURE GAUGE THEORIES ON THE LATTICE[1]

Richard W. Haymaker, Vandana Singh and Dana Browne
Department of Physics and Astronomy
Louisiana State University, Baton Rouge, Louisiana, 70808, USA

and

Jacek Wosiek
Chair of Computer Science
Jagellonian University, Institute of Physics, Reymonta 4 Cracow, Poland; and
Max-Planck-Institut für Physik - Werner-Heisenberg-Institute - P.O. Box 40 12 12,
Munich, Germany

ABSTRACT

We present results from simulations on two aspects of quark confinement in the pure gauge sector. First is the calculation of the profile of the flux tube connecting a static $q\bar{q}$ pair in $SU(2)$. By using the Michael sum rules as a constraint we give evidence that the energy density at the center of the flux tube goes to a constant as a function of quark separation. Slow variation of the width and energy density is not ruled out. Secondly in the confined phase of lattice $U(1)$ we calculate the curl of the magnetic monopole current and show that the dual London equation is satisfied and that the electric fluxoid is quantized.

1. Introduction

I would like to report on efforts of the Cracow-LSU collaboration to study the response of gauge fields to the presence of static sources. Confinement is the central issue which is seen as a consequence of the vacuum squeezing the field lines to form flux tubes connecting color charges. One is beginning to see considerable detail of the field distributions in the flux tube, e.g. its size and shape, the chromoelectric and chromomagnetic field components, and monopole currents that are responsible for squeezing the tube to form an Abrikosov vortex.

In the first part of this talk I will describe our results for SU(2) lattice gauge theory[1,2,3]. We examine the flux tube between a $q\bar{q}$ pair, calculating the six chromoelectric and chromomagnetic components to the energy density and action density. We parametrize the profile of the flux tube and study scaling. The results are subjected to the check provided by the Michael sum rules[4].

The second part of this talk concerns the mechanism that leads to flux tube formation. This is the role of the solenoidal magnetic monopole currents that surround

---

[1]Presented by R. Haymaker

the flux tube. In this work[5] we show in the confined phase of U(1) that the curl of the monopole current has a profile similar to the electric field and that the dual London equation is satisfied and electric fluxoid quantization occurs. We demonstrate that the the flux tube is precisely the dual of the Abrikosov vortex in Type II superconducting materials.

2. Flux Tubes in SU(2) [2]

2.1. Background

In this calculation we measure the field energy densities by correlating the small plaquette with the Wilson loop. Full details of the simulation are given in Ref. [2], which also contains further references. Further details of the flux profiles will appear in a companion paper Ref. [3]. By fixing our attention on the middle time slice of the Wilson loop, the time-like segments form world lines that approximate a static $q\bar{q}$ pair. The 3 space-space plaquettes measure the magnetic component of the energy density and similarly the 3 space-time plaquettes measure the electric components.

Before defining the flux calculation in more detail, we point out that the Wilson loops themselves are used to extract the transfer matrix eigenvalues which give the static quark potential and are further used to extrapolate the flux measurements to infinite time extent of the Wilson loops. Specifically we determine the eigenvalues of the transfer matrix by fitting the Wilson loops to the exponentials as described in Ref.[2].

$$< W(R,T) >= \sum_i A_i e^{-E_i(R)T}. \tag{1}$$

$E_0(R)$ is of special interest since it contains the static quark potential:

$$E_0(R) = -\frac{\alpha}{R} + \sigma R + \frac{c(\beta)}{a(\beta)}. \tag{2}$$

The term independent of R is the self energy of the two quarks which does not scale but diverges as $a \to 0$. Since Wilson loops can be calculated quite accurately, the static potential is a useful physical quantity to check scaling and thereby determine the lattice spacing $a(\beta)$. All our data is consistent with standard values $a(2.3) = 0.171$ fm, $a(2.4) = 0.128$ fm and $a(2.5) = 0.089$ fm.

2.2. Flux Tube Profiles

The lattice observable needed to measure the flux is the following[6,1,2].

$$\begin{aligned}
f^{\mu\nu}(x) &= \frac{\beta}{a^4}\left(\frac{\langle W P_x^{\mu\nu}\rangle}{\langle W\rangle} - \langle P\rangle\right), \\
&\approx \frac{\beta}{a^4}\left(\frac{\langle W P_x^{\mu\nu} - W P_{x_R}^{\mu\nu}\rangle}{\langle W\rangle}\right), \tag{3}
\end{aligned}$$

---

[2]Haymaker, Singh and Wosiek

where W is the Wilson loop, $P_x^{\mu\nu}$ the plaquette located at x, $\beta = \frac{4}{g^2}$ and x_R is a distant reference point. In the classical continuum limit

$$f^{\mu\nu} \xrightarrow{a\to 0} -\frac{1}{2}\langle(F^{\mu\nu})^2\rangle_{q\bar{q}-vac}, \tag{4}$$

where the notation $\langle\cdots\rangle_{q\bar{q}-vac}$ means the difference of the average values in the $q\bar{q}$ and vacuum state. From now on we shall be using field components in Minkowski space and hence

$$f^{\mu\nu} \to \frac{1}{2}(-B_1^2, -B_2^2, -B_3^2; E_1^2, E_2^2, E_3^2). \tag{5}$$

Correspondence between various components and $f^{\mu\nu}$ is standard: space-space plaquettes are magnetic, space-time plaquettes are electric. The energy and action densities are respectively

$$\epsilon = \frac{1}{2}(E^2 + B^2),$$
$$\gamma = \frac{1}{2}(E^2 - B^2). \tag{6}$$

Since the magnetic contribution turns out to be negative, there is a strong cancellation between the two terms in the energy, whereas they are enhanced in the action.

Figure 1 gives the flux profiles. The cancellation which suppresses the energy density in the flux tube is evident. However notice that the self energy of the quarks is not similarly suppressed. This follows because the self energy is primarily electric. We fitted the energy and action density in the plane at the midpoint between q and $\bar{q}$ using the function

$$f(r_\perp) = a\exp\left(-\sqrt{b^2 + (r_\perp/c)^2}\right). \tag{7}$$

The peak value and the width at half maximum were very well determined using a χ^2 fit for each of 70 cases of different loop sizes and values of β. For the third parameter we chose the decay length of the tail of this function and found it less well determined but with a value typically close to the width at half maximum. The details of the analysis will be given in a forthcoming paper[3]. Here we just give the results of the extrapolation to infinite Wilson loop time extent in Figs. 2 and 3.

The basic issue is whether the peak value of the energy density stabilizes to a constant or goes to zero with quark separation. This is not easy to settle as can be seen in Fig. 2. Roughly speaking we know from the linearly rising potential that the string tension $\sim$ (width)^2 $\times$ (peak value) should be constant. Both Figs. 2 and 3 show that we are marginally asymptotic in quark separation, R. The two curves are $\sim 1/R^4$ and $\sim 1/R$. A Coulomb field would fall like the former but since the string tension is constant the asymptotic width would have to grow like R^2 which clearly it does not. Therefore we can rule out a Coulomb field as expected. Interestingly for small separations, the eyeball fit to $\sim 1/R^4$ is quite good which may be due to a

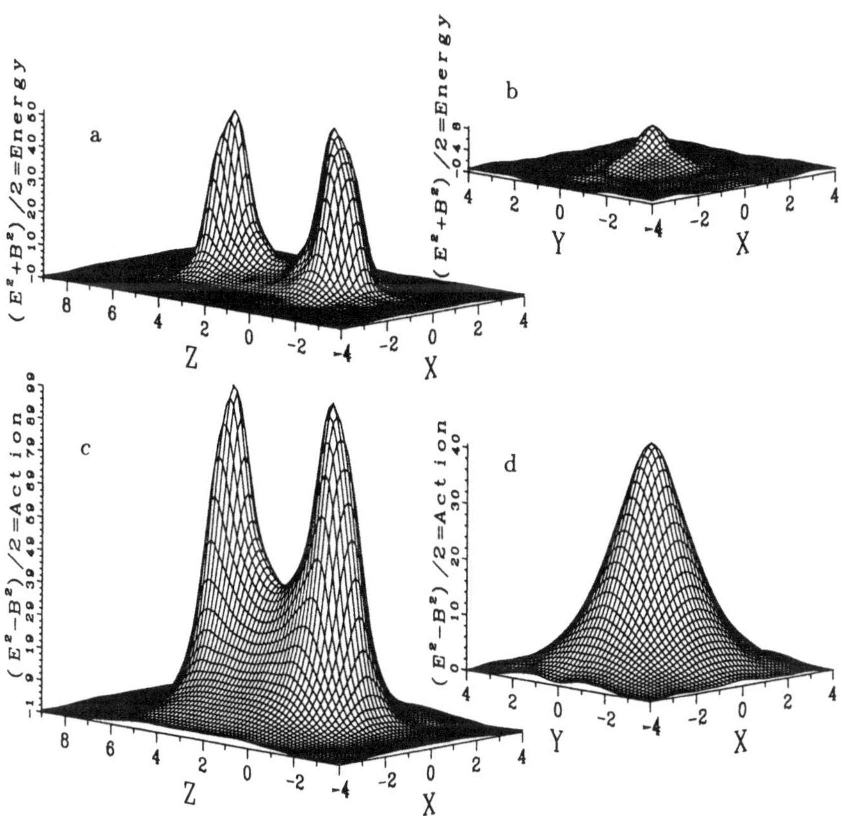

Figure 1: Energy and action profiles: (a) energy density in the plane containing $q\bar{q}$; (b) energy density in the plane midway between q and $\bar{q}$; (c) and (d) similarly for action density.

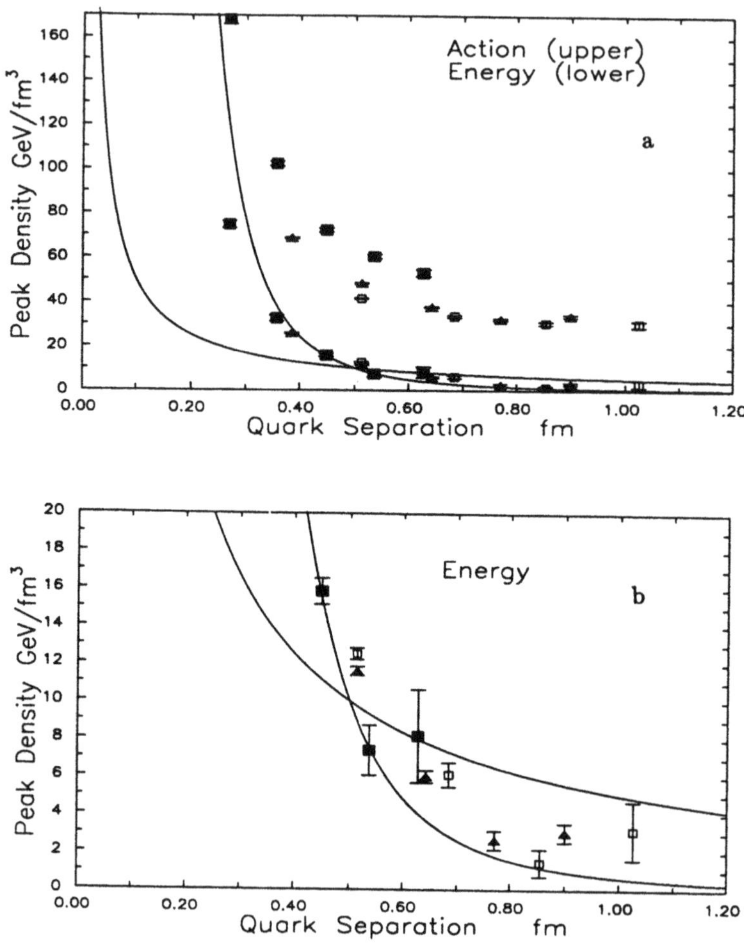

Figure 2: (a) Peak value of energy and action density; solid squares: $\beta = 2.5$; triangles: $\beta = 2.4$; open squares: $\beta = 2.3$. The two curves are $1/R$ and $1/R^4$ arbitrarily normalized; (b) blowup of (a).

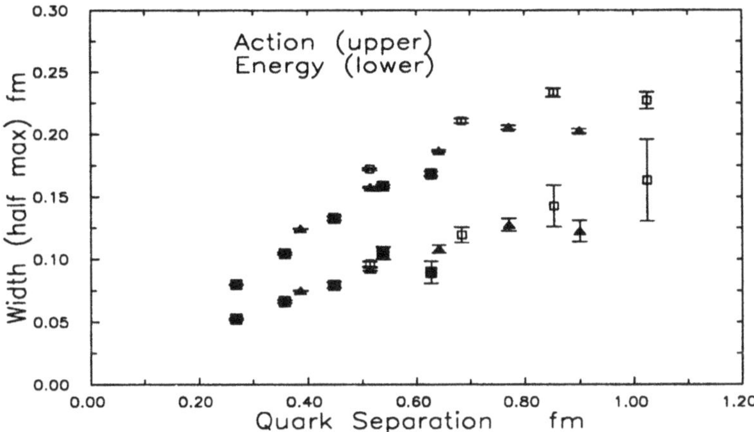

Figure 3: Width at half maximum for energy and action density.

Coulomb like behavior at small distances. (The above argument that the width must grow like R^2 does not apply because there is no string for small R.) The dielectric model[16] predicts the peak density $\sim 1/R$. This function (arbitrarily normalized) does not seem to fit the data very well. However such a behavior would imply the width $\sim \sqrt{R}$ which is certainly possible in our data. We can say that the peak energy density and width are consistent with a constant value for large quark separation but we can not rule out a slow variation. The issue can be tightened by making use of the Michael sum rules[4] as we mention in the next section.

Figure 4 illustrates a general feature of our data. The cluster of three points for each R and T correspond to the three quantities:

$$\epsilon = \frac{1}{2}(E_\parallel^2 + B_\parallel^2) + \frac{1}{2}(E_\perp^2 + B_\perp^2),$$

$$\epsilon_\parallel = \frac{1}{2}(E_\parallel^2 + B_\parallel^2),$$

$$\epsilon(T \leftrightarrow R) = \frac{1}{2}(E_\parallel^2 + B_\parallel^2) - \frac{1}{2}(E_\perp^2 + B_\perp^2). \tag{8}$$

If one turns the Wilson loop on its side, the $\parallel$ components are unchanged but the $\perp$ components of the electric and magnetic fields are reversed: $E_\perp^2 \leftrightarrow -B_\perp^2$. Hence there is a sign change in the third expression. The central points of the cluster are the $\parallel$ components only. *The clustering of the points implies that the $\perp$ components of the electric and magnetic contributions to energy density are approximately equal but of opposite sign and cancel.* The width of the peak is even less sensitive to the

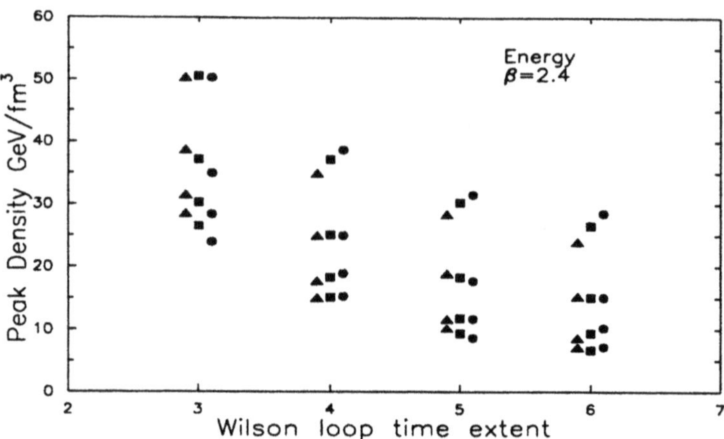

Figure 4: Peak density for $R \times T$ Wilson loop sizes $R = 3 - 6$, $T = 3 - 6$. For fixed T the points decrease monotonically with R. Triangles: energy density for $R \times T$ loop; circles: energy density for $T \times R$ loop; squares: $\parallel$ components of E^2 and B^2 only.

transverse components giving essentially the same value for all three points.

2.3. Sum Rules

A consistency check on the flux distributions can be obtained by using the Michael sum rules[4] for energy and action.

$$\frac{1}{2} \sum_{\vec{x}} (E(\vec{x})^2 + B(\vec{x})^2) = E_0(R),$$

$$\frac{1}{2} \sum_{\vec{x}} (E(\vec{x})^2 - B(\vec{x})^2) = -\beta \frac{\dot{a}}{a}[E_0(R) - \frac{c(\beta)}{a}] - \beta \frac{\dot{c}(\beta)}{a}. \tag{9}$$

Here $E_0(R)$ is given by Eqn.(1), and $(\cdot \equiv \frac{d}{d\beta})$. In ref.[2] we have shown that our data are essentially consistent with these sum rules. The one difficulty is the fact that the self energy, $c(\beta)/a(\beta)$, determined from the potential differs from the self energy determined from the the action sum rule. This may be due to an ambiguity in the definition of self energy or possibly due to our classical expressions for energy and action which ignores quantum corrections. By taking a derivative of these expressions with respect to the quark separation, R, this difficulty is avoided. This gives the relation

$$\sigma_A = -\beta \frac{\dot{a}}{a}\sigma; \quad \sigma_A \equiv \frac{1}{2} \sum_{\vec{x}_\perp} (E(\vec{x})^2 - B(\vec{x})^2); \quad \sigma \equiv \frac{1}{2} \sum_{\vec{x}_\perp} (E(\vec{x})^2 + B(\vec{x})^2). \tag{10}$$

The sums are now over the plane midway between the $q\bar{q}$ pair.

Using the sum rules we find that the β function $(-\beta\dot{a}/a) \approx 10.\pm2.$ compared to the current estimates $7.\pm1..$ The asymptotic value is $-51/121 + 3\pi^2\beta/11 = 6.0(\beta = 2.4)$. There is ample evidence from other measurements that although scaling works well, asymptotic scaling is violated[8] and hence we do not expect to get the asymptotic value.

An alternative approach is to assume the sum rules are correct and use them to infer information about the energy density from the action density which is far easier to measure since relative errors are down by an order of magnitude. As is clear from the sum rule, the action does not scale yet the variation over these values of β is very small. An examination of Fig. 2(a) shows that the action for each β seems to stabilize to a constant for increasing distance for the peak density and for the width. This is quite striking for $\beta = 2.3$ and 2.4. For $\beta = 2.5$ R appears to be too small to draw a conclusion. These data do not suggest that the peak value is tending to zero at all. We would like to use the sum rules to predict the behavior of the energy density. A constant peak energy density follows only if the widths of the energy and action peaks have the same behavior. Figure 3 shows that in fact they do. *From this and using the sum rules we conclude that the energy density stabilizes to a constant value also.* This conclusion is an argument against the dielectric model[16]. However we have little to say about logarithmic behavior of the flux tube width as predicted by Lüscher[7]. For more details see Ref.[3].

3. Mechanism for confinement in U(1)^3

3.1. Dual Superconductor Model of Confinement

We now turn to the mechanism for flux tube formation and present direct evidence that supercurrents of magnetic monopoles produce a dual Abrikosov vortex[9]. U(1) lattice gauge theory in 4 dimensions has both a confined phase at large charge and a weak coupling deconfined phase corresponding to continuum electrodynamics with a Coulomb interaction between static charges. Therefore confinement or its absence can be studied using U(1) lattice gauge theory as a prototype, before tackling the more complicated non-Abelian theories that actually describe quarks. Much evidence for the dual superconductor hypothesis has accumulated from studies[10,11,12] of lattice gauge theory. Polyakov[10] and Banks, Myerson and Kogut[11] showed that U(1) lattice gauge theory in the presence of a quark-antiquark pair could be approximately transformed into a model describing magnetic current loops (the monopoles) interacting with the electric current generated by the $q\bar{q}$ pair. DeGrand and Toussaint[12] demonstrated via a numerical simulation that the vacuum of U(1) lattice gauge theory was populated by monopole currents, copious in the confined phase and rare in the deconfined phase. This behavior has also been seen in non-Abelian models after gauge fixing[13]. Many studies of non-Abelian models using Dirac monopoles[13,14] or other topological excitations[15] support the dual superconductor mechanism, although other studies[17] dissent.

192

So far, studies of confinement have examined "bulk" properties such as the monopole density[12,13], and the behavior of the static quark potential[14]. In a recent paper[5] we presented the first direct evidence that the flux tube is a dual Abrikosov vortex. We further show that there are exact U(1) lattice gauge theory analogues of two key relations that lead to the Meissner effect in a superconductor; the London equation and the fluxoid quantization condition.

3.2. Electric Field Profiles

Our simulations were done on a Euclidean spacetime lattice of volume $9^3 \times 10$. The static charges are represented by a Wilson loop as in the previous section. We take a 3×3 loop in the $z - t$ plane and measure the fields in the $x - y$ plane at the midpoint between the charges. Because of the geometrical symmetry of the measurements only the z-components of $\langle \vec{\mathcal{E}} \rangle$ and $\langle \vec{\nabla} \times \vec{J}_M \rangle$ are nonzero. If the Wilson loop is removed, even the z-components average to zero, so the response is clearly induced by the presence of the static charges. Only the imaginary part of the Wilson loop contributes to the averages of these two quantities.

The plaquette measures flux passing through a unit square on the lattice[4]

$$\exp[iea^2 F_{\mu\nu}(\vec{r})] = \exp[i\theta_{\mu\nu}(\vec{r})] \equiv U_\mu(\vec{r})U_\nu(\vec{r}+\mu)U_\mu^\dagger(\vec{r}+\nu)U_\nu^\dagger(\vec{r}). \tag{11}$$

The electric flux in lattice variables is

$$\mathcal{E}_\mu(\vec{r}) = \mathrm{Im}\exp[i\theta_{\mu 4}(\vec{r})]. \tag{12}$$

Figure 5(a) shows the electric flux distribution for $\beta = 1.1$ where the vacuum is in the deconfined phase. The broad flux distribution seen is identical to the dipole field produced by placing two classical charges at the quark positions, except that the classical value of the flux on the $q\bar{q}$ axis is a factor of two smaller. We measure the total electric flux from one quark to the other, including not only the flux through the plane between the charges (0.8504 ± 0.0045) but also the flux (0.0951 ± 0.0028) that flows through the lattice boundary because of the periodic boundary conditions. This yields a total flux of (0.9453 ± 0.0053), close to the theoretical value $\Phi_e = e/\sqrt{\hbar c} = 1/\sqrt{\beta} = 0.9534$.

Figure 5(b) and 6(a) shows the electric flux in the confined phase ($\beta = 0.95$). In this case the flux is confined almost entirely within one lattice spacing of the axis and essentially no flux passes the long way around through the lattice boundary. The net flux is again equal to $1/\sqrt{\beta}$ within statistical error. This behavior is exactly what one would expect from the superconducting analogy, where the flux has been "squeezed" into a narrow tube.

3.3. Magnetic Monopole Supercurrents

The monopole currents are found by a prescription devised by DeGrand and Toussaint[12], which employs a lattice version of Gauss' Law to locate the Dirac

---

[4]The flux here means $E \times$ (area). We use the same term for E^2 or B^2 since it has become an accepted usage.

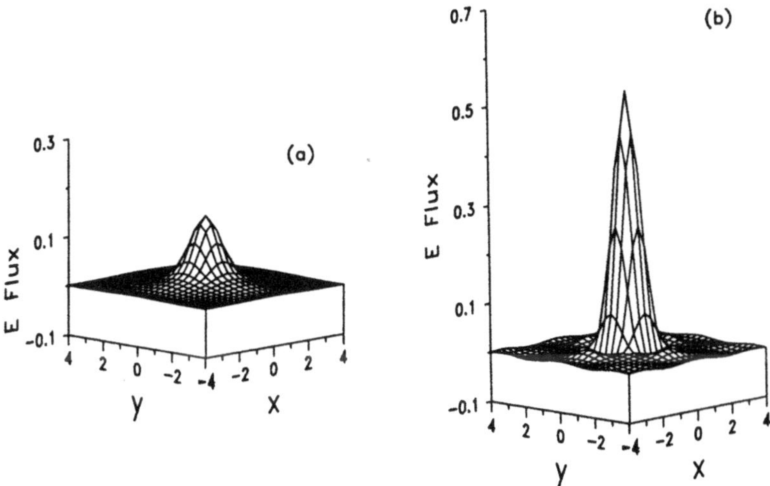

Figure 5: Surface plot of the electric flux through the xy plane midway between the $q\bar{q}$ pair when the system is in (a) the deconfined phase ($\beta = 1.1$) and (b) the confined phase ($\beta = 0.95$). The line joining the pair is located at (0,0).

string attached to the monopole. The net flux into each plaquette face is given by $(\theta_{\mu\nu}(\vec{r}) \bmod 2\pi)$. If the sum of the fluxes into the faces of a 3-volume at fixed time is nonzero, a monopole is located in the box. A non-zero net flux can occur only if a multiple of 2π arises from the mod operation. Therefore the net flux is

$$\sum_{6\,faces} a^2 F_{\mu\nu} = n(2\pi\hbar c/e) = ne_M \tag{13}$$

The net flux into the box at fixed time thus yields the monopole "charge" density, the time component of the monopole 4-current J_M. The spatial components are found similarly. The monopole currents form closed loops due to the conservation of magnetic charge. Finally the curl is calculated by the line integral of the current around a dual plaquette.

We show in Fig. 6(a) $\langle \vec{\mathcal{E}} \rangle$ and in Fig. 6(b) $-\langle \vec{\nabla} \times \vec{J}_M \rangle$ in the confined phase as a function of the distance from the $q\bar{q}$ axis. The data show that the spatial variation of the flux and the curl of the current are very similar, except for the point on the axis which will be discussed below.

3.4. London Equations, Fluxoid Quantization and the Abrikosov Vortex

In order to interpret this result we would like to review the ordinary London theory. In the next section we interchange electric and magnetic quantities to get the dual results.

194

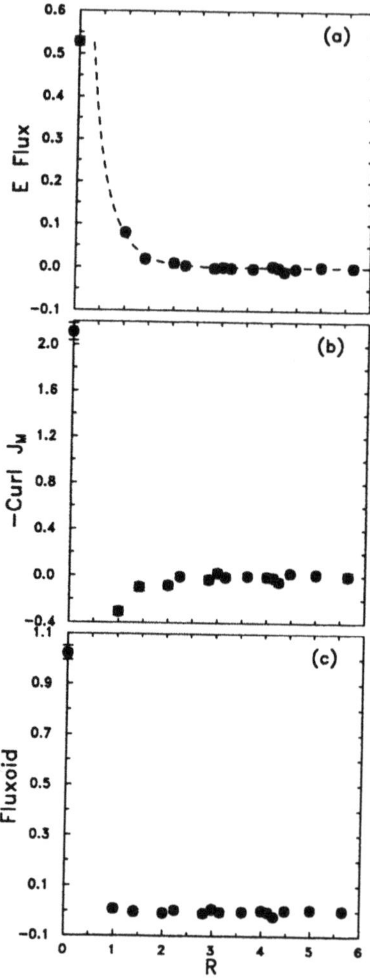

Figure 6: Behavior of (a) the electric flux, (b) the curl of the monopole current and (c) their sum, the fluxoid, in the confined phase ($\beta = 0.95$) as a function of the perpendicular distance $r_\perp$ from the $q\bar{q}$ axis. The solid line in (a) gives the flux for a pure Coulomb field .

A concise statement of the London theory is contained in the relation[5]

$$\vec{A} + \frac{\lambda^2}{c}\vec{J} = 0; \quad (\vec{\nabla} \cdot \vec{A} = 0). \tag{14}$$

If the charge density is zero, then in this gauge the electric field is given by $-\dot{A}/c$ and therefore $\vec{E} = \lambda^2 \dot{\vec{J}}/c$. This describes a perfect conductor and is just Newton's law for free carriers, $e\vec{E} = m\dot{\vec{v}}$. By taking the curl of Eqn.(14) we obtain the condition for a perfect diamagnet.

$$\vec{\nabla} \times \vec{J} = -\frac{c}{\lambda^2}\vec{B}. \tag{15}$$

This relation together with Ampere's law $\vec{\nabla} \times \vec{B} = \vec{J}/c$ gives $\nabla^2\vec{B} = \vec{B}/\lambda^2$ which implies that the magnetic field falls off in the interior of the superconductor with a skin depth λ.

Finally the fluxoid is given by the integral

$$\int_S (\vec{B} + \frac{\lambda^2}{c}\vec{\nabla} \times \vec{J}) \cdot \hat{n}da = \int_C (\vec{A} + \frac{\lambda^2}{c}\vec{J}) \cdot d\vec{l} = n\frac{\hbar c}{2e} = n\Phi_m \tag{16}$$

If the curve C is in a simply connected region of a superconductor, then $n = 0$. However if the curve encircles a hole in the material then n need not be zero but must be an integer. In an extreme type II Abrikosov vortex, a very small core is comprised of normal material. A single unit of magnetic flux of radius $\sim \lambda$ passes through the vortex. The fluxoid density $\vec{B} + \frac{\lambda^2}{c}\vec{\nabla} \times \vec{J}$ is zero everywhere except in the region of the normal material. In the limit in which the core is a delta function we obtain:

$$\vec{B} + \frac{\lambda^2}{c}\vec{\nabla} \times \vec{J} = \Phi_m \delta^2(x_\perp)\hat{n}_z. \tag{17}$$

Further if we use Ampere's law we can get an analytic expression for B_z.

$$B_z(r_\perp) - \lambda^2\nabla^2 B_z = \Phi_m \delta^2(x_\perp); \quad B_z = \frac{\Phi_m}{2\pi\lambda^2}K_0(r_\perp/\lambda). \tag{18}$$

3.5. Dual Superconductor

We interpret our results using the following relations which are the dual of the corrsponding relations in the previous section:

$$\vec{E} - \frac{\lambda^2}{c}\vec{\nabla} \times \vec{J}_m = \Phi_e \delta^2(x_\perp)\hat{n}_z. \tag{19}$$

where the unit of electric flux $\Phi_e = 1/\sqrt{\beta}$. Figure 5(c) shows the result of fitting our data to this relation. The London penetration depth, λ, is the only free parameter. We are able to determine a value of λ which gives zero away from the axis and a delta

---

[5]We use Heaviside-Lorentz units to be consistent with lattice gauge theory.

function with the correct coefficient on axis. Further we can check that the electric flux profile is given by

$$E_z = \frac{\Phi_e}{2\pi\lambda^2} K_0(r/\lambda) \tag{20}$$

This function has no free parameters. This curve is also shown in Fig. 6(a) showing excellent agreement. We find a value of $\lambda/a = 0.482 \pm 0.008$, which is consistent with the range of penetration of the electric flux in Fig. 6(a) and the thickness of the current sheet in Fig. 6(b). We expect that, as in a superconductor, the transition to the deconfined phase will be signalled by a divergence of the London penetration depth. We have therefore measured λ further from the deconfinement transition at $\beta = 0.90$, and find a smaller penetration depth of $\lambda/a = 0.32 \pm 0.02$. In the deconfined phase, $\beta = 1.1$ we find an almost insignificant value of $\langle \vec{\nabla} \times \vec{J}_M \rangle$ and fitted values of λ were larger than our lattice size.

In summary, one can find a value of the London penetration depth, λ, that satisfies Eqn.(19) off axis. One then finds that for the point on axis, the same value of λ gives one quantum of electric flux as predicted by Eqn.(19). Finally the same value of λ gives a good fit to the profile using Eqn.(20). Hence considerable detail of the dual Abrikosov is verified. It is perhaps surprising that a nonlinear, strongly interacting, model such as U(1) lattice gauge theory could be described by such a simple model as the linear London equations but our results indicate that the operators $\langle \vec{\mathcal{E}} \rangle$ and $\langle \vec{\nabla} \times \vec{J}_M \rangle$ when measured in the presence of source of external flux like a Wilson loop, give an unambiguous indication of the confinement of electric flux by a monopole current distribution. The simulation yields a large signal even with modesnt amounts of computer time on a Sun workstation. Although the Meissner effect itself requires only that Eq. (20) hold off axis, our data also support the more restrictive fluxoid quantization relation on axis. This additional relation reflects the single-valued nature of the order parameter in a Ginzburg-Landau description of the monopole condensate. Because the monopoles appear pointlike in our simulations, lattice gauge theory looks like an extreme type-II superconductor.

4. Acknowledgements

Two of us (R.W.H. and V.S.) would like to thank A. Kotanski, Y. Peng, G. Schierholz and T. Suzuki for many fruitful discussions on this problem. V.S. thanks A. Kronfeld and R. Wensley for useful conversations. R.W.H and V.S. are supported by the DOE under grant DE-FG05-91ER40617 and D.A.B. is supported in part by the National Science Foundation under Grant No. NSF-DMR-9020310 J.W. is supported in part by the Polish Government under Grants Nos. CPBP 01.01 and CPBP 01.09.

5. References

1. J. Wosiek and R. Haymaker, Phys. Rev. Rapid Commun. **D 36**, 3297 (1987); J. Wosiek, Nucl. Phys. **B** (Proc. Suppl.) **4**, 52 (1988); R. W. Haymaker and J.

Wosiek, Acta Physica Polonica **B 21**, 403 (1990). R. W. Haymaker, Y. Peng, V. Singh and J. Wosiek, Nucl.Phys. **B** (Proc. Suppl.) **17**, 558 (1990).

2. R. W. Haymaker and J. Wosiek, Phys. Rev. **D 43**, 2676, (1991)

3. R. W. Haymaker, V. Singh and J. Wosiek, In preparation

4. C. Michael, Nucl. Phys. **B 280 [FS18]** 13, (1987).

5. V. Singh, R. W. Haymaker, and D. Browne, LSU Preprint LSUHEP - 01 (1992).

6. M. Fukugita and T. Niuya, Phys. Lett. **B 132**, 374 (1983); R. Sommer, Nucl. Phys. **B 291**, 673 (1987); **B 306**, 180 (1988); I. H. Jorysz and C. Michael, Nucl. Phys **B 302**, 448 (1987); C. Michael and M. Teper, Nucl. Phys. **B 305**, 453 (1988); N. A. Campbell, A. Huntley and C. Michael, Nucl. Phys. **B 206**, 51 (1988); S. Perantonis, A. Huntley and C. Michael, Nucl. Phys. **B326**, 544 (1989); W. Feilmair and H. Markum, Nucl. Phys. **B 370**, 299 (1992); W. Burger, M. Faber, W. Feilmair, H. Markum and M Muller, Nucl. Phys. **B20** (Proc. Suppl.), 203, (1991);

7. M. Lüscher, Nucl. Phys.**B 180** [FS2], 317 (1981); M. Lüscher, G. Münster and P. Weisz, Nucl. Phys. **B180[FS2]**, 1 (1981).

8. C. Michael, Phys. Lett. **B283**, 103, (1992).

9. H. B. Nielsen and P. Olesen, Nucl. Phys. **B61**, 45 (1973); J. Kogut and L. Susskind, Phys. Rev. **D 9**, 3501 (1974); S. Mandelstam, Phys. Rep. **23C**, 245 (1976); G. 't Hooft, in *High Energy Physics*, ed. A. Zichichi (Editrice Compositori, Bologna, 1976), Nucl. Phys. **B190[FS3]**, 455 (1981).

10. A. M. Polyakov, Phys. Lett. **59B**, 82 (1975).

11. T. Banks, R. J. Myerson, and J. Kogut, Nucl. Phys. **B129**, 493 (1977).

12. T. A. DeGrand and D. Toussaint, Phys. Rev. **D 22**, 2478 (1980).

13. A. S. Kronfeld, M. L. Laursen, G. Schierholz, and U.-J. Wiese, Phys. Lett. **198B**, 516 (1987); Nucl. Phys. **B293**, 461 (1987); F. Brandstaeter, G. Schierholz, and U.-J. Wiese, Phys. Lett. **272B**, 319 (1991); M. Teper, Phys. Lett. **171B**, 86 (1986); R. J. Wensley and J. D. Stack, Phys. Rev. Lett. **63**, 1764 (1989).

14. T. Suzuki and I. Yotsuyanagi, Phys. Rev. **D 42**, 4257 (1990).

15. J. Smit and A. J. van der Sijs, Nucl. Phys. **B355**, 603 (1991); T. L. Ivanenko, A. V. Pochinsky, and M. I. Polikarpov, Phys. Lett. **252B**, 631 (1990).

16. S. L. Adler, Nucl. Phys. **B 217**, 381 (1983).

17. J. Greensite and J. Winchester, Phys. Rev. **D 40**, 4167 (1989); A. Di Giacomo, M. Maggiore, and S. Olejnik, Nucl. Phys. **B347**, 441 (1990).

WHAT DO LATTICE SIMULATIONS
TELL US ABOUT THE QCD VACUUM?[*]

W. BÜRGER, M. FABER, H. MARKUM, M. MÜLLER,
W. SAKULER, M. SCHALER

Institut für Kernphysik, Technische Universität Wien
A-1040 Vienna, Austria

ABSTRACT

A review is given over diverse physical observables calculated on space-time lattices which reflect the vacuum structure of QCD. We analyse the gluon string distribution in a static meson. It turns out that the energy density is restricted to a thin flux tube of constant diameter. We demonstrate flux tube breaking when sea quarks are considered. We further investigate the local chiral condensate and charge density around static quarks. A restoration of chiral symmetry in the vicinity of color sources as a consequence of gluon string formation is observed. It is shown that virtual antiquarks screen static quarks as expected from QED.

1. Introduction

Today there exists a vast variety of explorations of the mechanism for quark confinement in QCD. The most direct evidence results from lattice QCD where the computation of gluon exchange in a static quark-antiquark pair yields a linear potential.[1] As a consequence, two further questions raise immediately: i) What is the detailed mechanism being responsible for confinement. ii) What happens when the effect of sea quarks is taken into account.

For a deeper study of the confinement mechanism one has to calculate additional observables beyond the static potential. One observable of interest is the chromoelectric and chromomagnetic field distribution in a quark-antiquark system. We present high quality results for the flux tube formation between quark sources for the pure $SU(3)$ gluon theory. By computing the thickness of the flux tube we can give some statements about the relation between confinement in QCD and string theories. Further, this is the first study of the breaking of the flux tube in full QCD with dynamical quarks.

A great challenge for a fundamental quantum field theory starting from first principles like lattice QCD is to show effects of the vacuum polarization. So far screening effects from the quark sea on the linearly rising confining potential have been studied. We investigate the influence of light dynamical quarks by analyz-

---

[*]Supported in part by "Fonds zur Förderung der wissenschaftlichen Forschung" under Contract No. P7510-TEC.

ing the charge density of the polarization cloud around hadrons directly. It is expected that the charge of a valence quark is screened by virtual anticharges similar to QED. The chiral condensate around static quarks and mesons gives us information about the behavior of chiral symmetry. The idea is that due to gluon string formation the chiral condensate is suppressed around color sources which means that chiral symmetry is restored in this region. Since there are several ways to put fermions on a space-time lattice it is important to show that the physical results are independent of a specific lattice discretization method. Thus, we employ both Kogut-Susskind and Wilson fermions in our calculations and compare these regularization schemes.

2. Theory

The meson is constructed from a static quark and antiquark which are expressed by Polyakov loops $L(\vec{r}_1)$ and $L^\dagger(\vec{r}_2)$, respectively. The Polyakov loop[2]

$$L(\vec{r}) = \frac{1}{N} \, \mathrm{Tr} \prod_{i=1}^{N_t} U_{\mu=4}(\vec{r}, t_i) \, , \tag{1}$$

where N is the number of colors and N_t the time extension of the lattice, represents the propagator in the time direction for a heavy quark at position $\vec{r}$ in a gluon field $U_\mu(x)$ and satisfies the static Dirac equation. The finite time extension corresponds to a finite temperature T of the system.

We begin with the color electromagnetic observables. In the lattice formulation the components of the electromagnetic field reduce to plaquette operators[3]

$$P_{\mu\nu}(\vec{r}, t_i) = \prod_{\sigma \in \square} U_\sigma(\vec{r}, t_i) \, , \tag{2}$$

corresponding to the gauge field on the smallest unit squares $\square$ of length a on the fourdimensional space-time grid. The space-time like plaquettes yield the electric energy densities and the pure space like objects the magnetic energy densities in a gluon field U

$$e_i^2(\vec{r}) = \frac{2\beta}{N N_t} \sum_{l=1}^{N_t} \mathrm{Re} \, \mathrm{Tr}(P_{i4}(\vec{r}, t_l) - 1) \, , \quad b_k^2(\vec{r}) = \frac{2\beta}{N N_t} \sum_{l=1}^{N_t} \mathrm{Re} \, \mathrm{Tr}(1 - P_{ij}(\vec{r}, t_l)) \, , \tag{3}$$

with i, j, k cyclic.

The chiral condensate $\bar{\psi}\psi$ is an order parameter for chiral symmetry breaking and a measure for the sum of virtual quark and antiquark densities. The color charge density $\psi^\dagger\psi$ is related to the difference of virtual quark and antiquark densities. The charge density can be constructed similar as the chiral condensate by inserting a γ_4 matrix between $\bar{\psi}$ and ψ

$$\psi^\dagger(x)\psi(x) \; = \; \bar{\psi}(x)\gamma_4\psi(x) \, . \tag{4}$$

To obtain the expectation value of a physical observable O in a static quark-antiquark system we have to calculate the corresponding path integral of the correlation function

$$< L(\vec{r}_1)L^\dagger(\vec{r}_2)O(\vec{r}) > = \frac{1}{Z} \int \mathcal{D}[U]\mathcal{D}[\bar{\psi},\psi] \; L(\vec{r}_1)L^\dagger(\vec{r}_2)O(\vec{r}) \; e^{-S[U,\bar{\psi},\psi]} \; . \tag{5}$$

The functional integration extends over all degrees of freedom of the gauge field U and of the fermionic fields $\bar{\psi}$ and ψ in euclidean space-time. For the gluonic action we use Wilson's plaquette action. As far as the fermionic action is concerned we employ the Kogut-Susskind formulation[4]

$$S_{KS} = \frac{1}{2} \sum_{x,\mu} \Gamma_\mu(x)\big[\bar{\psi}(x)U_\mu(x)\psi(x+\hat{\mu}) - \bar{\psi}(x+\hat{\mu})U_\mu^\dagger(x)\psi(x)\big] + m \sum_x \bar{\psi}(x)\psi(x) \; , \tag{6}$$

where the fields $\bar{\psi}$ and ψ reduce to scalars in Dirac space and the phases Γ_μ play the role of the Dirac matrices, and alternatively the Wilson formulation[5]

$$S_W = \kappa \sum_{x,\mu} \big[\bar{\psi}(x)U_\mu(x)(\gamma_\mu-1)\psi(x+\hat{\mu}) - \bar{\psi}(x+\hat{\mu})U_\mu^\dagger(x)(\gamma_\mu+1)\psi(x)\big] + \sum_x \bar{\psi}(x)\psi(x) \; , \tag{7}$$

with the hopping parameter $\kappa = 1/(8 + 2m_0a)$.

Both formulations serve to remove the fermion doubling problem on the lattice. In the Kogut-Susskind scheme the four Dirac spinors are distributed over a hypercube. This method retains the chiral invariance of the lagrangian for dynamical quark mass $m = 0$. But it reduces the degeneracy only from 2^4 to 4 fermion modes which survive in the continuum. One tries to cure this by taking $n_f/4$ in order to represent only n_f flavors. The Wilson scheme works quite different. An additional term is introduced giving the 16 fermions different masses and thus removing the degeneracy completely. This method has the correct continuum limit. But for finite lattice constants it explicitly breaks chiral symmetry. This means that calculating $\bar{\psi}\psi$ we always measure an unphysical part stemming from that additional term.

In the continuum limit both regularization schemes should give the same results. Since we are away from this limit we have to find a relation between the two methods. For comparison the quark-antiquark potentials $V(r)$ given by

$$V(r) = -\frac{1}{N_t a} \ln < L(0)L^\dagger(r) > \tag{8}$$

can be determined. By computing the static quark potentials and the virtual quark densities one can investigate relations between the bare Kogut-Susskind mass m and the Wilson hopping parameter κ.

3. Results

In the underlying Monte Carlo simulations of QCD a hypercubic lattice with 8 points in space direction and 4 points in time direction is used for all cases. The results in Sec. 3.1 were produced applying the Metropolis algorithm and those of Sec. 3.2 to Sec. 3.4 with the Hybrid Monte Carlo method. Dynamical quarks are always treated as pseudofermions.[6]

3.1. Energy Densities and Flux Tubes

We start with the gluonic energy density distributions and the flux tube profiles between a static quark-antiquark pair where we compare QCD with and without dynamical fermions. The corresponding inverse gluon couplings β are choosen to give roughly the same physical lattice constant. The parallel electric flux with respect to the quark-antiquark dipole of length $d = |\vec{r}_2 - \vec{r}_1|$ lying in the z direction is computed by the connected normalized three-point function[3]

$$E_{\parallel}^2(\vec{r}_1, \vec{r}_2, \vec{r}) = \frac{2\beta}{N} \frac{< L(\vec{r}_1)L^{\dagger}(\vec{r}_2)P_{34}(\vec{r}) >}{< L(\vec{r}_1)L^{\dagger}(\vec{r}_2) >} - < P_{34} > . \tag{9}$$

Here the time averaged plaquette operator $P_{\mu\nu}(\vec{r}) = 1/N_t \sum_{l=1}^{N_t} \mathrm{Re}\, \mathrm{Tr} P_{\mu\nu}(\vec{r}, t_l)$ is introduced. Analog expressions hold for $E_{\perp}^2, B_{\parallel}^2$ and $B_{\perp}^2$. The total energy density ϵ is given by the sum of the electric and the magnetic energy densities, whereas the total action density σ is defined as the difference between these quantities

$$\epsilon = \frac{1}{2}(\vec{E}^2 + \vec{B}^2) , \quad \sigma = \frac{1}{2}(\vec{E}^2 - \vec{B}^2) . \tag{10}$$

In Fig. 1 the normalized gluonic energy density correlations $< L(\vec{r}_1)L^{\dagger}(\vec{r}_2)\,\epsilon(\vec{r}) >$ for a static quark-antiquark system of separation $d = |\vec{r}_2 - \vec{r}_1| = 1, 2, 3, 4$ are illustrated.[7] The creation of excess energy around the sources is seen. In the pure gluonic case ($\beta = 5.6$) the formation of a flux tube is observed which does not break with increasing distance. The situation changes in the presence of dynamical quarks ($\beta = 5.2$). Here the breaking of the flux tube sets in but is not completed at maximum separation $d = 4$.

To obtain a further idea of the gluon string in different vacua we present in Fig. 2 flux tube profiles $< L(\vec{r}_1)L^{\dagger}(\vec{r}_2)P_{\mu\nu}(\vec{r}_{\perp}) >$ normalized as in Eq. (9). The static meson is cut by the symmetry plane at $d/2$ between the two charges and the observables are taken at radial separation $r_{\perp} \leq 4$. We compute all spatial components of the color electromagnetic field densities $E_{\parallel}^2, E_{\perp}^2, B_{\perp}^2, B_{\parallel}^2$ by evaluating the corresponding space-time like and space-space like plaquette operators $P_{\mu\nu}$ according to (9). The parallel components $E_{\parallel}, B_{\parallel}$ measure in the direction of the quark-antiquark pair and the normal components $E_{\perp}, B_{\perp}$ perpendicular to it. In the pure gluonic case we see that all field components have the tendency to spread out with increasing meson elongation and obey the relation

$$E_{\parallel}^2 > E_{\perp}^2 \approx |B_{\perp}^2| > |B_{\parallel}^2| . \tag{11}$$

The dominating contribution of $E_{\parallel}$ originates from the color-charge current transported from quark to antiquark via the gluon string. The non-vanishing contribution

202

ENERGY DENSITIES $\langle L(\vec{r}_1)L^\dagger(\vec{r}_2)\epsilon(\vec{r})\rangle$

no fermions $(\beta = 5.6)$

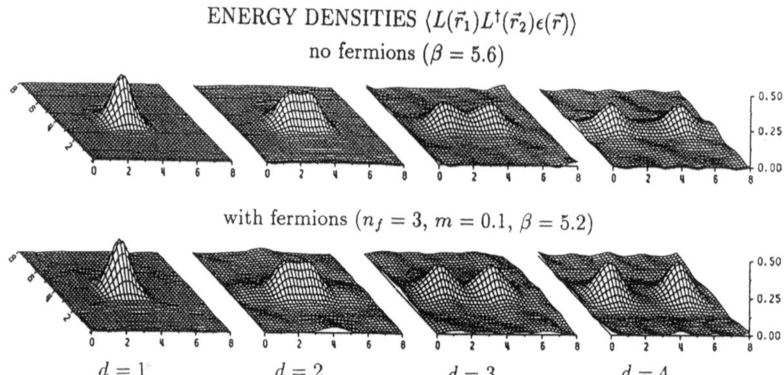

with fermions $(n_f = 3,\ m = 0.1,\ \beta = 5.2)$

$d = 1 \qquad d = 2 \qquad d = 3 \qquad d = 4$

Fig. 1: Energy density distributions $< L(\vec{r}_1)L^\dagger(\vec{r}_2)\ \epsilon(\vec{r}) >$ for a static quark-antiquark pair with elongation $d = |\vec{r}_2 - \vec{r}_1|$ with and without dynamical quarks. The meson position is kept fixed and the observable $\epsilon(\vec{r})$ is moving around the plane spanned by the three objects. In the pure $SU(3)$ theory a gluon string is formed which breaks if dynamical fermions are included.

FLUX TUBE PROFILES $\langle L(\vec{r}_1)L^\dagger(\vec{r}_2)P_{\mu\nu}(\vec{r}_\perp)\rangle$

no fermions $(\beta = 5.6)$

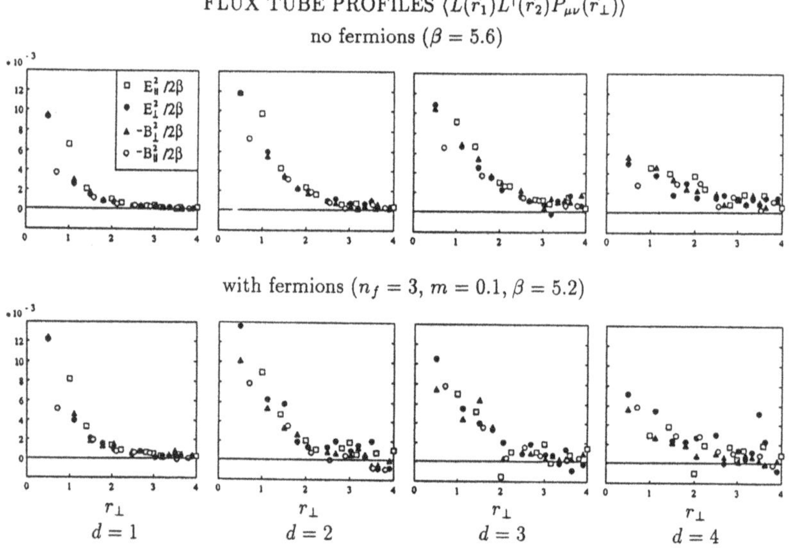

Fig. 2: Flux tube profiles $< L(\vec{r}_1)L^\dagger(\vec{r}_2)\ P_{\mu\nu}(\vec{r}_\perp) >$ in a static meson with length $d = |\vec{r}_2 - \vec{r}_1|$ in the presence of dynamical quarks. The plaquette operator $P_{\mu\nu}(\vec{r}_\perp)$ corresponding to the color electromagnetic energy densities $E_\parallel^2$, $E_\perp^2$, $B_\perp^2$, $B_\parallel^2$ is measured at $d/2$ orthogonal to the two charges as a function of radial separation $r_\perp$. All field components spread out with increasing meson length both with and without dynamical fermions.

of $E_\perp$ can be interpreted to arise from the current of the charged gluons of the vibrating string. The negative magnetic net energies for $B_\perp$ and $B_\parallel$ might reflect the dilution of the monopole condensate inside the electric flux tube.[8] Taking dynamical fermions into account we see again a spreading of the field components with increasing meson elongation. The above relation (11) is also valid but the magnitude of all components has a stronger tendency to shrink with growing meson length which can be attributed to the screening of the static sources by sea quarks.

A lively discussed problem is the thickness of the gluon string.[9] As a convenient measure the root mean square radius of the flux tube profile can be taken. We tried to compute this observable and end up with the important result that the gluon string is very thin with a width of about one lattice unit and practically independent of the meson length.

3.2. Potentials

We turn to the potential between a static quark and antiquark.[10] Fig. 3a shows both pure gluonic and full QCD with sea quark masses $m = 0.1, 0.025$ and 0.0125 in lattice units. The latter represents the lowest mass currently accessible in lattice simulations and corresponds to roughly 8 MeV. To get an idea of the scale the lattice constant is about 0.3 fm.

The potential mediated by pure gluon exchange has a Coulomb plus linear form which is characteristic for confinement. The linear part can be interpreted as the energy of a string of gluons between the separated quark and antiquark being built to locally neutralize the color of the system.

Switching on the quark sea the linearly rising potential becomes screened due to the production of virtual quark-antiquark pairs.[11] The amount of energy needed to create a $q\bar{q}$ pair decreases with smaller quark mass. Thus, the screening effect is stronger for light quarks as can be seen in Fig. 3a.

3.3. Chiral Condensates and Charge Densities

In Fig. 3b we display the chiral condensate $\bar\psi\psi$ around a static quark as a function of distance r for various quark masses.[12] The data has been normalized to the cluster value $< L >< \bar\psi\psi >$ to get rid of renormalization constants. We see that the chiral condensate is lowered in the vicinity of color sources. Taking $\bar\psi\psi$ as an order parameter of chiral symmetry breaking we find that chiral symmetry is locally restored in the vicinity of static color sources which is in agreement with the basic ideas of chiral bag models where a perturbative vacuum is assumed inside the bag. Interpreting $\bar\psi\psi$ as a measure for the sum of virtual quark and antiquark densities we observe that the production of virtual particles is reduced in the vicinity of a static color source which supports the picture of the suppression of the pion cloud in the bag of valence quarks. The suppression is a consequence of the increasing gluon density near an external charge which reflects the antiscreening mechanism of QCD. Looking at the dependence on the dynamical quark mass we find that the effect is more pronounced for lighter quarks. This indicates that a heavy quark sea can reach close to an external charge and might give an explanation for a rather strong strange sea inside a hadron

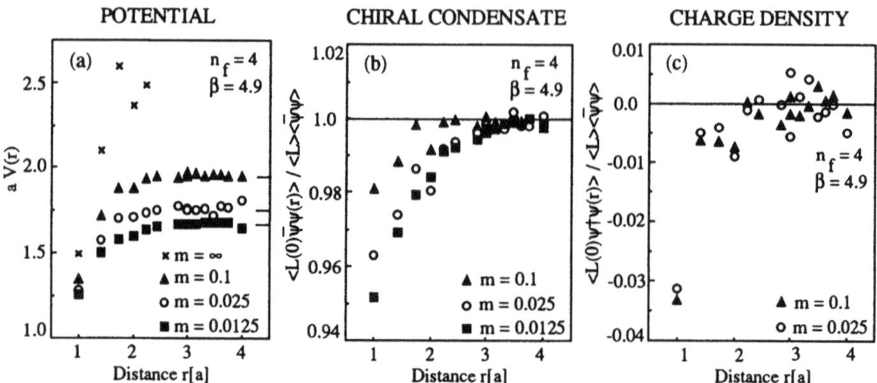

Fig. 3: (a) Screening of the linearly rising potential $V(r)$ in a static quark-antiquark system when sea quarks are taken into account. (b) Restoration of chiral symmetry $\bar{\psi}\psi(r)$ around a static quark charge. (c) Enhancement of the virtual charge density $\psi^\dagger\psi(r)$ with opposite sign around an external quark.

sheding some light on the proton spin crisis.[13]

The charge density $\psi^\dagger\psi$ around a static quark is shown in Fig. 3c for different quark masses.[14] The data has been normalized to the cluster value of the condensate. We find a net charge excess of opposite sign around a source. It goes to zero with increasing distance from the source approaching the undisturbed vacuum. The screening is in contrast to the antiscreening from the gluons expected for a non-abelian theory. Concerning the mass dependence the correlations agree within error bars. This might indicate that a single color charge is screened on average by exactly one anticharge.

We now turn to the case of a static quark-antiquark pair separated by distances $d = |\vec{r}_2 - \vec{r}_1| = 1, 2, 3$ in lattice units. In Fig. 4 the chiral condensates normalized to their cluster values are presented for quark masses $m = 0.1$ (first row) and $m = 0.025$ (second row). One observes the formation of a single quark and a single antiquark with increasing distance. Again, the restoration of chiral symmetry is stronger for lighter dynamical fermions.

Fig. 5 illustrates analogously the charge density for $m = 0.1$ and $m = 0.025$ in a static meson. With increasing elongation the polarization of the quark and antiquark by virtual particles of opposite charge becomes more pronounced. Our plots show practically no mass dependence.

3.4. Discretization Schemes

The results described so far were obtained using the Kogut-Susskind discretization of the fermionic degrees of freedom. Continuum physics should not depend on the regularization technique and thus one expects employing dynamical Wilson fermions

CHIRAL CONDENSATE IN MESON $\langle L(\vec{r}_1)L^\dagger(\vec{r}_2)\bar{\psi}\psi(\vec{r})\rangle$

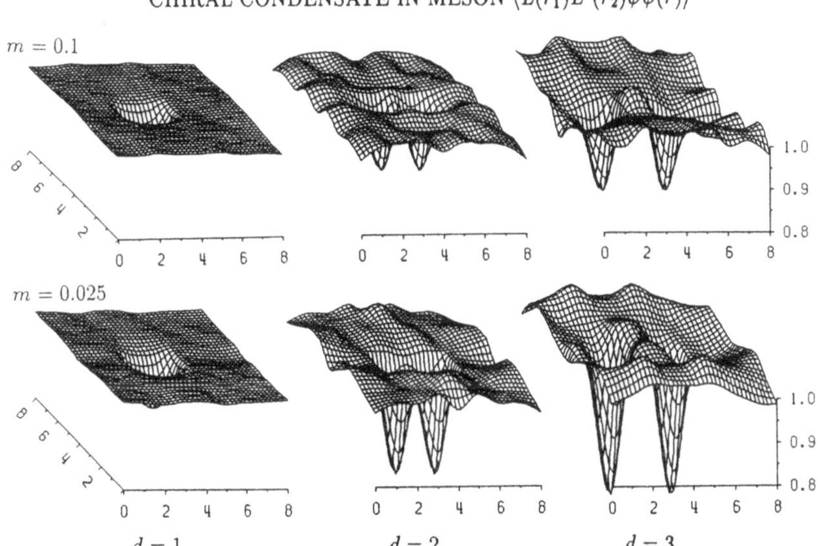

Fig. 4: Chiral condensates around a static quark-antiquark system separated by distances $d = |\vec{r}_2 - \vec{r}_1|$ for virtual quark masses $m = 0.1$ and $m = 0.025$. With increasing distance the formation of an isolated quark and an isolated antiquark each suppressing the condensate takes place. The effect is stronger for lighter sea quarks.

CHARGE DENSITY IN MESON $\langle L(\vec{r}_1)L^\dagger(\vec{r}_2)\psi^\dagger\psi(\vec{r})\rangle$

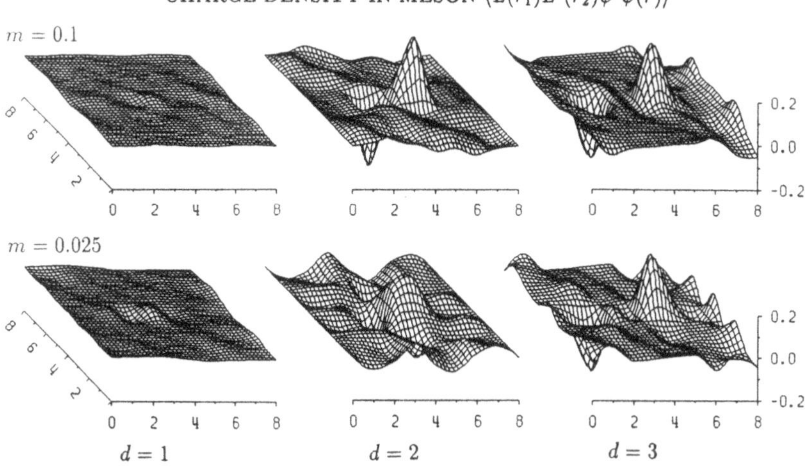

Fig. 5: Charge densities in a static meson. Around the valence quarks the virtual quark density of opposite sign is clearly enhanced. Hadronization becomes visible with increasing elongation.

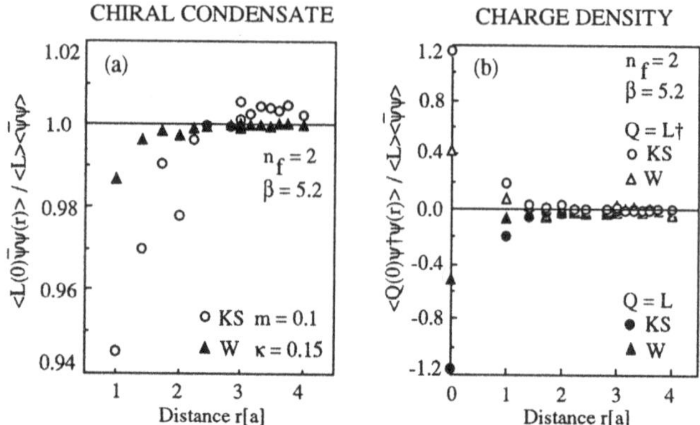

Fig. 6: Chiral condensates and charge densities around a static quark for Kogut-Susskind and Wilson fermions. The chiral condensate (a) is lowered around external sources reflecting gluonic antiscreening. The virtual charge density (b) with opposite sign increases around an external charge in analogy to QED. The effect is less pronounced for Wilson fermions on both observables in contrary to the potential which may reflect the different discretization schemes. For better visibility the Wilson data of the charge density is multiplied by a factor of 100.

to yield qualitatively the same results as Kogut-Susskind fermions. We compare both discretization schemes by calculating the chiral condensate and the charge density around a static quark. For Kogut-Susskind fermions we use a bare mass $m = 0.1$ and for Wilson fermions we work with a hopping parameter $\kappa = 0.15$ which leads to a stronger screened potential in the Wilson case.[15] An exact formula converting κ to an equivalent Kogut-Susskind mass m is not known.

The chiral condensate $\bar{\psi}\psi$ around a static color source is shown in Fig. 6a for both formulations. The restoration of chiral symmetry is observed for both definitions. However, the screening effect on the chiral condensate is less pronounced for Wilson fermions. A possible explanation might be that $\bar{\psi}\psi$ which is the order parameter of chiral symmetry is more sensitive to the additional chiral symmetry breaking term of the Wilson action.

In Fig. 6b the charge density around a static quark is compared for both definitions. The data has been normalized to the cluster value of the condensate which is known in the Wilson case only up to a constant proportional to the renormalized mass. The screening effect can be seen for both definitions but is again smaller for Wilson fermions. In explaining the mechanism the same arguments as for the condensate hold true.

4. Summary

We have reviewed what is known about the structure of the QCD vacuum by analyzing flux tubes, potentials, chiral condensates and charge densities in numerical simulations of lattice QCD with dynamical fermions. We could clearly demonstrate the formation of a flux tube for a non-abelian $SU(3)$ gauge theory and its breaking by dynamical fermions. It was shown that the linear part of the confining potential is screened by sea quarks. We observed that the chiral symmetry is restored in the vicinity of static color sources and that the effect is stronger for lighter dynamical quarks. Our results support the picture of the pion cloud suppression in bag models. We further found that virtual anticharges screen external color charges of QCD in analogy to QED. Employing two completely different fermion discretization schemes it turned out that there is qualitative agreement between the polarization observables. To learn more about the QCD vacuum we recommend for future investigations to analyze the monopole condensate and the instanton distribution in hadrons with lattice techniques.

6. References

1. H.-Q. Ding, *Phys. Lett.* **B200** (1988) 133.

2. L. D. McLerran and B. Svetitsky, *Phys. Rev.* **D24** (1981) 450.

3. R. Sommer, *Nucl. Phys.* **B291** (1987) 673;
   J. Wosiek and R. W. Haymaker, *Phys. Rev.* **D36** (1987) 3297;
   R. Sommer, *Nucl. Phys.* **B306** (1988) 181.

4. J. Kogut and L. Susskind, *Phys. Rev.* **D11** (1975) 395.

5. K. Wilson, *Phys. Rev.* **D10** (1974) 2445.

6. W. Hamber, E. Marinari, G. Parisi and C. Rebbi, *Phys. Lett.* **B124** (1983) 99.

7. W. Feilmair and H. Markum, *Nucl. Phys.* **B370** (1992) 299.

8. A. S. Kronfeld, G. Schierholz and U.-J. Wiese, *Nucl. Phys.* **B293** (1987) 461.

9. M. Lüscher, G. Münster and P. Weisz, *Nucl. Phys.* **B180** (1981) 1;
   M. Lüscher, *Nucl. Phys.* **B180** (1981) 317.

10. K. D. Born, R. Altmeyer, W. Ibes, E. Laermann, R. Sommer, T. F. Walsh and P. M. Zerwas, *Nucl. Phys.* **B** (Proc. Suppl.) **20** (1991) 394.

11. M. Faber, P. de Forcrand, H. Markum, M. Meinhart and I. Stamatescu, *Phys. Lett.* **B200** (1988) 348;
    E. Gerstenmayer, M. Faber, W. Feilmair, H. Markum and M. Müller, *Phys. Lett.* **B231** (1989) 453.

12. W. Feilmair, M. Faber and H. Markum, *Phys. Rev.* **D39** (1989) 1409;
    Phys. Lett. **B221** (1989) 363.

13. M. Karliner, *Nucl. Phys.* **B** (Proc. Suppl.) **23B** (1991) 98.

14. M. Müller, M. Faber, W. Feilmair and H. Markum, *Nucl. Phys.* **B335** (1990) 502.

15. W. Sakuler, W. Bürger, M. Faber, H. Markum, M. Müller, P. De Forcrand, A. Nakamura and I. O. Stamatescu, *Phys. Lett.* **B276** (1992) 155.

Introduction to Lattice Gaugefixing
and Effective Quark and Gluon Masses

Claude Bernard
Physics Dept., Washington University, St. Louis, MO 63130, USA

Amarjit Soni and Ken Yee[1]
Physics Dept., Brookhaven National Laboratory, Upton, NY 11973, USA

ABSTRACT

We report on the status of quark and gluon propagators in quenched, gaugefixed lattice QCD. In Landau gauge we find that the effective quark mass in the chiral limit is $M_q \sim 350(40) MeV$. Quark and gluon propagators, the slope of the quark dispersion relation, and effective masses all appear to depend on gauge. A link-chain picture of lattice gaugefixing in the color $N \to \infty$ and strong coupling limit, where the system becomes almost solvable, supports the gauge variance of these numerical results.

1. Motivation and Overview

Lattice gaugefixing has drawn considerable attention in recent years. Gluon [1], quark and photon propagators and effective masses have been evaluated on the lattice in Landau, Coulomb and related "λ-gauges," $\lambda \partial_0 A^0 + \nabla \cdot \vec{A} = 0$. Wavefunctions of valence quarks inside mesons have been probed with gauge-variant extended sources. Furthermore, lattice gaugefixing is routinely performed nowadays in order to use extended gauge-variant sources as interpolating fields in the calculation of gauge-*invariant* quantities. While beyond the scope of this talk (see Refs. [2]-[4] for some references), the numerical work has spawned analytical and computational studies of longitudinal and topological gaugefixing ambiguities and their effects on gluon, quark and photon correlation functions and operator product expansion coefficients determined from gauge covariant matching conditions.

For perspective let us focus momentarily on the solvable Schwinger model, where quarks(electrons) are confined but the photon is physical and has a gauge-invariant nonzero mass from the $U_A(1)$ anomaly. Coulomb gauge—the unitary gauge for the Schwinger model—has no unphysical modes and the Coulomb gauge quark propagator is the physical amplitude for quark propagation. Since in $1 + 1$ dimensional empty space a quark's electric field $\vec{E}$ is constant and cannot die away, $\vec{E}$ produces a constant flux of quark-antiquark pairs moving away from the original quark at the speed of light. Consequently the quark propagator—the amplitude to have just one quark at $x \neq 0$ starting with an $x = 0$ quark—vanishes and the effective *Coulomb* gauge quark mass diverges. On the other hand, the *covariant* gauge quark propagator is not the physical amplitude for quark propagation. In fact the Landau gauge quark mass vanishes. This discrepancy with Coulomb gauge is due to unphysical modes (or,

---

[1]Presented by K. Yee, present address: Dept. of Physics and Astronomy, LSU, Baton Rouge, LA 70803-4001, USA; kyee@rouge.phys.lsu.edu

alternatively, Gupta-Bleuler physical-state conditions on the Hilbert space) which ruin the identification of the covariant gauge quark propagator with the physical propagation amplitude. Thus, the effective quark mass can be shown to *vary* with covariant gauge parameter [4]. The photon mass, being physical, is gauge *invariant*.

The point is, in association with confinement, the effective quark mass is ambiguous because it varies with gauge—at least in the Schwinger model. Like Schwinger model quarks, QCD quarks and gluons are confined and unphysical and nothing guarantees their masses be gauge-invariant. If what is not forbidden occurs (as usually happens), quark and gluon masses would be gauge-variant.

2. Numerical Lattice QCD Results

Numerical quark and gluon propagators at large times—as large as our lattices permit—look approximately, but not exactly, like free particle propagators.[2] In λ-gauges M_q and M_g are evaluated by matching gaugefixed $\vec{p} = 0$ propagators to free particle propagators at large Euclidean time t. In particular, the zero momentum λ-gauge $M_q^{(\lambda)}$ is obtained by comparing the numerical quark propagator to $Z_q(\frac{1+\gamma^0}{2})e^{-M_q^{(\lambda)}t}$. The "$(\lambda)$" superscript allows for λ dependence. Since the scalar "1" part of the numerical quark propagator equals the vector "γ^0" part at large enough times, we can focus on the scalar part without loss of generality. Correlated χ^2-fits of the large time scalar part of lattice quark propagators to $Z_q e^{-M_q^{(\lambda)}t}$ give unacceptably large χ^2s. Quark propagators fit well to trial functions like $At^B e^{-m_B t}$.

Rather than use an unmotivated but better-fitting trial function, in the rest of this talk we will ignore the large χ^2's and assume free particle form $Z_q e^{-M_q^{(\lambda)}t}$. Then the following features are observed: (i)The Dirac scalar component of the $\vec{p} = 0$ numerical quark propagator equals the vector component at large times. (ii)As depicted in Figure 1, quark and gluon effective masses vary with gauge parameter λ. They decrease with increasing λ such that most of the change occurs when $0 < \lambda \leq 2$. As $\lambda \to \infty$, the masses stabilize to a nonzero value. (iii)Motivated by the chiral perturbation theory relation $M_\pi^2 \propto m_q$, a fit to

$$M_q^{(\lambda)} = b^{(\lambda)} M_\pi^2 + M_c^{(\lambda)} \tag{1}$$

in the chiral regime yields in Landau gauge at $\beta = 6.0$ that $b^{(1)} \sim 2.7(.3) \times 10^{-4}/\text{MeV}$ and $M_c^{(1)} \sim 350(40)\text{MeV}$. In general, M_q is about one-half the ρ mass at the same hopping constant K, the lattice bare mass parameter, independently of K. (iv)M_g is about 70% larger than $M_c^{(1)}$. (v)The slope of the dependence of quark energy square, E_q^2, on spatial momentum square, $\vec{p}^2$, varies with λ so that not all gauges can be consistent with Lorentz invariance.

3. Lattice Gaugefixing and $\beta = 0$, $N \to \infty$ Model

The λ dependences of the effective quark mass and dispersion relation are quali-

---

[2]The numerical calculation is on $\beta = 5.7$ and 6.0 quenched Wilson lattices. For noncogniscenti, β is a measure of physical lattice spacing; the larger β is, the smaller the lattice spacing and the closer one is to the continuum limit. $\beta = 6.0$ correspond to a lattice spacing of $\sim 1.7 GeV^{-1}$.

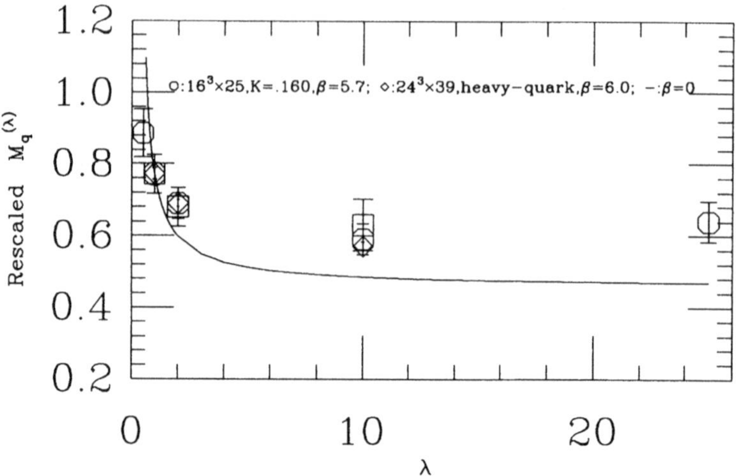

FIG. 1. Comparison of $M_q^{(\lambda)}$ to Strong Coupling λ-Dependence

tatively reproduced in a $\beta = 0$, color $N \to \infty$ analytical solution of λ-gauge lattice QCD. Figure 1 compares the chiral quark mass $M_c^{(\lambda)}$ at strong coupling to the numerical effective quark masses rescaled so that they all agree at $\lambda = 1$. The strong coupling calculation was done in the following way. λ-gauge vacuum expectation value $\langle \Theta \rangle$ in lattice QCD is defined as

$$\langle \Theta \rangle \equiv \left[\, [\, [\Theta]_f]_v \, \right]_u, \quad [\Theta]_\theta \equiv z_\theta^{-1} \int [d\theta] e^{-S^\theta} \Theta, \quad z_\theta \equiv \int [d\theta'] e^{-S^{\theta'}}, \tag{2}$$

where labels $\theta = f$, v or u refer respectively to quark, gauge transformation, and lattice gauge "link" fields. $[\Theta]_\theta$ is the average of the operator (or combination of operators) Θ over the fields $\theta = f$, v, or u. f stands for quark fields ψ; v, for fields V which transform a background gauge field into λ-gauge; and u, for the lattice gauge fields U. S^f and S^u are the usual lattice Wilson fermion and plaquette actions. The lattice gaugefixing action S^v is

$$S^v \equiv -\sum_{x,\mu} \frac{J_n}{2} \, \mathrm{Retr}(V_x U_{x,\mu} V_{x+\hat\mu}^\dagger), \quad J_\mu \equiv \frac{\lambda_\mu}{\xi} \tag{3}$$

where $\lambda_0 \equiv \lambda$ and $\lambda_i \equiv 1$. Quarks are quenched by choice; consistency with the Fadeev-Popov method requires $\{V_x\}$ to be quenched. S^v is minimized site-by-site with respect to $\{V_x\}$ in numerical simulations to achieve λ-gauge. Similarly, the $\xi \to 0$ limit corresponds to λ-gauge.

Since this system is not solvable, we take the λ-gauge, $N \to \infty$ limit [3]. Identifying $\xi \equiv 1/(2N)$, taking $N \to \infty$ and integrating out gaugefixing fields $\{V_x\}$ yields

$$\langle V_y(_y\mathbf{A}_w)V_w^\dagger \rangle_{\alpha j} = \delta_{\alpha j} \sum_{w \tilde{\mathbf{a}}_y} A_{w\tilde{\mathbf{a}}_y}(J) \, \mathrm{tr}[_y\mathbf{A}_w \, _w\tilde{\mathbf{A}}_y]_u, \quad A_{w\tilde{\mathbf{a}}_y} = \frac{1}{N^2} \mathrm{tr} \langle V_{yy}\tilde{\mathbf{A}}_w V_w^\dagger \rangle_{|\beta=0}. \tag{4}$$

α and j are color indices ranging from 1 to N and $_y\!A_w$ is a link chain from w to y. Eq. (4) says the gaugefixed expectation value of $_y\!A_w$ is the weighted sum over all Wilson loops made by joining to $_y\!A_w$ a selfavoiding link segment $_w\tilde{A}_y$. (4) identifies weights $A_{_w\tilde{A}_y}$ with the $\beta = 0$ gaugefixed expectation value of $_w\tilde{A}_y$. All of the λ dependence is in these weights.

Hopping expanding the quark fields, applying (4), and resumming the hopping expansion leads to an expression for the quark propagator. In $D = 3 + 1$ dimensions in the color $N \rightarrow \infty$ and $\beta = 0$ limit, and in an orthogonal trace approximation described in Ref. [3], the quark propagator dispersion relation is

$$E_q^{(\lambda)2} = M_q^{(\lambda)2} + \sum_i \bar{g}_i^2 p_i^2, \qquad \bar{g}_i(\lambda) = \left\{ \begin{array}{ll} \frac{3}{4\lambda} & \lambda \leq \frac{1}{2}; \\ \frac{3\lambda}{4\lambda-1} & \lambda \geq \frac{1}{2}. \end{array} \right. \tag{5}$$

Since slope $\bar{g}_i(\lambda \neq 1) \neq 1$, the dispersion relation is not free-particle form except in Landau gauge. At $\vec{p} = 0$ $\bar{g}_i$ dependence drops out, but $M_q^{(\lambda)}$ remains λ-dependent. The quark propagator in this limit (with Wilson parameter $r = 1$) has the free particle exponential form.

While we do not explain Wilson parameter r here, let us note that in strong coupling the $r = 0$ point (where backtracking and hence spontaneous chiral symmetry breaking is allowed) is more similar than $r = 1$ to continuum QCD. At $r = 0$ $M_q^{(\lambda)}$ varies with λ, diverges in Coulomb gauge, and is analytic and linear in m_q near $m_q \rightarrow 0$. Following (1), the Taylor expansion $M_q^{(\lambda)} = M_c^{(\lambda)} + B^{(\lambda)}m_q + \mathcal{O}(m_q^2)$ yields

$$M_c^{(\lambda)} = \frac{\sqrt{7}}{242\lambda} \left\{ \begin{array}{ll} 67 - 25\lambda^2 & \lambda \leq \frac{1}{2}; \\ \frac{-25+200\lambda+672\lambda^2}{4(4\lambda-1)} & \lambda \geq \frac{1}{2}; \end{array} \right. \qquad B^{(\lambda)} = \left\{ \begin{array}{ll} \frac{20203+4753\lambda^2}{29282\lambda} & \lambda \leq \frac{1}{2}; \\ \frac{4753-38024\lambda+399296\lambda^2}{117128\lambda(4\lambda-1)} & \lambda \geq \frac{1}{2}. \end{array} \right. \tag{6}$$

4. Acknowledgements

Computing was at NERSC (partially within "Grand Challenge") and SDSC. AS and KY are partially supported by DOE contract $DE - AC02 - 76CH00016$; CB partially by DOE grant $DE2 - FG02 - 91ER40628$.

1. J. Mandula and M. Ogilvie, *Phys. Lett.* B185 (1987) 127; M. Ogilvie, *Phys. Lett.* B231 (1989) 161.

2. C. Bernard, D. Murphy, A. Soni, K. Yee, *Nucl. Phys.* B(Proc. Suppl.) 17 (1990) 593; C. Bernard, A. Soni, K. Yee, *Nucl. Phys.* B (Proc. Suppl.) 20 (1991) 410; C. Bernard, A. Soni, K. Yee, *Gaugefixed Lattice QCD: Masses and Matching Coefficients*, Wash. U. preprint HEP/92-33, to be published.

3. K. Yee, *Gauge Dependence of Effective Quark Mass and Matrix Elements in Gaugefixed Large N Strong Coupling Lattice QCD*, BNL preprint #47712 (1992), to be published; K. Yee, *Central Charge of the Parallelogram Lattice Strong Coupling Schwinger Model*, BNL preprint #47604 (1992), to be published.

4. A. Casher, J. Kogut and L. Susskind, *Phys. Rev.* D10 (1974) 732; K. Yee, *Phys. Rev.* D45 (1992) 4644.

LIGHT-CONE QUANTIZED QCD

Kent Hornbostel
Department of Physics, Ohio State University
Columbus, OH 43210

ABSTRACT

This talk introduces the fundamentals of light-cone quantization as a tool for nonperturbative QCD. Numerical results from 1+1 dimensional QCD provide an illustration. The vacuum is discussed in detail, showing in what sense it may be considered trivial and how effects of a complicated vacuum may be naturally incorporated.

1. Introduction

In this talk I will present an introduction to light-cone (LC) quantization, focusing on nonperturbative calculations in QCD. In particular, I will present arguments which suggest that the perturbative, or trivial, vacuum is also an eigenstate of the full theory. If true, this would provide an enormous simplification and justify some of the recent interest in this approach. As an illustration, and to demonstrate what can be accomplished with a few minutes of CPU, I will review its application to computing the spectrum and wavefunctions of hadrons in $1 + 1$ dimensional QCD. Finally, in light of the emphasis of this workshop, I will return to the LC vacuum to discuss in what sense it may be considered trivial, when kinematic arguments which suggest this are overly simplistic and must be modified, and how effects from a nontrivial vacuum may be naturally incorporated.

2. Light-Cone Quantization[1]

LC quantization is a Hamiltonian formalism which, apart from the choice of initial surface, is very similar to standard equal-time (ET) canonical quantization. Given a Lagrangian, the problem is to solve the equations of motion to determine the fields ϕ^i at all x^μ. To accomplish this, one first chooses initial conditions for the ϕ^i on an appropriate surface, then uses a Hamiltonian to evolve these off the surface. For a quantum system, setting initial conditions for independent fields involves fixing appropriate commutation relations.

As the name suggests, in the ET formalism $t = 0$ defines the initial surface. There commutation relations may be realized by equating $\phi(0, \mathbf{x})$ to a free field expanded in creation and annihilation operators. The Poincaré generators which translate, rotate and boost states are constructed from these fields. The rotation and translation generators $\mathbf{J}$ and $\mathbf{P}$ leave the initial surface invariant and are simple to implement, acting as if on free fields, while the Hamiltonian P_0 and boosts $\mathbf{K}$ evolve the system off $t = 0$. They therefore involve the full dynamics and are complicated. By diagonalizing P_0 in a Fock basis constructed with creation operators acting on the vacuum, one can determine the evolution of states in t, which is equivalent to solving the equations of motion. The perturbative or free vacuum, defined as the state with no particles, will not generally be the ground state of the interacting theory.

The same procedure can be followed on the LC, but the initial surface is defined by zero LC time, $x^+ \equiv (x^0 + x^3)/\sqrt{2} = 0$. The LC longitudinal and transverse spatial coordinates, $x^- \equiv (x^0 - x^3)/\sqrt{2}$, $\mathbf{x}_\perp \equiv (x^1, x^2)$, parametrize this surface. The next step is to determine equal-x^+ commutation relations among independent fields. This is often more involved than for ET systems; for example, new constraint equations often appear. The fields can then be expanded in creation and annihilation operators. The LC generators of translations, boosts, and J_z preserve $x^+ = 0$. As a result, it is simple to boost states, but J_x and J_y involve interactions.

In LC variables, the scalar product

$$P_\mu x^\mu = P_+ x^+ + P_- x^- - \mathbf{P}_\perp \cdot \mathbf{x}_\perp \tag{1}$$

reveals that $(P_-, \mathbf{P}_\perp)$ are conjugate to spatial variables $(x^-, \mathbf{x}_\perp)$, and so act as three-momenta, while P_+, conjugate to time x^+, is the Hamiltonian. Diagonalizing P_+ in the Fock basis determines evolution in x^+. (Note that $g_{\mu\nu}$ is off-diagonal, with $g_{+-} = g^{-+} = 1$, $g_{++} = g_{--} = 0$ and $P_\pm = P^\mp$. To avoid confusion, I will use upper indices for coordinates $x^\pm$ and lower for momenta $P_\pm$.)

Free particles, which satisfy $k^2 = m^2$, have LC energy $k_+ = (m^2 + k_\perp^2)/2k_-$, in contrast to the ET energy $k_0 = \pm(\mathbf{k}^2 + m^2)^{1/2}$. Requiring particles to have positive energy k_+ means that their momentum k_- is positive. This is sensible, given that $k_\pm \equiv (k_0 \pm k_3)/\sqrt{2}$. That particles carry positive k_-, while P_- and $\mathbf{P}_\perp$ are conserved in interactions leads to substantial simplifications. Vertices in which particles appear from or disappear into the vacuum are excluded, since momentum conservation would require the sum of their (positive) k_- to vanish. This is only possible if all k_- are zero and all LC energies k_+ infinite.

Consequently, the large number of diagrams in LC time-ordered perturbation theory with such vertices vanish, making it competitive with covariant perturbation theory.[2] Also, imposing a minimum k_- to exclude infinite-k_+ states restricts the number of quanta in any state of finite P_-. Finally, and most importantly for any attempt to compute a spectrum, the perturbative vacuum is apparently an eigenstate of the full Hamiltonian with eigenvalue zero. No vertex can couple the vacuum to another Fock state while conserving P_-. Furthermore, any attempt to build up a nontrivial Lorentz-invariant vacuum would seem to require particles of zero k_- and infinite k_+. (Note however that particles with zero mass and transverse momentum $\mathbf{k}_\perp$ could have zero k_- but finite energy. Here I will consider $m \neq 0$. The $k_- = 0$ region will be discussed in some detail later.)

To the extent that this picture coincides with reality, this provides an enormous advantage, since the ground state is already determined by simple kinematic arguments. Furthermore, attempts at computing hadrons are not burdened by the need to recreate an involved vacuum with processes occurring at completely unrelated locations and energy scales. All particles in a hadron's wavefunction are specifically related to that hadron. This also makes interpretation of wavefunctions straightforward: all the quanta in the proton wavefunction belong to the proton; none are spent building the vacuum. In addition, the eigenstate wavefunctions in this scheme are boost invariant, which makes it well-suited for studying processes with large momentum transfer. If I solve for the pion in the frame $(P_-, \mathbf{P}_\perp) = (1, 0)$,

214

$$|\pi(P^+ = 1, \mathbf{P}_\perp = 0)\rangle$$

$$= \sum_{\lambda_{1,2}} \int_0^1 \frac{dx}{4\pi[x(1-x)]^{1/2}} \int_{-\infty}^{\infty} \frac{d^2k_\perp}{(2\pi)^2} \, \psi_{q\bar{q}}(x, \mathbf{k}_\perp, \lambda_i) b^\dagger(x, \mathbf{k}_\perp, \lambda_1) d^\dagger(1-x, -\mathbf{k}_\perp, \lambda_2) |0\rangle$$

$$+ \sum \int \psi_{q\bar{q}g} \, b^\dagger d^\dagger a^\dagger |0\rangle + \dots . \tag{2}$$

then the wavefunctions $\psi_{q\bar{q}}, \psi_{q\bar{q}g}, \cdots$ are the same for any pion momentum. As a result, if I am interested in computing the pion form factor

$$F^\mu(q) = \langle \pi(p+q)| J^\mu(0) |\pi(p)\rangle , \tag{3}$$

I only need solve for the pion in one frame to know $F^\mu(q)$ for all q. Conversely, if I measure $F^\mu(q)$, it gives me information related to a single (multi-Fock) wavefunction, so that LC wavefunctions are closely tied to such measurable quantities.

3. $SU(N)_{1+1}$

To show how this works in a concrete example, I will present some general results obtained by applying it to numerical calculations of $1+1$ dimensional QCD.[3] These will include various numbers of color and one flavor, although the program has since been extended to allow various flavors as well.

$SU(N)$ gauge theories restricted to one spatial dimension and time have been studied extensively, both analytically and numerically, predominantly for the case when N is large.[4] There are some special properties of these theories peculiar to $1+1$ dimensions which should be mentioned for the sake of orientation. Because there are no transverse directions, the gluons are not dynamical, and (in $A^+ = 0$ gauge) their presence is felt only by the constraint equation they leave behind. Likewise the quarks carry no spin. The fermion field is a two-component spinor, $(\psi_L(x)_c, \psi_R(x)_c)$; chirality for massless fermions identifies only the direction of motion. The coupling constant g carries the dimension of mass, and for one quark flavor of mass m, the relevant parameter is g/m. The theory is superenormalizable. Finally, restriction to one spatial dimension produces confinement automatically, even for QED_{1+1}. The electric field is unable to spread out and the energy of a nonsinglet state diverges as the length of the system.

The Lagrangian is the usual

$$\mathcal{L} = -\tfrac{1}{4} F^a_{\mu\nu} F^{\mu\nu a} + \bar{\psi}(i \not{D} - m)\psi \tag{4}$$

with $F_{\mu\nu} = \partial_\mu A^a_\nu - \partial_\nu A^a_\mu - g f^{abc} A^b_\mu A^c_\nu$ and $iD_\mu = i\partial_\mu - g A^a_\mu T^a$. A useful gauge choice is $A^+ = 0$. In this gauge there are neither ghosts nor negatively normed gauge bosons, so the Fock space quanta, and therefore the wavefunction constituents, are physical and positively normed. Also, in $1+1$ dimensions, this gauge choice is Lorentz (but not parity) invariant. The equations of motion are then

$$i\partial_-\psi_L = \tfrac{1}{2} m\psi_R \tag{5}$$

$$-\partial_-^2 A^{-a} = g\psi_R^\dagger T^a \psi_R \equiv \tfrac{1}{2} g j^{+a} \tag{6}$$

$$i\partial_+ \psi_R = \tfrac{1}{2} g A^{-a} T^a \psi_R + \tfrac{1}{2} m \psi_L \tag{7}$$

$$\partial_+ \partial_- A^{-a} = g\psi_L^\dagger T^a \psi_L - \tfrac{1}{2} g f^{abc} \partial_- A^{-b} A^{-c} \equiv \tfrac{1}{2} g j^{-a}. \tag{8}$$

Only Eqs. (7, 8) are dynamical; these will be generated by the Hamiltonian P_+. Eqs. (5, 6) are constraints, as they involve only spatial derivatives.

To quantize the model, one solves for the dependent fields ψ_L and A^{-a} by inverting these derivatives, and then assigns an appropriate commutation relation to the only remaining dynamical degree of freedom, ψ_R,

$$\{\psi_R(x)_{c_1}, \psi_R^\dagger(y)^{c_2}\}_{x^+=y^+} = \delta_{c_1}^{c_2} \delta(x^- - y^-), \tag{9}$$

implemented at $x^+ = 0$ by expanding in terms of creation and annihilation operators:

$$\psi_R(x^-)_c = \frac{1}{\sqrt{2L}} \sum_{n=\frac{1}{2},\frac{3}{2},\cdots}^{\infty} \left[b_{n,c} e^{-ik_{n-}x^-} + d_{n,c}^\dagger e^{ik_{n-}x^-} \right]. \tag{10}$$

Here $k_{n_-} = n\pi/L$. Then

$$\begin{aligned}
\psi_L(x^-)_c &= \frac{m}{2}\frac{1}{i\partial_-}\psi_R = -\frac{im}{4} \int dy^- \epsilon(x^- - y^-)\psi_R(y^-) \\
&= \frac{1}{\sqrt{2L}} \sum_{n=\frac{1}{2},\frac{3}{2},\cdots}^{\infty} \left(\frac{m}{2k_{n-}}\right) \left[b_{n,c} e^{-ik_{n-}x^-} - d_{n,c}^\dagger e^{ik_{n-}x^-} \right].
\end{aligned} \tag{11}$$

For future reference, ψ_L apparently disappears when $m = 0$ and is singular at $k_- = 0$. Also

$$A^{-a} = -\frac{1}{\partial_-^2} j^{+a} = -\frac{g}{2} \int dy^- |x^- - y^-| : \psi_R^\dagger T^a \psi_R(y^-): . \tag{12}$$

The generators for translations, computed from the energy-momentum tensor, include the momentum

$$P_- = \int dx^- \Theta^+{}_- = \sum_{n>0} (k_{n-}) \left[b_n^\dagger b_n + d_n^\dagger d_n \right], \tag{13}$$

which is diagonal, and the Hamiltonian $P_+ = \int dx^- \Theta^+{}_+$, which divides into three parts. The first, $LQ^a Q^a/4$, enforces confinement, restricting finite energy solutions to singlets. The second is the usual Hamiltonian for free fermions, P_+^0, which assigns an energy $m^2/2k_-$, while the third contains the Coulomb interaction,

$$P_+^I = -\frac{g^2}{4} \int dx^- dy^- j^{+a}(x^-) |x^- - y^-| j^{+a}(y^-). \tag{14}$$

It includes vertices in which an instantaneous gluon is exchanged between quarks with a propagator of $1/k_-^2$. These can change fermion number by at most two. There are, however, no vertices in which four fermions emerge or disappear from the vacuum.

The only approximation is the imposition of antiperiodic boundary conditions in a box of length $2L$, so that the momenta are discrete and denumerable. The total momentum

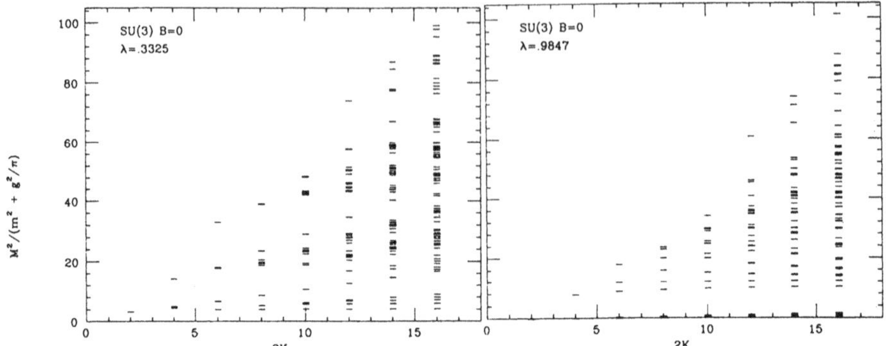

Figure 1: Meson spectra for weak $(m/g = 1.6)$ and strong $(m/g = .1)$ coupling vs. K.

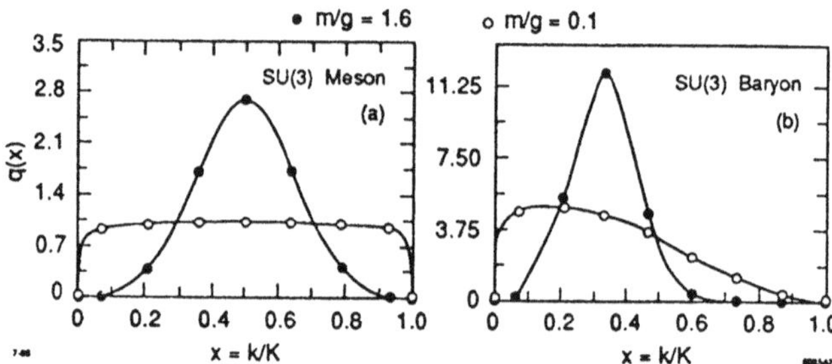

Figure 2: Valence structure functions for lightest meson, baryon.

$P_- = K(\pi/L)$, with K the sum of the half-integer momenta carried by the constituents. By momentum conservation, K sets the upper limit on particle number in a state. In the continuum limit, P_- is fixed while L and K go to infinity.

The program takes as input the number of colors N, baryon number B, total momentum K, and coupling over mass g/m. It generates the complete set of Fock states consistent with these quantum numbers, computes and diagonalizes P_+ in this basis, and returns the full spectrum of hadrons and their wavefunctions.

Fig. 1 shows the meson $(B = 0)$ spectrum of $SU(3)$ for weak $(m/g = 1.6)$ and strong $(m/g = .1)$ coupling as K is increased and the continuum approached. In the weakly coupled case, the spectrum breaks into a set of low-lying radial excitations, plus a continuum of states beginning with a pair of the lightest mesons at rest. Structures such as resonances are embedded in the continuum. The strongly coupled case is less simple to interpret. One feature worth noting is the cluster of very light mesons; their mass may be shown analytically to vanish as $m/g \rightarrow 0$.[3] The number of basis states generated in this example range from 1 to 117 at $K = 8$, and the typical required CPU on an IBM 3081 ranged from a couple seconds to 2.5 minutes.

Fig. 2 displays the valence wavefunctions of the lightest $SU(3)$ meson and baryon by

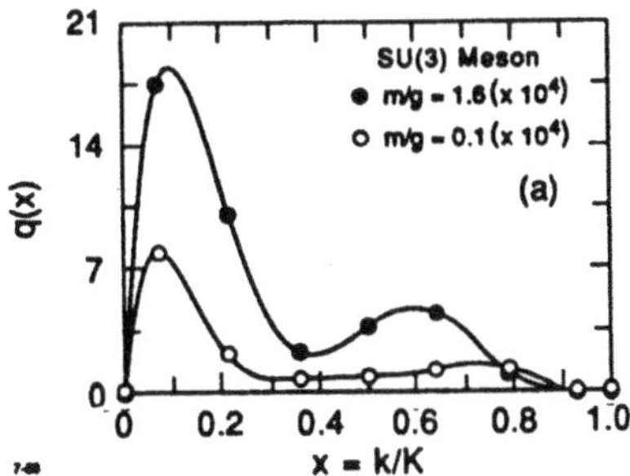

Figure 3: Four-quark structure function for lightest meson.

means of their structure functions. By boost invariance, the wavefunctions contain all the information about these hadrons at any momentum. The momentum k_- carried by a quark over the total P_- is denoted x. At weak coupling, the meson and baryon wavefunctions peak strongly at 1/2 and 1/3 respectively. There quarks are heavy, and the LC momenta are dominated by sums of masses. At strong coupling, the distribution spreads, and at $m/g \to 0$ a quark is equally likely to have any allowed momentum. Finally, the endpoints are tied to zero because the kinetic energy $m^2/2k_-$ blows up when any quark has zero momentum.

No approximation was made to explicitly restrict the number of quarks, so hadrons would be expected to also contain higher-Fock wavefunctions with extra $q\bar{q}$ pairs. Fig. 3 isolates the four-quark contribution to the lightest $SU(3)$ meson. It illustrates several features common to most higher-Fock components. As expected, the distribution is pushed to lower x since the momentum is shared among more constituents. Its magnitude is unexpectedly small, however, even for strong coupling; almost all of the meson is in the valence state. It can be shown analytically that at extremely strong coupling it is entirely valence.[3] This is only true for the lightest meson; states higher in the spectrum can have large higher-Fock components. Finally, the double bump structure is general and can be understood roughly in terms of the splitting of valence quarks. This structure is probably not specific to two dimensions and may be observable.

Diagonalizing P_+ also produces excited states. Fig. 4 shows the first three states in the meson spectrum at weak coupling. These are obviously the lightest meson and its first two radial excitations, all of which are predominantly valence. They are simple to interpret because the quanta are associated only with the meson and need not be extracted from a complicated ground state. The fourth plot is taken from the continuum. Its dominant component has two $q\bar{q}$ pairs with a shape the same as the first state but peaked at 1/4. It therefore represents a pair of the lightest mesons, and has the appropriate mass. That its shape is the same is a consequence of boost invariance.

Because of the peculiarities of LC quantization, comparisons with available results from other methods are valuable. Hamer[5] used a Hamiltonian lattice to compute the low-lying $SU(2)$ meson masses, and his lightest mass is compared in Fig. 5 to the LC results for $N = 2$ to 4 as a function of m/g. Hamer's error bars are displayed; the LC errors were comparable.

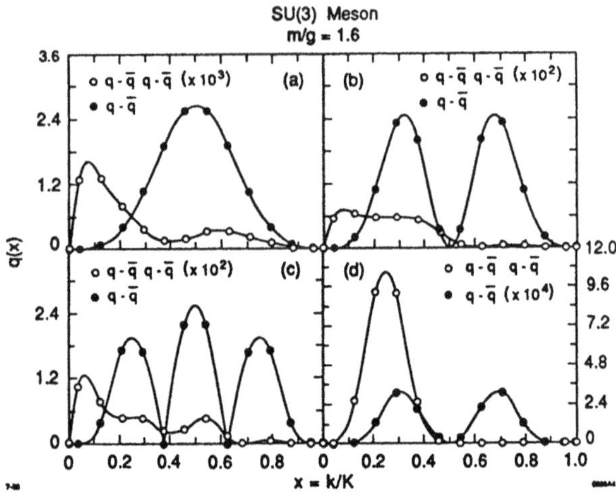

Figure 4: (a-c) First three meson structure functions. (d) Continuum meson pair.

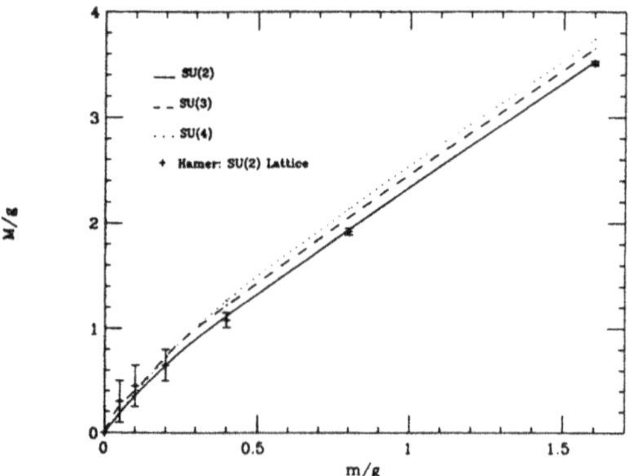

Figure 5: Lightest $N = 2, 3, 4$ meson mass vs. lattice result.

That these agree is not trivial, as the calculations differed in quantization surface ($t = 0$ vs. $x^+ = 0$), gauge ($A^0 = 0$ vs. $A^+ = 0$), space (position vs. momentum), continuum limit (box size $\to \infty$ and lattice spacing $\to 0$ vs. $K \to \infty$), and vacuum (complex vs. trivial).

Before leaving QCD$_{1+1}$ I will point out a few curious results, mainly analytic, which I will return to at the end. Although $\psi_L = (m/2i\partial_-)\psi_R$, ψ_L doesn't disappear as $m \to 0$; also, current conservation holds; finally, the correct axial anomaly is reproduced. But these are only true when measured inside the matrix elements of eigenstates.[6] It is also possible to reproduce the derivation of a nonzero vacuum expectation value for $\bar{\psi}\psi$ (at large N) even though the vacuum remains perturbative.[7] Again, all intermediate steps in the derivation involve hadronic matrix elements.

4. The LC Vacuum

There are some things about this discussion of the LC vacuum that might be bothering you. On kinematic grounds the perturbative vacuum is alleged to also be the full vacuum. However, the QCD vacuum is believed to be quite involved, and responsible for chiral symmetry breaking. If we are interested in solving for the spectrum of QCD we had better get the ground state right. Other vacuum effects, such as the Higgs mechanism in the Standard Model, seem precluded. Finally, for $1 + 1$ dimensional fermions, $\psi_L \propto m\psi_R$ and appears to vanish as $m \to 0$. In QED$_{1+1}$, for example, the only difference between the gauge and axial currents is due to ψ_L. If it vanishes they are identical, although the first is conserved and the second anomalous, and it is not clear what is going on.

The source of the problem is straightforward, especially in $1 + 1$ dimensions. The initial surface $x^+ = 0$ is insufficient to specify a wave equation;[8] left-moving massless particles, for example, never intercept it. This can be rectified by specifying initial data on the perpendicular surface $x^- = const$,[9] although I will discuss a different approach here.

To study these questions I will present some simple 1+1 dimensional systems in which LC quantization will be defined as a limit of a more conventional procedure.[10] This is in the spirit of the old infinite-momentum frame, but without the choice of a particular frame. At each step the quantization will be performed and the Hamiltonian constructed on a spacelike surface; the quantization will be canonical and unambiguous. As a result, some of the formal simplicity of LC quantization will be sacrificed for the sake of better control over quantities such as the vacuum.

I will define the initial surface to interpolate from $x^0 = 0$ to $x^0 + x^1 = 0$. The angle which the initial or quantization surface makes relative to $x^0 = 0$ will be left as a parameter. Lorentz-invariant quantities such as masses must in the end be independent of this angle, while in intermediate stages it may be chosen for convenience. Specifically, the time coordinate x^+ and space coordinate x^- are defined to be

$$\begin{bmatrix} x^+ \\ x^- \end{bmatrix} \equiv \begin{bmatrix} \sin\theta/2 & \cos\theta/2 \\ \cos\theta/2 & -\sin\theta/2 \end{bmatrix} \begin{bmatrix} x^0 \\ x^1 \end{bmatrix} , \tag{15}$$

with $(\pi - \theta)/2$ the angle between the quantization surface and $x^0 = 0$. The ET limit is at $\theta = \pi$, while $\theta = \pi/2$ is the LC limit. Because I will focus on the LC, I have chosen to retain the conventional LC notation: x^+ for time, x^- for space, with p_+ and p_- their conjugate momenta. It should be kept in mind that what these mean now depends on this

220

angle. In the ET limit $x^+ \to x^0$ and $x^- \to -x^1$; in the LC limit, $x^+ \to (x^0 + x^1)/\sqrt{2}$ and $x^- \to (x^0 - x^1)/\sqrt{2}$.

The metric

$$g^{\mu\nu} = g_{\mu\nu} = \begin{bmatrix} -\cos\theta & \sin\theta \\ \sin\theta & \cos\theta \end{bmatrix} \tag{16}$$

also depends on θ, and is seen to interpolate between the usual ET ($g_{00} = -g_{11} = 1$) and LC ($g_{+-} = g_{-+} = 1$) metrics. It has both on-diagonal and off-diagonal elements, so lowering or raising indices is somewhat involved. As before, I will use upper indices for coordinates and lower indices for their conjugate momenta. Finally, for the simplicity of later discussion, define $c \equiv -\cos\theta$, $s \equiv \sin\theta$. In the LC limit, $c \to 0$ and $s \to 1$.

4.1. Scalars

At this point, some simple systems with nontrivial but calculable vacua can be studied at arbitrary θ to see how they behave in the LC limit. A free massive scalar theory in 1+1 dimensions provides a good initial example to show how this works. The Lagrangian becomes

$$\mathcal{L} = \tfrac{1}{2}\partial_\mu\phi\,\partial^\mu\phi - \tfrac{1}{2}m^2\phi^2 = \tfrac{1}{2}c\left[(\partial_+\phi)^2 - (\partial_-\phi)^2\right] + s\,\partial_+\phi\,\partial_-\phi - \tfrac{1}{2}m^2\phi^2 \tag{17}$$

in these coordinates with the corresponding equation of motion

$$\left[\partial^2 + m^2\right]\phi = \left[c\left(\partial_+^2 - \partial_-^2\right) + 2s\partial_+\partial_- + m^2\right]\phi = 0 . \tag{18}$$

The conjugate momentum is, as usual, the variation of $\mathcal{L}$ with respect to the time derivative of ϕ,

$$\pi(x) = \frac{\delta\mathcal{L}}{\delta\left(\partial_+\phi(x)\right)} = \partial^+\phi = c\,\partial_+\phi + s\,\partial_-\phi . \tag{19}$$

For nonzero c the velocity $\partial_+\phi$ can be eliminated in favor of π by inverting Eq. (19),

$$\partial_+\phi = \frac{1}{c}\left(\pi - s\,\partial_-\phi\right) , \tag{20}$$

while in the LC limit, Eq. (20) evolves into the usual LC constraint equation $\pi = \partial_-\phi$. Away from that limit the system is canonical, as opposed to constrained, with the usual equal-x^+ commutation relation

$$[\pi(x), \phi(y)]_{x^+ = y^+} = -i\delta(x^- - y^-) . \tag{21}$$

Plane wave solutions of Eq. (18) have energy $p_+ = (\omega_p - sp_-)/c$, with $\omega_p \equiv (p_-^2 + cm^2)^{1/2}$. Expanded in these solutions,

$$\phi(x) = \int_{-\infty}^{\infty} \frac{dp_-}{(4\pi\omega_p)^{\frac{1}{2}}}\left[a(p_-)e^{-i(p_+x^+ + p_-x^-)} + a^\dagger(p_-)e^{i(p_+x^+ + p_-x^-)}\right] . \tag{22}$$

Imposing

$$\left[a(p_-), a^\dagger(q_-)\right] = \delta(p_- - q_-) \tag{23}$$

satisfies Eq. (21). The normalization in Eq. (23) is chosen to be independent of quantization angle, so that all dependence on c appears explicitly in coefficients in the fields, with no implicit dependence in a or $a^\dagger$. As a result it is simple to trace the c-dependence of states or fields.

The Hamiltonian in terms of these creation and annihilation operators is

$$P_+ = \int_{-\infty}^{\infty} dp_- \left[\frac{\omega_p - s\, p_-}{c} \right] a^\dagger(p_-)a(p_-) \,, \tag{24}$$

while the momentum

$$P_- = \int_{-\infty}^{\infty} dp_- \,[p_-]\, a^\dagger(p_-)a(p_-) \,. \tag{25}$$

For positive p_-, provided $|p_-| \gg c^{1/2}m$, the energy reduces to the conventional value

$$p_+ \sim \frac{m^2}{2p_-} + \mathcal{O}(c) \tag{26}$$

while negative p_- particles have divergent energy,

$$p_+ \sim \frac{2|p_-|}{c} + \mathcal{O}(1) \tag{27}$$

in the LC limit. At $p_- = 0$, $p_+ = m/c^{1/2}$. For the special case when $m = 0$, $p_+ = ((1+s)/c)|p_-|$ for $p_- < 0$, and $p_+ = ((1-s)/c)|p_-|$ for $p_- > 0$, with respective leading LC behaviors of $p_+ = (2/c)|p_-|$ and $p_+ = 0$.

A simple example which nevertheless contains most of the interesting results that appear in spontaneous symmetry breaking is that of a scalar field coupled to a constant source.[11] Specifically, the free Lagrangian $\mathcal{L}_0$ is symmetric under $\phi \to -\phi$. The mass term $m^2\phi^2/2$ may be thought of as a quadratic potential with minimum at $\phi = 0$, and a simple way to generate symmetry breaking is to shift the minimum to $-v$ by replacing $\phi(x)$ with $\phi(x)+v$. The effect is to add to $\mathcal{L}_0$ an interaction plus a constant piece,

$$\mathcal{L} = \mathcal{L}_0 - m^2 v\, \phi - \tfrac{1}{2}\, m^2 v^2 \,. \tag{28}$$

The sane approach, given this Lagrangian, would be to first shift ϕ back to its minimum and then quantize, as in treatments of the Higgs mechanism. The intention here is to use it as a model for more complicated theories such as QCD, where the solution is not known, so I will feign ignorance and quantize as is. However, this Lagrangian remains quadratic and soluble. If it is quantized first and the ground state solved, ϕ must find its way back to $-v$, and the vacuum must be complicated. Once obtained at arbitrary quantization angle, the object of this exercise is to see what happens near the LC.

Proceeding as usual, the constant term is dropped, and $\phi(x)$ is quantized and expanded as above. The Hamiltonian P_+ has the usual free part,

$$P_+^0 = \left(\frac{\pi}{L} \right) \sum_n \left[\frac{\omega_n - s\,n}{c} \right] a_n^\dagger a_n \tag{29}$$

plus an interaction

$$P_+^I = \int_{-L}^{L} dx^- m^2\, v\, \phi(x^-) = \left(\frac{\pi}{L} \right) \left(\frac{\hat{m}^{\frac{3}{2}} \pi^{\frac{1}{2}} v}{c^{\frac{1}{4}}} \right) \left[a_0 + a_0^\dagger \right] \,. \tag{30}$$

222

Here periodic boundary conditions have been imposed, with momentum n, mass $\hat{m}$, and ω_n in units of π/L.[12]

The result of solving $P_+ |\Omega\rangle = E_\Omega |\Omega\rangle$ is not surprising. The ground state

$$|\Omega\rangle = U |0\rangle = \exp\left\{-(c^{\frac{1}{2}}\pi\hat{m})^{\frac{1}{2}} v (a_0^\dagger - a_0)\right\} |0\rangle \tag{31}$$

is a coherent state of zero-momentum particles, and its energy

$$E_\Omega = -\tfrac{1}{2} m^2 v^2 (2L) \tag{32}$$

is just the discarded constant piece of $\mathcal{L}$ times the volume of space. Furthermore, the vacuum expectation value of $\phi(x)$ is also calculable. It involves only the constant part of $\phi(x)$, and

$$
\begin{aligned}
\langle\Omega| \phi(x) |\Omega\rangle &= \langle 0| \exp\left\{(c^{\frac{1}{2}}\pi\hat{m})^{\frac{1}{2}} v (a_0^\dagger - a_0)\right\} \left[\frac{(a_0 + a_0^\dagger)}{2(c^{\frac{1}{2}}\pi\hat{m})^{\frac{1}{2}}}\right] \exp\left\{-(c^{\frac{1}{2}}\pi\hat{m})^{\frac{1}{2}} v (a_0^\dagger - a_0)\right\} |0\rangle \\
&= -v ,
\end{aligned} \tag{33}
$$

as it must.

Because these results are valid at all quantization angles θ, it is possible to examine them near the LC; that is, as c vanishes. In particular, the full, nonperturbative vacuum given by Eq. (31),

$$|\Omega\rangle \sim \left[1 - \left(c^{\frac{1}{2}}\pi\hat{m}\right)^{\frac{1}{2}} v\, a_0^\dagger + \cdots\right] |0\rangle , \tag{34}$$

reduces to the perturbative vacuum $|0\rangle$ in this limit. This is in accord with the conventional result that in LC quantization, the perturbative vacuum is the full vacuum. Nevertheless, the ground state energy, $E_\Omega = -(1/2)m^2 v^2(2L)$ and the vacuum expectation value $\langle\Omega| \phi |\Omega\rangle$ are independent of c and so persist in this limit. While $|\Omega\rangle$ becomes trivial as $c \to 0$, the constant, zero-momentum part of ϕ is singular and diverges as $c^{-1/4}$. This is just sufficient to extract the leading correction to $|\Omega\rangle$ such that $\langle\Omega| \phi |\Omega\rangle$ is constant. E_Ω may be thought of similarly. P_+ diverges as $m/c^{1/2}$ near $p_- = 0$, and picks out the first relevant correction in $\langle\Omega| P_+ |\Omega\rangle$.

This singularity in c is the analog of the $1/k_-$ singularity which plagues the conventional LC approach. In this case, however, this singularity is fully controllable so long as c is not identically zero. One last observation from this simple model is that knowing only the first correction to $|\Omega\rangle$ in c is enough to determine $\langle\Omega| \phi |\Omega\rangle$ exactly as c vanishes. One might hope that in a real problem, such as QCD, this procedure might allow for other useful expansions in c.

4.2. Fermions

In this second simple example, based on free fermions, the full vacuum will be built of particles with all momenta rather than one of zero momentum as above.

The field $\psi = (\psi_L , \psi_R)$ in the free Lagrangian

$$\mathcal{L}_0 = \bar{\psi}(i\,\slashed{\partial} - m)\psi = \bar{\psi}(i\gamma^+\partial_+ + i\gamma^-\partial_- - m)\,\psi \tag{35}$$

has single components of left and right chirality. I will use the chiral representation, with γ_5 diagonal, where

$$\gamma^+ = \begin{bmatrix} 0 & (1+s)^{\frac{1}{2}} \\ (1-s)^{\frac{1}{2}} & 0 \end{bmatrix}, \quad \gamma^- = \begin{bmatrix} 0 & -(1-s)^{\frac{1}{2}} \\ (1+s)^{\frac{1}{2}} & 0 \end{bmatrix}. \tag{36}$$

These satisfy $\{\gamma^\mu, \gamma^\nu\} = 2g^{\mu\nu}$ for all θ and revert to the usual $\gamma^\pm$, with single, off-diagonal elements, in the LC limit. Quantization follows as usual, with

$$\pi(x) = \frac{\delta\mathcal{L}}{\delta\left(\partial_+\psi(x)\right)} = i\bar\psi\gamma^+ \tag{37}$$

and

$$\{\pi(x), \psi(y)\}_{x^+=y^+} = i\delta(x^- - y^-). \tag{38}$$

The free fermion spinor,

$$u_n = \frac{1}{(2\omega_n)^{\frac{1}{2}}} \begin{bmatrix} (1-s)^{-\frac{1}{4}}(\omega_n - n)^{\frac{1}{2}} \\ (1+s)^{-\frac{1}{4}}(\omega_n + n)^{\frac{1}{2}} \end{bmatrix}, \tag{39}$$

while v_n differs in the sign of the top entry. Throughout this example, antiperiodic boundary conditions are imposed and n is an odd half-integer. The upper component vanishes as $c^{1/2}$ for $n > 0$ and diverges as $c^{-1/2}$ for $n < 0$; this divergence prevents $n < 0$ particles from decoupling in the charge operator, for example.

The Hamiltonian is similar to that for the free scalar theory. Near the LC, eigenstates may be divided into those with positive $p_- = n\pi/L$ of energy $p_+ \sim m^2/2p_-$, and negative p_- with energy $p_+ \sim 2|p_-|/c$. As $c \to 0$, those with negative p_- possess infinite energy, and if states are restricted to those of finite energy these may be excluded by fiat, leaving the usual LC theory. Particles with $p_- \to 0$ have energy $m/c^{1/2}$ and so the usual LC energy singularity at $k_- = 0$ is cut off for finite c.

To modify $\mathcal{L}_0$ such that the theory is still trivially soluble but possesses vacuum condensation, add a term $\mathcal{L}_I = -\mu\,\bar\psi\psi$. This is simply a mass shift, and so the new eigenstates are obviously known. Here however it will be treated as though it were an interesting interaction, and its vacuum and eigenstates will be solved for in terms of those of $\mathcal{L}_0$.

The interacting Hamiltonian is

$$P_+^I = \mu \int dx^- :\bar\psi\psi(x): \tag{40}$$

and the full Hamiltonian

$$P_+^0 + P_+^I = \left(\frac{\pi}{L}\right) \sum_{n=\pm\frac{1}{2},\pm\frac{3}{2},\cdots} \left\{ \left[\frac{\omega_n - s\,n}{c} + \frac{\hat m\hat\mu}{\omega_n}\right] \left(b_n^\dagger b_n + d_n^\dagger d_n\right) \right.$$
$$\left. - \frac{\hat\mu n}{c^{\frac{1}{2}}\omega_n} \left(b_n^\dagger d_{-n}^\dagger + d_{-n}b_n\right) \right\}. \tag{41}$$

At this stage it is not possible to take a LC limit by letting c vanish and discarding states of infinite energy. The last term in P_+, which mixes particles of positive and negative

224

momenta, becomes infinitely strong in this limit. That is, states with divergent energy couple to finite-energy states strongly and cannot be ignored. Before infinite-energy states can be discarded, P_+ must be diagonalized, at least up to an appropriate power in c.

Of course, because $\mathcal{L}$ is free, P_+ can be diagonalized by a Bogoliubov transformation:

$$B_n \equiv G_+(n)\, b_n - \epsilon(n) G_-(n)\, d^\dagger_{-n}$$
$$D_n \equiv G_+(n)\, d_n - \epsilon(n) G_-(n)\, b^\dagger_{-n} \tag{42}$$

where

$$G_\pm(n) \equiv \frac{1}{(2\omega_n \tilde{\omega}_n)^{\frac{1}{2}}} \left[\omega_n(\tilde{\omega}_n \pm \omega_n) \pm c\hat{m}\hat{\mu} \right]^{\frac{1}{2}} \tag{43}$$

and

$$\omega_n \equiv \left(n^2 + c\hat{m}^2 \right)^{\frac{1}{2}}, \quad \tilde{\omega}_n \equiv \left(n^2 + c(\hat{m} + \hat{\mu})^2 \right)^{\frac{1}{2}}, \tag{44}$$

with $\epsilon(n)$ the antisymmetric step function. Near $c \to 0$, $G_+(n) \to 1$ and $G_-(n) \to c^{1/2}\hat{\mu}/2|n|$, provided $c^{1/2} \ll |n|/\hat{m}$. In terms of B_n and D_n,

$$P_+ = \left(\frac{\pi}{L} \right) \sum_{n=\pm\frac{1}{2},\pm\frac{3}{2},\cdots} \left[\frac{\tilde{\omega}_n - s\,n}{c} \right] \left(B^\dagger_n B_n + D^\dagger_n D_n \right) + E_\Omega, \tag{45}$$

which is the free Hamiltonian for a fermion of mass $m + \mu$.

The new vacuum should satisfy $B_n |\Omega\rangle = D_n |\Omega\rangle = 0$, and so is given by

$$|\Omega\rangle = \left\{ \prod_{n>0} B_n B_{-n} D_n D_{-n} \right\} |0\rangle . \tag{46}$$

For a particular momentum n, using Eq. (42),

$$B_n B_{-n} D_n D_{-n} |0\rangle = \left[G_+^2(n) + \epsilon(n) G_+(n) G_-(n) \left(d^\dagger_n b^\dagger_{-n} - d^\dagger_{-n} b^\dagger_n \right) \right.$$
$$\left. - G_-^2(n) d^\dagger_{-n} d^\dagger_n b^\dagger_{-n} b^\dagger_n \right] |0\rangle . \tag{47}$$

Near the LC, with $c^{1/2} \ll |n|/\hat{m}$, this reduces to $|0\rangle$ as

$$\left[1 + \frac{c^{\frac{1}{2}}\hat{\mu}}{2n} \left(d^\dagger_n b^\dagger_{-n} - d^\dagger_{-n} b^\dagger_n \right) - \frac{c\hat{\mu}^2}{4n^2} d^\dagger_{-n} d^\dagger_n b^\dagger_{-n} b^\dagger_n \right] |0\rangle . \tag{48}$$

The vacuum energy of $|\Omega\rangle$ relative to $|0\rangle$ is

$$E_\Omega = -\left(\frac{\pi}{L} \right) \sum_n \left[\frac{\tilde{\omega}_n - \omega_n}{c} - \frac{\hat{m}\hat{\mu}}{\omega_n} \right] , \tag{49}$$

which becomes

$$E_\Omega \sim -\left(\frac{\pi}{L} \right) \sum_n \frac{\hat{\mu}^2}{2|n|} = -\left(\frac{L}{\pi} \right) \sum_n \frac{\mu^2}{2|n|} \tag{50}$$

near the LC. The chiral condensate in this new vacuum is also calculable, with

$$\langle\Omega| :\bar{\psi}\psi: |\Omega\rangle = \frac{1}{2L}\sum_n \left[\frac{\hat{m}}{\omega_n} - \frac{\hat{m}+\hat{\mu}}{\tilde{\omega}_n}\right] \underset{c\to 0}{\sim} -\frac{1}{2L}\sum_n \frac{\hat{\mu}}{|n|} \,. \tag{51}$$

More precisely, in the continuum limit (and for $m = 0$)

$$\langle\bar{\psi}\psi\rangle = -\frac{\mu}{2\pi}\log\left(\frac{\Lambda^2}{\mu^2}\right)\,. \tag{52}$$

Just as in the scalar example, the operator $\bar{\psi}\psi$ is sufficiently singular to capture the vanishing corrections in Eq. (48). (This is reflected in the factor k_-^{-1} in ψ_L noted earlier.) As $c \to 0$, the support comes from a region increasingly focused around $k_- = 0$. In the continuum limit, the final result is c-independent, as it must be.[13]

Some more interesting cases, specifically the Gross-Neveu and Schwinger models, produce similar results.[10,14] The full vacuum evolves into the perturbative one as $c \to 0$ but quantities such as $\langle\bar{\psi}\psi\rangle$ are sustained. The Gross-Neveu model at large N can be solved simply by iterating the previous example. The Schwinger model is somewhat more complicated, but I will point out one tantalizing feature. The Schwinger boson wavefunction can be described in terms of a pair of effective or constituent quarks. At ET, these are complicated coherent objects composed of exponentiated fermion pairs. However, tracing these from ET to the LC limit, the complexity disappears and these constituent quarks evolve back to free quarks at $x^+ = 0$; the boson becomes just a $q\bar{q}$ pair. So constituent quarks are much more like LC than ET quarks in this model. It would be a great advantage if something similar turned out to be true in QCD. In particular, it would make LC calculations with a limited number of particles sensible.[15]

5. Effective LC Hamiltonian

Given the subtlety of the vacuum in the LC limit and the importance of negative-momentum, high-energy quanta in its construction, an obvious question is whether these effects can be incorporated into the conventional LC approach. Fortunately, a formalism exists which is ideally suited to this problem and in which these effects can be systematically computed.[16]

The conventional LC Hamiltonian includes interactions among quanta whose momenta are positive-definite, whereas the Hamiltonians in the previous sections included particles of all momenta. In the LC limit, however, the energies of particles with negative momentum typically become very large, and one might imagine that these are irrelevant to states of low energy in which positive momentum quanta should predominate. An exception occurs when these energetic particles have couplings to positive-momentum quanta which also become large near the LC. In this case, an effective Hamiltonian can be defined which acts only on positive momentum states but with extra interactions added to account for the effect of excluded states. Because these excluded states are very energetic, their interactions with low energy states occur over small (LC) times and these interactions will be local in x^+.

To see how these additional interactions can be computed, consider an operator $\mathcal{P}$ which projects out the subspace of states with only positive momentum particles, with $\mathcal{Q} = 1 - \mathcal{P}$

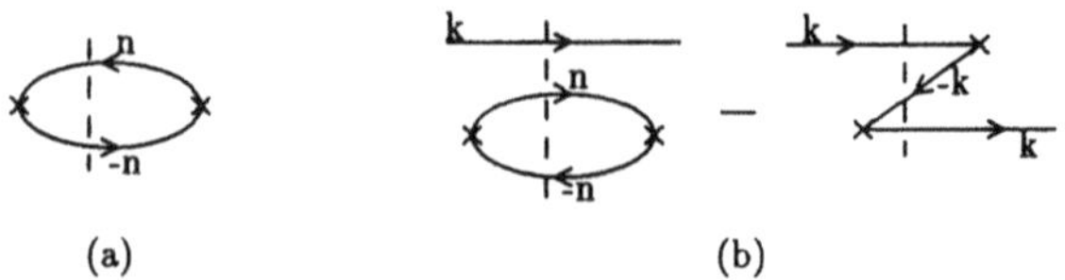

(a)   (b)

Figure 6: Diagrams contributing to H_{eff} on (a) the vacuum, and (b) the one particle state.

the complement. If, in the full state space, $H\,|\psi\rangle = E\,|\psi\rangle$, the effective Hamiltonian

$$H_{\text{eff}}(E) = H_{\mathcal{PP}} + H_{\mathcal{PQ}}\,\frac{1}{E - H_{\mathcal{QQ}}}\,H_{\mathcal{QP}} \tag{53}$$

satisfies $H_{\text{eff}}(E)\,\mathcal{P}\,|\psi\rangle = E\,\mathcal{P}\,|\psi\rangle$ within the projected subspace and may be thought of as a conventional LC Hamiltonian. Here $H_{\mathcal{PQ}} \equiv \mathcal{P} H \mathcal{Q}$, and so on.

If the states removed by $\mathcal{P}$ have energies much higher than the energy E for states of interest, the denominator in Eq. (53) may be expanded in $E/H_{\mathcal{QQ}}$. Then $H_{\text{eff}}(E)$ becomes a series of effective interactions of decreasing importance which are polynomial in E. Powers of E correspond to derivatives in time and so represent local interactions. This expansion works so long as $E/H_{\mathcal{QQ}}$ is small. However, in cases where $H_{\mathcal{PQ}}$ is large near the LC, H_{eff} may acquire terms in addition to the conventional LC Hamiltonian $H_{\mathcal{PP}}$.

Consider the example of fermions with a mass shift when c becomes small. The dominant terms are

$$
\begin{aligned}
H_{\mathcal{PP}} &= \left(\frac{\pi}{L}\right)\sum_{n>0}\left[\frac{\hat{m}^2}{2n} + \frac{\hat{m}\hat{\mu}}{n}\right]\left(b_n^\dagger b_n + d_n^\dagger d_n\right)\,, \\
H_{\mathcal{PQ}} &\sim \mathcal{P}\left(\frac{\pi}{L}\right)\sum_{n}\left[\frac{-\hat{\mu}\epsilon(n)}{c^{\frac{1}{2}}}\right]\left(b_n^\dagger d_{-n}^\dagger + d_{-n}b_n\right)\mathcal{Q}\,, \\
H_{\mathcal{QQ}} &\sim \left(\frac{\pi}{L}\right)\sum_{n<0}\left[\frac{2|n|}{c}\right]\left(b_n^\dagger b_n + d_n^\dagger d_n\right)\,.
\end{aligned}
\tag{54}
$$

Because the interactions $H_{\mathcal{PQ}}$ which couple states in $\mathcal{Q}$ to $\mathcal{P}$ are of order $c^{-1/2}$, while those of $H_{\mathcal{QQ}}$ are order c^{-1},

$$H_{\text{eff}} \sim H_{\mathcal{PP}} - H_{\mathcal{PQ}}\,\frac{1}{H_{\mathcal{QQ}}}\,H_{\mathcal{QP}} + \mathcal{O}(c)\,. \tag{55}$$

The correction to $H_{\mathcal{PP}}$ in the subspace $\mathcal{P}$ is simple to evaluate. Each vertex in $H_{\mathcal{PQ}}$ contributes a factor $-\hat{\mu}\epsilon(n)\pi/c^{1/2}L$ with vacuum pair production or annihilation, while the denominator from $H_{\mathcal{QQ}}$ contributes, to leading order, an energy $2\pi|n|/cL$.

Acting on the perturbative vacuum, which is included in $\mathcal{P}$, with H_{eff} gives zero for $H_{\mathcal{PP}}$ while the correction, shown in Fig. 6(a), is

$$H_{\text{eff}}\,|0\rangle \sim -H_{\mathcal{PQ}}\,\frac{1}{H_{\mathcal{QQ}}}\,H_{\mathcal{QP}}\,|0\rangle = \left[-\left(\frac{\pi}{L}\right)\frac{\hat{\mu}^2}{2}\sum_n\frac{1}{|n|}\right]|0\rangle = E_\Omega\,|0\rangle \tag{56}$$

as c vanishes. So $|0\rangle$ is an eigenstate of H_{eff} but with the same vacuum energy computed by diagonalizing H in the full space. On the one particle state of momentum $k_- > 0$, H_{PP} assigns an energy $(\pi/L)[\hat{m}^2/2k + \hat{m}\hat{\mu}/k]$, while the correction reproduces both the vacuum energy and a term $(\pi/L)[\hat{\mu}^2/2k]$. This is illustrated in Fig. 6(b). As a result

$$H_{\text{eff}}\, b_k^\dagger |0\rangle = \left[E_\Omega + \frac{\pi}{L}\frac{(\hat{m} + \hat{\mu})^2}{2k} \right] b_k^\dagger |0\rangle \ . \tag{57}$$

So $b_k^\dagger |0\rangle$ is an eigenstate of H_{eff}, but with both the vacuum energy and the mass shift $m \to m + \mu$ correctly accounted for.

This is a typical renormalization group result. High energy states are removed through a cutoff, but their effect, which can be large, can be entirely incorporated into a series of extra terms in the Hamiltonian. An alternative to computing these extra terms would be to enumerate the possible interactions consistent with the symmetries of the QCD LC Hamiltonian and match them to experiment.[17]

6. Conclusions

Due to its simple vacuum and boosts, LC quantization is potentially an efficient framework for studying QCD bound states. This proved true in $1+1$ dimensions, where more or less complete solutions were generated at little CPU expense. Of course, for $3+1$ dimensions, important issues such as renormalization must be worked out.

The LC vacuum can be studied using a limit of spacelike quantizations. As usual kinematic arguments suggest, the full vacuum becomes trivial. The entire effect of a nontrivial vacuum is pushed into the $k_- \sim 0$ region. Operators such as $\bar{\psi}\psi$ can maintain a nonzero vacuum expectation value because they are sufficiently singular in this region as $c \to 0$ to pick out the nontrivial corrections. Whether the LC vacuum is trivial then depends on what is being measured. For example, one could have $\langle \bar{\psi}\psi \rangle \neq 0$ but with the low energy spectrum unaffected. This should also occur in $3 + 1$ dimensions. The connection between short times in x^+ and long distances in x^- associated with the vacuum suggest that an effective, or cut-off, Hamiltonian is an appropriate formalism in which to incorporate the effects of a nontrivial vacuum while preserving the simplicity of the LC.

At the end of the discussion of QCD_{1+1} I listed some curiousities. Certain operator properties, such as current conservation, could only be demonstrated within matrix elements of finite-energy states. These operators made sense because the wavefunctions sufficiently suppressed large k_+ states from the $k_- \sim 0$ region. So this model, and LC quantization in general, has long been treated as if it were an effective theory.

7. Acknowledgements

I would like to thank the organizers for a diverse and lively conference. This work was supported in part by the Department of Energy and Ohio State University.

8. References

1. D. E. Soper, SLAC Report No. 137, Ph.D Thesis (1971) provides a good introduction to LC quantization, and the LC Poincaré representation in particular. See also T.-M. Yan, Phys.

Rev. **D7**, 1780 (1972), and references therein; S. J. Brodsky and G. P. Lepage, in *Perturbative Quantum Chromodynamics*, edited by A. H. Mueller (World Scientific, Singapore, 1989)). An expanded version of this section is in K. Hornbostel, in *From Fundamental Fields to Nuclear Phenomena*, edited by J. A. McNeil and C. E. Price, (World Scientific, Singapore, 1991).

2. S. Weinberg, Phys. Rev. **150**, 1313 (1966).

3. These results are from K. Hornbostel, SLAC Report No. 333, Ph.D Thesis (1988); K. Hornbostel, S. J. Brodsky, H. C. Pauli, Phys. Rev. **D41**, 3814 (1990). For other recent work, see M. Burkardt, Nucl. Phys. **A504**, 762 (1989).

4. This model was introduced in G. 't Hooft, Nucl. Phys. **B75**, 461 (1974), and has been developed extensively. See Ref. 5 for a useful summary, and also Refs. 1, 3.

5. C. J. Hamer, Nucl. Phys. **B195**, 503 (1982).

6. See, for example, C. G. Callan, N. Coote, and D. J. Gross, Phys. Rev. **D13**, 1649 (1976).

7. A. R. Zhitnitskii, Phys. Lett. **165B**, 405 (1985).

8. See for example F. Rohrlich, Acta Phys. Austr. **32**, 87 (1970); P. Steinhardt, Ann. Phys. **128**, 425 (1980).

9. G. McCartor, Z. Phys. **C41**, 271 (1988). See also Ref. 3.

10. Y. Frishman, C. T. Sachrajda, H. Abarbanel, and R. Blankenbecler, Phys. Rev. **D15**, 2275 (1977). A similar approach appears in T. W. Chen, Phys. Rev. **D3**, 1989 (1971); E. Elizalde and J. Gomis, Nuov. Cim. **35A**, 367 (1976); E. V. Prokhvatilov and V. A. Franke, Yad. Fiz. **49**, 1109 (1989); F. Lenz, M. Thies, S. Levit, and K. Yazaki, Ann. Phys. **208**, 1 (1991). This discussion follows K. Hornbostel, Phys. Rev. **D45**, 3781 (1992).

11. I thank P. Mulders for discussions on this exercise.

12. Periodic conditions simplify, but are not essential for, this result.

13. I have been somewhat sloppy with limits for simplicity. The correct order is the continuum limit ($L \rightarrow \infty$) prior to the LC limit ($c \rightarrow 0$). Otherwise, important contributions from the region $p_- < c^{1/2}m$ can be lost. See Ref. 10.

14. G. McCartor, Z. Phys. **C52**, 611 (1991).

15. I thank G. McCartor, R. Perry, and K. Wilson for discussions on this point.

16. For a good discussion of effective Hamiltonians see G. P. Lepage, in *From Actions to Answers*, Proceedings of the TASI-89 Summer School, T. DeGrand and D. Toussiant, eds. (World Scientific, Teaneck, N. J., (1990)).

17. Such an approach is advocated, for example, in R. J. Perry, A. Harindranath, and K. G. Wilson, Phys. Rev. Lett. **65**, 2959 (1990).

LIGHT FRONT TAMM-DANCOFF METHOD HYDROGEN ATOM AND 2D YUKAWA MODEL

OSAMU ABE

Asahikawa College, Hokkaido University of Education
9 Hokumoncho, Asahikawa 070, JAPAN

and

K. TANAKA and K.G. WILSON

Department of Physics, The Ohio State University
174 West 18th Avenue, Columbus, OH 43210, USA

Abstract

The light front frame and Tamm-Dancoff method is reviewed briefly and
the light front Tamm-Dancoff method is discussed. The method is applied
to the hydrogen atom and the 2D Yukawa model.

The light front Tamm-Dancoff (LFTD) method is introduced as an alternative
approach to lattice gauge theory to understand relativistic bound states.[1,2]
To alleviate the case the coupling parameter is not small, Tamm and Dancoff[3]
(TD) independently considered the possibility of expanding in terms of amplitude that
represent a finite number of particles, and solve the coupled set of integral equations
of a small number of these amplitudes. This method was applied to a variety of
problems in strong interactions, but it was unsuccessful.

In the light front (LF) or null plane formalism, a physical state is given by the
evolution variable that plays the role of time $x^+ = (x^0 + x^3)/\sqrt{2} = \text{const}$, and
$x^- = (x^0 - x^3)/\sqrt{2}$, $x_\perp = (x^1, x^2)$ are spatial variables. There are seven generators
P^1, P^2, P^+, M_{12}, M_{1+}, M_{2+}, and M_{-+} that leave the LF plane invariant.[4]
The momentum of a particle of mass m is

$$p^\mu = \left(p^+, \ p^- = \left(m^2 + p_\perp^2\right)/2p^+, \ p_\perp = \left(p^1, p^2\right)\right) \quad \mu = +, -, 1, 2$$

where $p^\pm = (p^0 \pm p^3)/\sqrt{2}$ and $p^2 = m^2 + p_\perp^2 - p_\perp^2 = m^2$, where

$$g_{\mu\nu} = \begin{pmatrix} 0 & 1 & 0 & 0 \\ 1 & 0 & 0 & 0 \\ 0 & 0 & -1 & 0 \\ 0 & 0 & 0 & -1 \end{pmatrix}.$$

The p^+ and $p_\perp$ are conserved at each vertex and particles are on mass shell. We require spectrum of p^μ be contained in the forward cone

$$p^0 > 0, \quad \text{so } p^+ = \left(\sqrt{m^2 + p^{3^2} + p_\perp^2} + p^3\right)/\sqrt{2} > 0\,.$$

The advantages of the LF formalism is that[5] (1) If the LF wave function of a system is known at rest, it can be obtained at other momentum by boost operators. (2) The property $p^+ > 0$, and its conservation reduces the number of Fock space states needed to generate a covariant result. (3) The bare vacuum is equal to the physical vacuum.

But these advantages are accompanied by some disadvantages, (1) It is necessary to require for renormalization, mass and counter terms that depend on the sector of Fock space within which they act. (2) Note if the momentum p^3 is cut off at Λ, $p^+ = (p^0 + p^3)/\sqrt{2} = (2\Lambda + \dfrac{m^2}{2\Lambda})/\sqrt{2}$. There is an infrared divergence as $p^+ \to$ small. So ultraviolet and infrared divergences have to be treated in a consistent manner. (3) Renormalization 'constants' Z_1 and Z_2 are momentum dependent.[6] (4) How to handle zero modes and gluons in QCD is not worked out.
Consider $H_I = g\phi^3$ as an example.[7]

If one restricts the TD amplitudes to less than two particles, then the second Z diagram is dropped. Both are needed for covariance in the equal time frame. In the LF frame the Z diagram does not exist, as the leg going backward has negative energy and $p^+ = (p^0 + p^3)/\sqrt{2}$ cannot be positive. But in the LF the first diagram alone is covariant

$$\begin{aligned}
p_1 &= px + p_{1\perp}\\
p_2 &= p(1-x) + p_{2\perp}\\
p &= p_1 + p_2 \qquad \therefore p_{1\perp} + p_{2\perp} = p_\perp = 0
\end{aligned}$$

$$M = g^2 \frac{1}{p^+} \frac{1}{2\left(p_{in}^- - p_{int}^-\right)} \qquad p_1^- = \frac{m^2 + p_{1\perp}^2}{2p_1^+}$$

$$M = \lim_{p \to \infty} g^2 \left[\frac{1}{p^+ \left(\frac{m^2 + p_{1\perp}^2}{p_1^+} \right) + p^+ \left(\frac{m^2 + p_{2\perp}^2}{p_2^+} \right) - p^+ \left(\frac{m^2}{p^+} \right)} \right]$$

$$= g^2 \left[\frac{1}{(m^2 + p_{1\perp}^2)/x + (m^2 + p_{2\perp}^2)/(1-x) - m^2} \right] = \frac{g^2}{s^2 - m^2} \tag{1}$$

We begin with Einstein's equation to consider the mass renormalization of the electron.

$$\left(M^2 + P_\perp^2 - 2P^+ P_0^- \right) |\Psi > = 2P^+ P_I^- |\Psi > \tag{2}$$

We outline the calculation[8] for the hydrogen atom of state $|H >$

$$|H > = C_0 |Pe > + C_1 |Pe\gamma > \tag{3}$$

Project out C_0 amplitude from Eqs. (2) and (3).

$$FC_0 = e \int L_{01} C_1 + e^2 \int L_{00} C_0 , \tag{4}$$

where L_{00} represents the instantaneous photon exchange and

$$F = \left[M^2 + (p_\perp + k_\perp)^2 - (p^+ + k^+) \left\{ \frac{k_\perp^2 + M_{e0}^2}{k^+} + \frac{p_\perp^2 + M_{P0}^2}{p^+} \right\} \right] .$$

Project out C_1 from Eqs. (2) and (3), and obtain

$$HC_1 = eL_{10} C_0 + e^2 \int L_{11} C_1 \tag{5}$$

Eliminate C_1 from Eq. (4) with the aid of Eq. (5)

$$FC_0 = e^2 \int L_{01} H^{-1} L_{10} C_0 + e^2 \int L_{00} C_0 . \tag{6}$$

In the nonrelativistic limit, $|p|, |q| < \Lambda \ll M_e$, after mass renormalization we obtain, with $M^2 = (M_e + M_P - B)^2 = (M_P + M_e)^2 - 2(M_P + M_e)\overline{B}$

$$\left(\overline{B} + \frac{p^2}{2\mu} \right) \Phi(P) = \frac{\alpha}{2\pi^2} \int_{-\infty}^{\infty} d^3 k \, \frac{\Phi(k)}{(p-k)^2} \tag{7}$$

μ is the reduced mass $\alpha = e^2/4\pi$. This is the Lippman-Schwinger equation and its solution[9] is $\overline{B}_n = \frac{\alpha^2 \mu}{2n^2}$.

A poor person's solution when $p = 0$ can be found by using the trial function $\Phi(k) = C/(k^2 + a^2)^2$, where a is to be determined.

We find $\overline{B} = \dfrac{\alpha}{2}\, a$ where a has dimensions of mass. Put $a = \sqrt{2\overline{B}\mu}$, then

$$\overline{B} = \frac{\alpha^2 \mu}{2} = \frac{\alpha^2 m_e}{2} \quad \text{hydrogen} \quad n = 1$$

$$= \frac{\alpha^2 m_e}{4} \quad \text{positronium} \quad n = 1 \tag{8}$$

We consider next fermion-fermion interaction via meson exchange in two dimensions[10]. The fermions have equal mass, different flavor and are neutral. Expand the two fermion bound state $|\psi >$

$$|\psi > = C_0 \big| ff > + C_1 \big| ffb > \tag{9}$$

It is shown in Ref. 10 that the self-energy corrections can be ignored in the following approximation in the non-relativistic limit. We let the fractions of the momentum of the two fermions be x and $(1-x)$, so that $p^+/p_1^+ = 1/x$, $p^+/p_2^+ = 1/(1-x)$ and $p = p_1 + p_2$. We put $P_\perp = 0$, project out C_0 and C_1 from Eqs. (2) and (9), and obtain[10]

$$\left(M^2 - m^2/(1-x) - m^2/x\right) C_0 (1-x, x)$$

$$= \lambda m \left\{ \int_0^{1-x} \frac{dy}{\sqrt{4\pi y}} \left[1/(1-x) + 1/(1-x-y)\right] C_1 (1-x-y, x, y) \right.$$

$$\left. + \int_0^x \frac{dy}{\sqrt{4\pi y}} \left[1/x + 1/(x-y)\right] C_1 (1-x, x-y, y) \right\} \tag{10}$$

$$\left(M^2 - m^2/(1-x-y) - m^2/x - m_B^2/y\right) C_1 (1-x-y, x, y)$$

$$= \frac{\lambda m}{\sqrt{4\pi y}} \left\{ \left[1/(1-x-y) + 1/(1-x)\right] C_0 (1-x, x) \right.$$

$$\left. + \left[1/x + 1/(x+y)\right] C_0 (1-x-y, x+y) \right\} \tag{11}$$

Eliminate C_1 from Eqs. (10) and (11),

$$\left[M^2 - m^2/(1-x) - m^2/x\right] C_0 (1-x, x)$$

$$= \lambda^2 m^2 \left\{ \int_0^{1-x} \frac{dy}{4\pi y} \frac{\left[1/(1-x) + 1/(1-x-y)\right]\left[1/x + 1/(x+y)\right]}{M^2 - m^2/(1-x-y) - m^2/x - m_B^2/y} C_0 (1-x-y, x+y) \right.$$

$$\left. + \int_0^x \frac{dy}{4\pi y} \frac{\left[1/(1-x) + 1/(x-y)\right]\left[1/(1-x) + 1/(1-x+y)\right]}{M^2 - m^2/(1-x) - m^2(x-y) - m_B^2/y} C_0 (1-x+y, x-y) \right\} \tag{12}$$

Change variables in Eq. (12), $1 - x = x_1$, $x = x_2$, $1 - x - y = y_1$, $x + y = y_2$

$$x_{1,2} = \frac{1}{2}\left(1 \mp \frac{q}{\sqrt{m^2 + q^2}}\right) \qquad y_{1,2} = \frac{1}{2}\left(1 \mp \frac{\ell}{\sqrt{m^2 + \ell^2}}\right)$$

$$\phi(q) = 4\sqrt{\frac{m^2 + q^2}{m}}\, C_0(x_1 x_2) \qquad \phi(\ell) = 4\sqrt{\frac{m^2 + \ell^2}{m}}\, C_0(y_1 y_2)$$

and the resulting equation with $M = 2m - B$, $\overline{B} = B - (B^2/4m)$ is[10]

$$\left(\frac{q^2}{2\mu} + \overline{B}\right)\phi(q) = \frac{\lambda^2}{2\pi}\int d\ell \,\frac{1}{(q-\ell)^2 + m_B^2}\Phi(\ell) \tag{13}$$

This is remarkably similar to the hydrogen atom equation (7). The correspondence is $\dfrac{d\ell}{2\pi} \to \dfrac{d^3 k}{(2\pi)^3}$ and $m_B = 0$ as expected. To solve the 2D Eq. (13) at $q^2 = 0$, choose a trial wave function $\Phi(\ell) = \dfrac{C}{\sqrt{\ell^2 + a^2}}$ and minimize $\overline{B}$ with respect to a. The solution is $\overline{B} = \dfrac{\lambda^2}{\pi m_B}$ where λ has the dimension of mass in 2D.

One would like to regard the 2D Eq. (13) extended to 4D as a beginning equation to calculate the deuteron bound state energy. But the problem is much more complicated because of a repulsive core and D wave contributions. The relativistic limit of one boson exchange in the Yukawa model is discussed in Ref. 11.

For fun if one substitutes in Eq. (13)

$$\int \frac{d\ell}{2\pi} \to \int \frac{d^3\ell}{(2\pi)^3}$$

$$\phi = \frac{C}{\sqrt{k^2 + a^2}} \to \phi = \frac{C}{(k^2 + a^2)^{3/2}}$$

one obtains after minimizing with respect to a

$$\overline{B} = \frac{\lambda^2}{3\pi^2}\, m_B \;.$$

One of the authors (KT) would like to thank John Ellis for the hospitality at CERN. He also expresses his thanks to B. Webber, T.T. Wu and A. Harindranath for discussions. This work was supported in part by the U.S. Department of Energy under Contract No. EY-76-C-02-1415*00.

REFERENCES

1. K.G. Wilson, in *Lattice '89*, Proceedings of the International Symposium, Capri, Italy, 1989, edited by R. Petronzio *et al.* [Nucl. Phys. **B** (Proc. Suppl.) **17** (1990)].

2. R.J. Perry, A. Harindranath and K.G. Wilson, Phys. Rev. Lett. **65**, 2959 (1990). M. Krautgartner, H.C. Pauli and F. Wolz, preprint MPIH-V4-1991 Max-Planck Institute and references therein. This preprint focuses on numerical techniques of solving the positronium problem whereas we focus on divergence problems that arise in the bound states of the hydrogen atom. The problem of infrared singularities induced by light-cone gauge is also discussed.

3. I. Tamm, J. Phys. (USSR) **9**, 449 (1945); S.M. Dancoff, Phys. Rev. **78**, 382 (1950).

4. P.A.M. Dirac, Rev. Mod. Phys. **21**, 392 (1949).

5. H.C. Pauli and S.J. Brodsky, Phys. Rev. **D32**, 1993–2001 (1985). A. Harindranath and R.J. Perry, Phys. Rev. **D43**, 492 (1991).

6. D. Mustaki, S. Pinsky, J. Shigemitsu and K.G. Wilson, Phys. Rev. **D43**, 3411 (1991).

7. S. Weinberg, Phys. Rev. **150**, 1313 (1966).

8. O. Abe, K. Tanaka and K.G. Wilson, Ohio State University preprint DOE/ER/01545 564.

9. M. Sawicki, Phys. Rev. **D32**, 2666 (1985); V. Fock, Z. Phys. **98**, 145 (1935).

10. R.J. Perry and A. Harindranath, Phys. Rev. **D43**, 4051 (1991).

11. S. Glazek, A. Harindranath, S. Pinsky, J. Shigemitsu and K. Wilson, preprint OHSTPY-HEP-T-92-004.

CHIRAL SYMMETRY AND CONFINEMENT
IN MASSLESS TWO DIMENSIONAL
QUANTUM CHROMODYNAMICS

A. FERRANDO

Institut for Theoretical Physics, University of Bern, Sidlerstrasse 5
CH-3012 Bern, Switzerland

and

V. VENTO

Departament de Física Teòrica and I.F.I.C.
Universitat de València - C.S.I.C.
E-46100 Burjassot (València), Spain

ABSTRACT

Two dimensional massless Quantum Chromodynamics presents many features which resemble those of the true theory. Its spectrum consists of mesons and baryons arranged in flavor multiplets without parity doubling. By means of bosonization techniques we analyze the implications of chiral symmetry, which is not spontaneously broken in two dimensions, in the spectrum and in the quark condensate. We prove that a chiral phase transition is not possible in QCD_2. In the bosonized version of two dimensional theories, non-trivial boundary conditions (topology) play a crucial role. They are inevitable if one wants to describe non-singlet states. We show that confinement appears in QCD_2 due to the dynamical collapse of the topology associated with color.

1. Introduction

By bosonizing QCD_2 we have obtained an equivalent theory with a massless meson. This so-called chiral meson is completely decoupled from the remaining fields of the theory and is not affected therefore by the interactions. It is suggestive of a Goldstone boson, which however cannot be due to Colemans theorem [1]. The remaining meson spectrum is described by means of colored mesons, leading to a stable discrete and infinite spectrum with no parity doublets [2,3].

The interpretation of the baryonic spectrum relies on the existence of zero mass baryons. In the bosonic description they are a consequence of the non-triviality of the boundary conditions of the chiral meson field [4].

Chiral symmetry cannot be broken spontaneously in two dimensions, however the massless chiral meson is still associated with the long wave lenght behavior of the correlators through the Berezinskii-Kosterlitz-Thouless (BKT) phenomenon [5,6].

Confinement arises trivially in the fermionic description of the theory [7,8]. In the bosonic description it is intimately related to the lack of topology in color space [9].

Our aim in this presentation is to review and elaborate on some of these results.

2. Chiral symmetry

Since chiral symmetry is realized à la BKT, the correlation function falls off like a power law

$$< q\bar{q}(x)q\bar{q}(0) > \stackrel{x^2\to\infty}{\approx} < e^{i\sqrt{\frac{4\pi}{N}}\varphi(x)}e^{-i\sqrt{\frac{4\pi}{N}}\varphi(0)} > \approx \frac{1}{(2\pi k^2\mu^2 x^2)^{\frac{1}{N}}} \tag{1}$$

where N is the number of colors and φ the massless meson field. Thus the chiral meson, although not a Goldstone boson, governs the long distance behavior of the correlator, a property shared with its four dimensional analog. Moreover, the discontinuity of the vector current two point function, a signature of the anomaly [10], is saturated by this meson as occurs in spontaneously broken theories [4,11].

Despite the fact that chiral symmetry is not spontaneously broken, parity doublets do not appear. This feature, also in correspondence with the four dimensional situation, arises though, from a series of mechanisms very particular of the two dimensional world. In the zero mass sector, gauge interactions are absent. Since the dimensionality of space-time imposes

$$j_5^\mu = \varepsilon^{\mu\nu}j_\nu \tag{2}$$

it is immediate to see that a unique massless chiral meson exists without chiral partner.

The excited meson spectrum is associated to a minimally coupled Wess-Zumino-Wittten model [3,12]. Due to the gauge coupling the original $SU(N)\otimes SU(N)$ symmetry is explicitly broken to $SU(N)_{vector}$ (the bosonic remnant of the color anomaly). Since this latter symmetry can not be spontaneously broken, the vacuum is also invariant under it. Considering only the color degrees of freedom, due to confinement (the physical states have to be color singlets), there cannot exist degeneracy associated to color,i.e., if two singlet states were degenerate the symmetry would be larger than $SU(N)_{vector}$. Thus the degeneracy must be associated with flavor and space degrees of freedom. Eq.(2) eliminates the degeneracy of the former and the dimensionality of space that of the latter [4]. Thus surprisingly Coleman's theorem and no parity doubling are consequences of the same restriction: the dimensionality of space-time.

The cornerstone of the suspicion that a chiral phase transition might take place in QCD_2 is the singular behavior of the quark condensate $< q\bar{q} >$, since this expectation value is an order parameter for the breaking of axial flavor symmetry. Taking the behavior of the quark condensate in the large N limit, see Eq.(1), one can envisage a phase change from a situation in which axial symmetry is conserved ($< q\bar{q} >= 0$), to one in which is broken ($< q\bar{q} >\neq 0$) [13].

Using bosonization techniques one obtains for the condensate

$$< q\bar{q} >\sim< 0| : e^{i\sqrt{\frac{4\pi}{N}}\varphi} : |0 >< 0|tr(g)|0 > \tag{3}$$

In this expression the vacuum expectation value (vev) of $tr(g)$ is always different from zero, since the $SU(N) \otimes SU(N)$ symmetry is explicitly broken. Thus only the vev of the chiral operator $e^{i\sqrt{\frac{4\pi}{N}}\varphi}$ can be the cause of the vanishing of the condensate. The chiral operator changes under transformations of the $U(1)_{axial}$ group, defined by $\varphi \to \varphi + \alpha$ as

$$e^{i\sqrt{\frac{4\pi}{N}}\varphi} \to e^{i\sqrt{\frac{4\pi}{N}}(\varphi+\alpha)} \tag{4}$$

If the vacuum is invariant under $U(1) \otimes U(1)$ the vev must vanish (Coleman's theorem). This vev is therefore the order parameter in the bosonized theory for any phase transition associated with chiral symmetry.

In the large N limit the vev of the quark condensate does not vanish. Naively this would imply that our order parameter does not vanish and that the vacuum would break spontaneously the symmetry. This argument is incorrect. In the large N limit the fluctuations of the φ field disappear and $e^{i\sqrt{\frac{4\pi}{N}}\varphi}$ becomes the unit operator, so that the quark condensate becomes in this limit purely an $SU(N)$ object ($< tr(g) >$), which is different from zero due to the gauge interaction. Thus our order parameter, which was such because it transformed non trivially under the group, ceases to be an order parameter in the large N limit because it becomes the unit operator, which is trivially invariant under the group. In this way, the fact that the condensate becomes different from zero in the large N limit is independent of chiral dynamics and simply reflects the fact that $SU(N) \otimes SU(N)$ is explicitly broken [4].

3. Dynamical confinement

Quantum Chromodynamics in two dimensions is a confining theory. In the fermionic language this property is realized in a very simple manner [2,7,8]. We next analyze confinement in a bosonized version of the theory [9].

The bosonized action for a massless quark theory of one flavor and two colors using abelian bosonization is 14

$$S = \int d^2x \{\frac{1}{2}(\partial_\mu\varphi)^2 + \frac{1}{2}(\partial_\mu\eta)^2\} \tag{5}$$

This action shows explicitly only an $U(1)_F \otimes U(1)_C$ symmetry. In order for this simple action to realize the full global SU(2) color symmetry, the η field must possess non-trivial boundary conditions, which determine the structure of the vacuum and from it the quantum numbers of the physical states [15]. These can be represented by the (B, T_3) lattice

238

$$\{(n,m) \; ; \; n,m \in Z\} \bigcup \{(n+\frac{1}{2},m+\frac{1}{2}) \; ; \; n,m \in Z\} \qquad (6)$$

Baryon number, B, and color charge, T^3, are related by the bosonization formulae to the asymptotic conditions of the fields

$$B = \frac{1}{\sqrt{2\pi}}[\varphi(+\infty) - \varphi(-\infty)] \qquad (7)$$

$$T^3 = \frac{1}{\sqrt{2\pi}}[\eta(+\infty) - \eta(-\infty)] \qquad (8)$$

In QCD_2 the bosonized action acquires a term due to the quark-gluon interaction [16]

$$V(\eta) = \frac{e^2}{32\pi}\eta^2 + \sqrt{\pi}\Lambda^2\{1 - \frac{\sin\sqrt{2\pi}\eta}{\sqrt{2\pi}\eta}\} \qquad (9)$$

were Λ is a scale parameter coming from the renormalization through normal-ordering. This potential is positive definite and has only an absolute minimum $V(\eta) = 0$, which occurs for $\eta = 0$. Thefore the coupling of the gauge fields reduces the possible quantum numbers of the physical states to

$$\{(n,0) \; ; \; n \in Z\} \qquad (10)$$

One can show by calculating $< T^2 >$ that these states are color singlets.

This mechanism has a topological interpretation. The (B, T_3) charges generate the $U(1)_F \otimes U(1)_C$ explicit symmetry of the free lagrangian. The non-trivial boundary conditions produce the *winding number* associated with the homotopy classes of this group. The lattice of physical states corresponds to the first homotopy class

$$\Pi_1[U(1)_F \otimes U(1)_C] \approx \Pi_1[U(1)_F] \times \Pi_1[U(1)_C] \approx Z_F \otimes Z_C \qquad (11)$$

When the confinement interaction is added the vacuum structure collapses into a colorless subset of the lattice points. In topological language, the non-trivial topology induced by the mapping

$$\Pi_1[U(1)_C] \approx Z_C \qquad (12)$$

disappears, since the confining interaction forces the physical states to have $T_3 = 0$. The topology associated with color becomes trivial.

4. Conclusion

We have used abelian and non-abelian bosonization techniques to study the properties of two-dimensional QCD. The bosonization rules have provided a well defined connection between the Green functions of the bosonized models and those of their fermionic counterparts. The crucial importance of the apparently naive chiral sector has been unveiled, both for mesons and baryons, and the role of the gauge dynamics in the formation of the spectrum has been clarified.

The zero mass baryon is essential to describe the excited baryons. Its existence is guaranteed by the presence of a topological chiral sector. We have constructed its current operator and described the properties of the baryonic spectrum as obtained from the four point correlators and its absence from the chiral anomaly [3,4].

The chiral meson is instrumental in describing the flavor symmetry à la BKT. This mechanism converts the impossibility of manifesting itself as the Goldstone boson of a spontaneously broken symmetry, into a dominance of the long range behavior of the correlators.

The resonant mesonic spectrum has its origin in the dynamics of color. The non-existence of parity multiplets is due to the explicit breaking of the chiral color symmetry in a confining environment in such a way as to preserve parity. In bosonized QCD_2 this breaking appears as an explicit term in the lagrangian, while in its fermionic equivalent version it is due to the anomaly, and therefore a consequence of the non-invariance of the fermionic measure under the chiral color symmetry.

We have analyzed the possibility of a phase transition in QCD_2. The cornerstone of this proposal is the behavior of the chiral condensate in the large N limit. We have proven that only one phase exists and that the apparent anomalous behavior of the chiral order parameter is a consequence of the trivialization of the chiral sector in that limit. We have shown that the physics becomes independent of the chiral dynamics. We therefore must conclude that all argumentation aiming at justifying the existence of a phase transition in QCD_2, based on the spontaneous symmetry breaking of the $U(1) \otimes U(1)$ symmetry is spurious and interprets incorrectly the mechanisms regarding the symmetries of QCD_2.

The fermionic description indicates in a qualitative manner that QCD_2 is a confining theory beyond leading order in the $\frac{1}{N}$ expansion. The spectrum, as well as, the vanishing of quark creation amplitudes, corroborate the non existence of *free* color non siglet states. The quark self-energy shows a confining behavior.

In the bosonized version of fermionic two dimensional theories topology plays an important role. States with baryon number and color charge are described by solitons. The properties of the vacuum, which give rise to non-trivial boundary conditions, determine the quantum number structure of the Fock space. We have analyzed initially the rich spectrum of non confining theories by discussing the role of the boundary conditions. It is appealing that in the bosonized version of the theory , this discussion can be carried out purely at the classical level.

Once the dynamics of QCD_2 is incorporated the Fock space collapses to the subset of color singlet states and we recover the fermionic result independent of any large N assumption. The mechanism can be cast in a more mathematical language by invoquing homotopy groups, but one should not avoid the very naive dynamical statement, namely that color solitons are given infinite energy.

5. Acknowledgements

We have mantained illuminating discussions with R. Brockmann, D. Espriu, H. Leutwyler, H. Minkowski, A. Smilga and U.J. Wiese. V. Vento would like to thank the members of the Institut für Physik der Universität Mainz for their hospitality. Our work was supported by Dgicyt grant # PB88-0064 and Cicyt grant # AEN90-0040. A. Ferrando is a post-doctoral fellow of Cicyt- Plan Nacional de Altas Energías.

6. References

1. S. Coleman, Comm. Math. Phys. **31** (1973) 259.

2. G. 't Hooft, Nucl. Phys. **B72** 1974 461; G. 't Hooft, Nucl. Phys. **B75** 1974 461;

3. A. Ferrando and V. Vento, Phys. Lett. **B256** (1991) 503; Phys. Lett. **B265** (1991) 153; Ph. D. Thesis, Universitat de València (1991).

4. A. Ferrando and V. Vento, Preprint FTUV/92-2

5. V.L. Berezinskii, Sov. Phys. JETP **32** (1970) 493; ibid 34 (1971) 610; J.M. Kosterlitz and D.J. Thouless, J. Phys. **C6** (1973) 1181.

6. E. Witten, Nucl. Phys. **B145** (1978) 110.

7. C.G. Callan, N. Coote and D.J. Gross, Phys. Rev. **D13** (1976) 1649.

8. M.B. Einhorn, Phys. Rev. **D12** (1976) 3451.

9. A. Ferrando and V. Vento, Preprint FTUV/92-3

10. G. 't Hooft in *Cargese Lectures* 1979; Y. Frishman, A. Schwimmer, T. Banks and S. Yankielowicz, Nucl. Phys. **B177** (1981) 157.

11. W. Buchmüller, S.T. Love and R.D. Peccei, Phys. Lett. **B108** (1982) 426; D. Amati, K.-C. Chou and S. Yankielowicz, Phys. Lett. **B110** (1982) 309.

12. E Witten, Comm. Math Phys. **92** (1984) 455; D Gepner, Nucl. Phys. **B252** (1985) 481; I. Affleck, Nucl. Phys. **B265** (1986) 448.

13. A.R. Zhitnitsky, Phys. Lett. **B165** (1985) 405.

14. M.B. Halpern, Phys. Rev. **D12** (1975) 1684; Phys. Rev. **D13** (1976) 337.

15. S. Elitzur, Y. Frishman and E. Rabinovici, Phys. Lett. **B106** (1981) 403.

16. V. Baluni, Phys. Lett. **90B** (1980) 407.

LARGE N_C QCD
SOLITONS AND THEIR INTERACTIONS

DIMITRI KALAFATIS

Division de Physique Théorique, IPN, Orsay
Orsay, 91406, France

and

Laboratoire de Physique Théorique des Particules Elémentaires, Université P. et M. Curie
4 Place Jussieu,Paris Cedex 05, 75272, France

ABSTRACT

In this talk the large colour number limit of QCD, where baryons are considered to be solitons of non-linear meson theories is discussed. The application of the $1/N_C$ expansion in baryon phenomenology will be sketched, with special emphasis on its consistency with the semi-classical approximation. It will be finally argued that for the purpose of this consistency, the large N_C effective QCD Lagrangian should include the physics of low-lying meson resonances, especially vector and scalar mesons.

1. Introduction

The relevance of the $1/N$ expansion to the low-energy hadron physics was first recognized by t' Hooft in his pioneering work on planar diagram theory[1]. The qualitative outcome of his investigation is that any $SU(N)$ Yang-Mills theory is equivalent to a weakly coupled meson theory when N becomes very large. This property, when applied to QCD, yields a formidable set of predictions about meson physics; for example the absence of exotic multiquark meson states, a natural feature of large N_C diagrammatics.

But how baryons fit to this picture? Witten[2] made the observation that like in some weakly coupled theories, there should be in large N_c QCD classical states with masses inversely proportional to the coupling constant $(1/N_c)$. These states are usually refered in the literature as solitons (or lumps). The conjecture that the large colour QCD monopoles are indeed the baryons observed in Nature was advanced. If one believes the arguments of Witten, the temptation to build a unified semiclassical theory of mesons and baryons is difficult to refute. The trouble is that we are not yet able to sum all the QCD planar diagrams, and the large colour chromodynamical action is still unknown. If one intends to describe baryons from mesons semiclassically, one is left with the task to construct an effective Lagrangian possessing non-trivial soliton solutions. The latter will be designed to incorporate the maximum of QCD's general properties like chiral symmetry, its anomalous scaling behaviour at the quantum level[3] etc., and all this in terms of meson fields. However, symmetries alone are not sufficient to constrain the specific form of the Lagrangian density. My purpose here is to illustrate how one can reduce this degree of arbitrariness in the construction of the effective Lagrangian, i) by simply examining the consistency of the semi-classical expansion for soliton observables and ii) including some basic phenomenological requirements of low-energy nucleon physics related to the conventional meson exchange. My discussion of point i) will be concentrated on the original Skyrmion. We will see in section 3 how its generalization is related to point ii).

2. The semi-classical expansion of the soliton mass

The fundamental building block for low-energy meson physics is the non-linear σ model which is of order two in the chiral expansion. In order to generate stable soliton solutions one has to add at least a fourth order term . One such term is the Skyrme term, introduced first long ago[4], and more recently by Adkins et al.[5]. For the Lagrangian density:

$$\mathcal{L}_{2+4a} = -\frac{f_\pi^2}{4}\mathrm{Tr}(L_\mu L^\mu) + \frac{1}{32e^2}\mathrm{Tr}([L_\mu, L_\nu][L^\mu, L^\nu]) \tag{1}$$

where $L_\mu = \partial_\mu U U^\dagger$, U being an SU(2) unitary matrix carrying one unit of winding number, the latter authors calculated the first and third terms of an expansion of the baryon mass $\mathcal{M}$ in decreasing powers of N_c:

$$\mathcal{M} = N_c\mathcal{M}_0 + \mathcal{M}_1 + \frac{1}{N_c}\mathcal{M}_2 \tag{2}$$

It is obviously very important to estimate the zero order contribution $\mathcal{M}_1$ to eq. (2) . To do so one has to evaluate the role of quantum fluctuations around the classical solution $U_0(\vec{x})$. For that purpose we write $U = U_0 \exp(i\vec{\tau}\vec{\alpha})$ and consider up to the 2nd order the expansion in the parameter $\alpha(\vec{x},t)$: $\int d^4x\mathcal{L}' = \int d^4x(\mathcal{L} + \alpha_a H^{ab}\alpha_b)$ where H^{ab} is a three dimensional operator[6]. $\mathcal{M}_1$ is then given by:

$$\mathcal{M}_1 = -\frac{1}{T}(S_{eff}[L_\mu^0] - S_{eff}[0]) \tag{3}$$

This form is not very convenient for practical purposes, but it has the merit to display the physical meaning of what we call the Casimir energy[6]: it is the shift in the energy of the vacuum fluctuations brought in by the presence of the soliton. Eq. (3) tells us also that this energy is ultraviolet divergent, since the effective action can be expanded as a sum over one-loop diagrams in perturbation theory. The spectrum of H_{ab} being continuous, the difference in eq. (3) can be expressed as an integral over its phase shifts[7]:

$$\mathcal{M}_1 = -\frac{1}{2\pi}\int_0^\infty dp\left\{ \sum_0^\infty (2j+1)\delta_j^H(p) - a_1 p - \frac{a_2}{\sqrt{p^2+\mu^2}} - c(\mu)\right\} \tag{4}$$

where the high energy behaviour of the phase shifts is subtracted away and the scale dependence introduced in eq. (4) is cancelled by suitable counterterms[8]. The results for the phase shifts[6] show that "soft"pions dominate the Casimir energy as the integrand decreases smoothly from $\approx 6\pi$ to zero at $p \approx 0.5$ GeV. The scale of that decrease is related to the size of the soliton. The behaviour at the origin is related to a more general feature of the classical solution, namely its symmetries under certain transformations. Upon quantization these classical symmetries (space translation, space and isospin rotations) are broken and this gives rise to the existence of zero modes. In the phase shift representation of the Casimir effect, these $j = 1$ modes will obviously dominate the integrand in eq. (4) since only the low-energy behaviour matters. All the other modes vanish at the origin. These heuristic arguments, are more transparent in 1+1 dimensional theories where the UV divergences can be treated by simple normal ordering[9], and the Casimir effect is exactly given by a sum over the bound state phase shifts.

Computing now the ratio of the zero order term $\mathcal{M}_1$ to the leading one for the Skyrme model one finds that it is equal to 0.9, which is rather high. This means that the Skyrme soliton is too small to be well described at leading classical order. Quantum effects are thus definitely important for the original Skyrme model. The situation is improved if one adds a sixth order term in the Lagrangian of eq. (1), accounting for ω meson "exchange"[10]:

$$\mathcal{L}_{2+4a+6} = \mathcal{L}_{2+4a} + \frac{b}{2f_\pi^2}\left\{\frac{\epsilon^{\mu\nu\alpha\beta}}{24\pi^2}\mathrm{Tr}(L_\nu L_\alpha L_\beta)\right\}^2 \tag{5}$$

In that case the ratio $\dfrac{\mathcal{M}_1}{N_c\mathcal{M}_0}$ is equal to 0.4, leading to a well defined hierarchy in the various contributions to the soliton mass, which is what one would expect if the large N_c expansion truly makes sense in practice to describe the nucleon[11]. All this give support to generalize the original Skyrme model.

3. More mesons

We will shortly see how the predictions of the nucleon physics from soliton models can be substantially improved at the semi-classical level. For this, one has to go beyond the local approximation in powers of the derivatives of the pion field (eq. (1),(5)), and consider finite mass meson resonances, say the basic isovector and isoscalar ones: $\vec{\rho}^{\mu}, \vec{A}_1^{\mu}, \omega^{\mu}, \epsilon$. The Lagrangian dynamics should account for the low-energy properties of these mesons, such as decay widths, physical masses, electromagnetic properties etc. The isovector mesons, in the simplest approach are introduced as massive Yang-Mills fields or as "hidden"gauge fields, the two approaches being equivalent[12]. The omega field, coupled to the U(1) baryon current ensures the soliton stability. One example of such a Lagrangian, based on the nonlinear σ model was shown to yield reasonable agreement with phenomenology[13].

A proper extension of this work, taking into account the physics of scalar isoscalar resonances, was shown to yield (besides a simultaneous fit of meson and single baryon observables) attractive forces in the nucleon-nucleon potential[14], and this at the correct range (≈ 1 fm). Let me only remind to you that besides the problem of the mass, the original Skyrme model was shown to be inappropriate for the description of the NN interaction[15-16]. In these papers it was found that at interdistances of ≈ 1 fm, the Skyrme model yields a repulsion of ≈ 400 MeV between nucleons. It was also advocated that this drawback can be amended by the inclusion of scalar fields. In order to investigate the role of the scalar degrees of freedom we include them in a meson Lagrangian based on the (gauged) linear σ model rather than the non-linear one:

$$\mathcal{L}_{\pi\xi\rho A_1\omega} = \frac{1}{2}\partial_\mu\xi\partial^\mu\xi + \frac{\xi^2}{4}\mathrm{Tr}(D_\mu U D^\mu U^\dagger) - \lambda(\xi^2 - \Gamma^2)^2 + \mathcal{L}_{MYM}(X_\mu, Y_\mu) + \mathcal{L}_\omega + \mathcal{L}_{WZ} \qquad (6)$$

The isovector mesons related to the $\vec{\rho}^\mu, \vec{A}_1^\mu$ resonances are introduced as massive SU(2) Yang-Mills fields through the covariant derivative $D_\mu U = \partial_\mu U + ig(X_\mu U - UY_\mu)$ in eq. (6). Physical masses for the X_μ, Y_μ fields are used as inputs in $\mathcal{L}_{MYM}$, and full account of the meson sector anomalies is taken in the effective Wess-Zumino term $\mathcal{L}_{WZ}$[14].

The baryon number one sector is then constructed by searching for classical solutions possessing the hedgehog symmetry, and the NN potential is calculated in the product approximation. The inclusion of the scalar field plays a crucial role in reducing the range of the classical omega solution through a highly non-linear mechanism. We believe that this mechanism[14] is responsible for the relative agreement we find with phenomenology, knowing that the repulsive part of the NN interaction is mainly due to the ω field[17].

4. Some issues

I thus would like to conclude by saying that there are strong indications to believe that it is unreasonable to demand the original Skyrme model to fit in details to baryon phenomenology, since quantum corrections are as large as the leading order.

A proper extension of the Skyrme model taking into account the low mass mesons seems to yield substantial agreement with phenomenology, and this at the classical level. I would like to mention here, that because of asymptotic freedom, in SU(N) Yang-Mills theory there is an infinity of meson states. There is no a priori reason to consider only the pion.

Finally, concerning the so-called $1/N_c$ expansion, we do not yet know whether the relevant expansion parameter for QCD is $1/N_c$, or $\frac{1}{8\pi}1/N_c$, or even $\frac{1}{100\pi}1/N_c$. In that latter case, it would be possible even for the "real world" QCD to have a semi-classical limit.

5. Acknowledgements

I would like to express my deep gratitude to Prof. R. Vinh Mau who initiated me to the soliton models of the nucleon. The ideas I exposed here are the result of discussions we had during the last two years. I would like also to thank Dr. B. Moussallam for explaining to me the physics of the the Casimir effect. I finally acknowledge many stimulating discussions with Drs. B. Loiseau, M. Lacombe and W. N. Cottingham.

6. References

1. G.'t Hooft, *Nucl. Phys.* **B72** (1974) 461; **B75** (1974) 461

2. E. Witten, *Nucl. Phys.* **B160** (1979) 57

3. J. Schechter, *Phys. Rev.* **D34** (1986) 868

4. T.H.R. Skyrme, *Proc. Roy. Soc.* **A260** (1961) 127

5. G.S. Adkins, C.R. Nappi and E. Witten, *Nucl. Phys.* **B228** (1983) 552

6. B. Moussallam and D. Kalafatis, *Phys. Lett.* **B272** (1991) 196

7. J. Schwinger, *Phys. Rev.* **94** (1951) 1362

8. J. Gasser and H. Leutwyller, *Ann. Phys.* (N.Y.) **158** (1984) 142

9. K. Cahill, A. Comtet and R. J. Glauber, *Phys. Lett.* **B64** (1976) 283

10. G.S. Adkins and C.R. Nappi, *Phys. Lett.* **137B** (1984) 251

11. In some respects, my discussion of the ratio $\dfrac{\mathcal{M}_1}{N_c \mathcal{M}_0}$, could be as silly as if I was asking whether the photon is a wave or a particle. But the key argument is the magnitude of the coupling. My point here is that in view of the recent results[6] on the Casimir energy, the Skyrme soliton seems to be a strong coupling soliton, rather than a weak one. Thus there is probably no hope to obtain a reliable approximation for baryon physics using this particular model. Instead of going through higher and higher quantum corrections (which is beyond the scope of the effective Lagrangian approach) it is more economical to build a weak coupling approximation, where the description of low-energy hadron physics is possible at leading order.

12. H. Forkel, A. D. Jackson and C. Weiss, *Nucl. Phys.* **A526** (1991) 453

13. M. Lacombe, B. Loiseau, R. Vinh Mau and W. N. Cottingham, *Phys. Rev.* **D38** (1988) 1491

14. D. Kalafatis and R. Vinh Mau, *Phys. Lett.* **B283** (1992) 13

15. A. Jackson, A. D. Jackson and V. Pasquier, *Nucl. Phys.* **A432** (1985) 567

16. M. Lacombe, B. Loiseau, R. Vinh Mau, W. N. Cottingham and P. Lisboa, *Phys. Lett.* **B150** (1985) 259

17. The effect on the ω field is non-trivial, because this field is only implicitly coupled to the scalar field. The reduction of his range is no more effective, if one switches off the isovector fields X_μ, Y_μ by taking $g = 0$. This shows that it is important to keep the masses finite for the mesons.

HOW CAN QED$_3$ HELP US UNDERSTAND QCD$_4$?

C. J. BURDEN

Department of Theoretical Physics, Institute of Advanced Studies,
Australian National University,
Canberra, ACT 2601, Australia

ABSTRACT

Quantum electrodynamics in (2+1) dimensions (QED3) has properties in common with QCD which make it an ideal toy model for studying nonperturbative field theory, namely confinement and a global 'chiral'-like symmetry which gives rise to goldstone 'pions'. We study the Schwinger-Dyson equations of QED3 by modelling the photon-fermion vertex with an ansatz designed to respect the symmetries of the theory. Our results are compared with those of lattice calculations.

1. Introduction

Quantum electrodynamics in two spatial and one temporal dimensions (QED3) is an ideal field theoretical model for testing ideas that may be relevant to quantum chromodynamics (QCD). It shares with QCD the property of confinement, albeit for the relatively simple reason that classical electrodynamics in (2+1) dimensions has a logarithmic potential. Our analysis of the photon polarisation tensor below indicates that this feature is maintained in the full quantum field theory.

In order to study chiral symmetry breaking we study the massless, 4-component version of QED3[1], with Euclidean action

$$S[\psi, \overline{\psi}, A_\mu] = \int d^3x \left[\frac{1}{4} F_{\mu\nu} F_{\mu\nu} + \frac{1}{2\xi} (\partial_\mu A_\mu)^2 + \overline{\psi} \gamma_\mu (\partial_\mu + ie A_\mu) \psi \right], \qquad (1)$$

where $\mu = 1, 2, 3$, the γ_μ's are 4×4 and ξ is the usual gauge fixing parameter ($\xi = 0$: Landau gauge, $\xi = 1$: Feynman gauge). The global $U(2)$ symmetry generated by $\{I, \gamma_4, \gamma_5, i\gamma_4\gamma_5\}$ is broken in the full quantum theory by the generation of a dynamical fermion mass to a $U(1) \times U(1)$ symmetry generated by $\{I, i\gamma_4\gamma_5\}$.

2. Schwinger Dyson Equations

In our analysis we study the truncated subset of the Schwinger Dyson (SD) equations consisting of the fermion and photon self energy equations:

$$S^{-1}(p) = i\gamma \cdot p + e^2 \int \frac{d^3q}{(2\pi)^3} \Gamma_\mu(p, q) S(q) \gamma_\nu D_{\mu\nu}(p - q), \qquad (2)$$

$$\Pi_{\mu\nu}(k) = -e^2 \int \frac{d^3q}{(2\pi)^3} \text{tr} \left[\gamma_\mu S(q + \tfrac{1}{2}k) \Gamma_\nu(q + \tfrac{1}{2}k, q - \tfrac{1}{2}k) S(q - \tfrac{1}{2}k) \right]. \qquad (3)$$

Here $S(q)$ and $D_{\mu\nu}(p)$ are the dressed fermion and photon propagators and $\Pi_{\mu\nu}$ the photon polarisation tensor. The dressed fermion-photon vertex $\Gamma_\mu(p,q)$ is determined by the remaining infinite tower of SD equations involving higher n-point functions. Solving this infinite tower is clearly not possible, and instead we adopt an ansatz for Γ_μ. We are guided by the restrictions that the vertex (a) must satisfy the Ward-Takahashi identity $i(p-q)_\mu \Gamma_\mu(p,q) = S^{-1}(p) - S^{-1}(q)$, (b) must be free of kinematic singularities, (c) must have the same transformation properties as the bare vertex γ_μ under C, P and T, and (d) must reduce to the bare vertex in the free field limit $S^{-1}(p) \to i\gamma \cdot p$.

Ideally one should also demand that local gauge invariance be respected. A well defined set of transformation laws for the fermion and photon propagators and for the vertex under an arbitrary gauge transformation are given in an early paper by Landau and Khalatnikov[2] (LK). These laws leave the SD equations and the Ward-Takahashi identity invariant. One can therefore, in principle, ensure gauge covariance of the solutions of the SD equations by specifying the ansatz for Γ_μ at one particular value of the gauge parameter, $\xi = \xi_0$, and then define the ansatz at other values via the action of the LK transformations. This is the procedure we shall adopt here, with the cautionary note that, by doing so, we are forced to accept a vertex ansatz with an explicit gauge parameter dependence. We also mention that the LK transformation rule for the vertex is quite complicated. However, there is no need to state explicitly the vertex ansatz in all gauges; in practice we need only solve the SD equations at ξ_0, and apply the transformation rules to the solutions obtained for the propagators if they are required in other gauges.

The following set of vertices, parameterised by the gauge parameter ξ_0, satify the above requirements:

$$\Gamma_\mu(p,q;\xi_0) = \frac{1}{2}\left(A(p^2) + A(q^2)\right)\gamma_\mu$$
$$+\frac{(p+q)_\mu}{p^2 - q^2}\left[\left(A(p^2) - A(q^2)\right)\frac{\gamma\cdot(p+q)}{2} - i\left(B(p^2) - B(q^2)\right)\right]. \qquad (4)$$

This form was initially proposed by Ball and Chiu[3]. In Eq. (4) the functions A and B come from our form of the dressed fermion propagator:

$$S^{-1}(p) = i\gamma \cdot p\, A(p^2) + B(p^2). \qquad (5)$$

Furthermore gauge invariance tells us that the photon polarisation tensor must be transverse. We therefore write $\Pi_{\mu\nu}(k) = (\delta_{\mu\nu}k^2 - k_\mu k_\nu)\Pi(k^2)$. With the ansatz Eq. (4), solving the SD equations Eqs. (2) and (3) amounts to solving three coupled integral equations for the three real scalar functions A, B and Π.

3. Numerical Solutions

3.1. Quenched Results

We concentrate first on the case $N_f = 0$, that is, we ignore the feedback of the photon polarisation tensor into the fermion equation Eq. (2)[4]. This is equivalent to

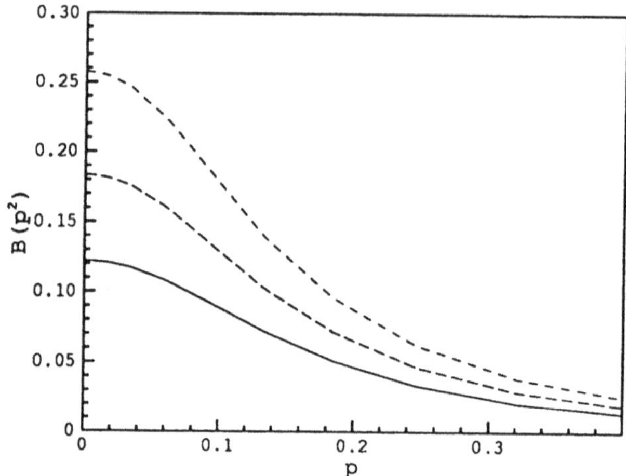

Figure 1: The function B in the fermion propagator obtained from the quenched SD equation with $\xi_0 = 0.0$ (solid curve), 0.5 (long dashes) and 1.0 (short dashes).

the quenched approximation of lattice gauge theory. The solution for the function B in the fermion propagator is shown in fig. 1 for a range of values of the specified gauge parameter ξ_0. We also find that the function A is of order 1, and asymptotes towards 1 as $p \to \infty$. We see that the fermion propagator asymptotes towards the free propagator in the space-like UV limit, reminiscent of asymptotic freedom in QCD. Chiral symmetry breaking is signalled by the function B being non-zero.

These functions are strongly dependent on the input gauge parameter ξ_0, as indeed they should be. In order to make comparisons between different ansätze for the vertex specified by different input values of ξ_0, it is necessary to gauge transform each solution back to the Landau gauge using the LK transformations[5]. This is most easily done in coordinate space where the (2+1)-dimensional Euclidean fermion propagator undergoes the simple transformation $S(x,\xi) \to S(x,\xi') = e^{(\xi'-\xi)|x|/8\pi}S(x,\xi)$. Fig. 2 illustrates the effect of the transformation to Landau gauge of the scalar part $Y(x^2)$ of the fermion propagator, defined by

$$S(x) = \gamma \cdot x X(x^2) + Y(x^2). \tag{6}$$

We see that the agreement between the different solutions is quite good. Similar agreement is found for the vector part $X(x^2)$.

The intercept on the left hand side of fig. 2 is essentially the chiral condensate, the value of which can be compared with that obtained from lattice simulations. In three dimensions, staggered lattice fermions naturally model four component fermions in their naive continuum limit, making QED3 an ideal field theory for lattice simulations.

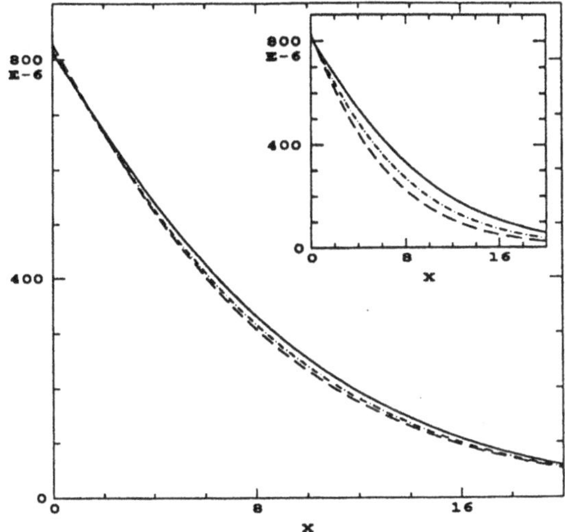

Figure 2: In the inset we plot $Y(x^2)$ for $\xi_0 = 0$ (solid curve), 0.5 (dash-dot curve) and 1.0 (dashed curve). In the main body of the figure we plot the LK transforms of these functions back to Landau ($\xi = 0$) gauge.

In fig. 3 are plotted the quenched lattice results of Dagotto et al.[6] on 8^3 and 10^3 lattices, together with our SD equation result. The agreement with the scaling window exhibited by the 10^3 results is very promising.

3.2. Dynamical Fermions

We present here preliminary results for the coupled system of fermion and photon SD equations Eq. (2) and (3) for $N_f = 2$ flavours of 4-component fermions. The ansatz of Eq. (4) is used for the fermion-photon vertex with $\xi_0 = 0$. The function $B(p^2)$ is plotted in fig. 4 together with the equivalent quenched case. We note that the scale of the dynamical symmetry breaking is considerably reduced as the number of flavours increases, an observation previously made in both SD equation[7] and lattice gauge theory[6] calculations. There is some debate in the literature about the possible existence of a critical number of flavours[6,7,8] above which there is no chiral symmetry breaking. Until our calculations are extended to higher values of N_f we have nothing to say on this matter. The function $A(p^2)$ is once again of $O(1)$.

In fig. 5 we plot the vacuum polarisation scalar $\Pi(k^2)$ [5] 'per number of flavours' for $N_f = 0$ and 2, together with the 1-loop result for chiral QED3 [9], $\Pi(k^2) = e^2/8k$. The effect of the 1-loop correction is to soften the logarithmic static potential associated with the free photon propagator to a non-confining $1/r$ potential. However, dynamical generation of an effective fermion mass at the infared end of the fermion propagator

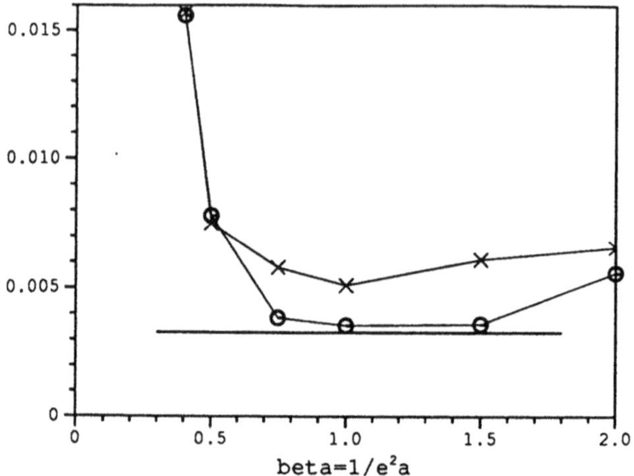

Figure 3: The chiral condensate from the quenched lattice results of Dagotto et al.[6] for 10^3 (circles) and 8^3 (crosses) lattices. The thick line across the bottom covers the range of values obtained by solving the quenched SD equation with ξ_0 between 0 and 1.

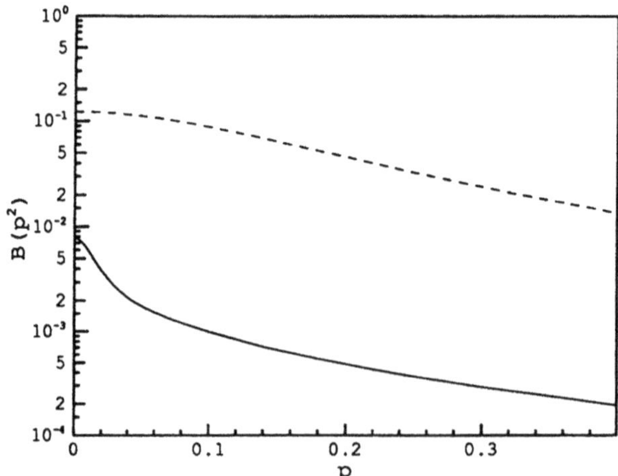

Figure 4: The function $B(p^2)$ for $N_f = 0$ (dashed curve) and $N_f = 2$ (solid curve) flavours of fermions at $\xi_0 = 0$.

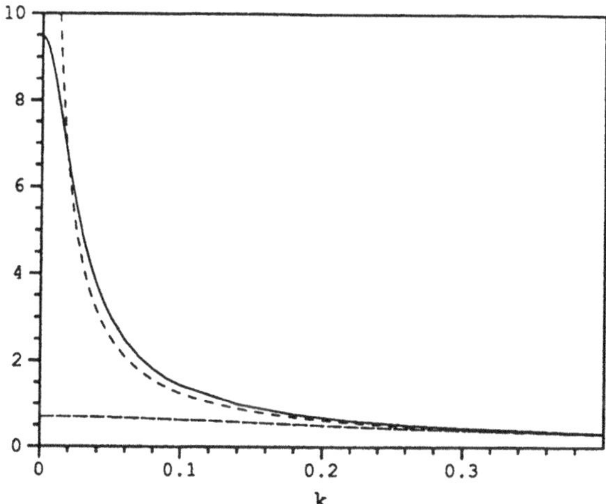

Figure 5: The vacuum polarisation scalar $\Pi(k^2)$ for $N_f = 0$ (long dashes) and $N_f = 2$ (solid curve) together with the 1-loop result (short dashes).

ensures that $\Pi(0)$ remains finite, even though the 1-loop result is approached as the number of fermions increases. A consequence of this is that the static potential associated with the dressed photon propagator maintains its confining logarithmic nature at large r.

References

1. R. D. Pisarski, *Phys. Rev.* **D29** (1984) 2423.
2. L. D. Landau and I. M. Khalatnikov, *Sov. Phys. JETP* **2** (1956) 69.
3. J. S. Ball and T.-W. Chiu, *Phys. Rev.* **D22** (1980) 2542.
4. C. J. Burden and C. D. Roberts, *Phys. Rev.* **D44** (1991) 540.
5. C. J. Burden, J. Praschifka and C. D. Roberts, *Photon polarisation tensor and gauge dependence in three dimensional quantum electrodynamics*, to appear in *Phys. Rev.* **D.**
6. E. Dagotto, A. Kocic and J. B. Kogut, *Nucl. Phys.* **B334** (1990) 270.
7. M. R. Pennington and D. Walsh, *Phys. Lett.* **B253** (1991) 246.
8. D. Nash, *Phys. Rev. Lett.* **62** (1989) 3024.
9. T. W. Appelquist et al., *Phys. Rev.* **D33** (1986) 3074.

ZETA FUNCTION REGULARIZATION OF THE VACUUM ENERGY

E. ELIZALDE

Department E.C.M., Faculty of Physics, University of Barcelona
Diagonal 647, E-08028 Barcelona, Spain

ABSTRACT

A compendium of the formulas resulting from the zeta-regularization techniques, developed by the author and collaborators during the last three years, is given. It is mainly intended as a table for practical use by the reader —the derivations and arguments involved can be found in the accompanying bibliography. In particular, useful expressions are provided for the analytic continuation of Riemann, Hurwitz and Epstein zeta functions and generalizations of them, for their asymptotic expansions (including those for derivatives of Hurwitz's ζ), integrals of Γ's, the zeta-function regularization theorem —and its use for multiple zeta-functions with arbitrary exponents– and, finally, some applications of the zeta-function regularization procedure to the definition of the vacuum energy and to the Casimir effect, with hints about the modern experiments employing liquid helium. Some mistakes which regretfully appeared here and there in the formulas of the original papers have been corrected.

1. Asymptotic expansion of Hurwitz's ζ and its derivatives

The Hurwitz (or generalized Riemann) zeta function is

$$\zeta(z,a) = \sum_{n=0}^{\infty} (n+a)^{-z}, \quad Re\ z > 1, \quad a \neq 0, -1, -2, \ldots \tag{1}$$

The ordinary Riemann zeta function is the particular case $a = 1$.

1.1. First derivative

We look for the asymptotic expansion for the first derivative

$$\zeta'(-n,a) \equiv \frac{\partial}{\partial z}\zeta(z,a)\bigg|_{z=-n}, \quad n = 0, 1, 2, \ldots, \tag{2}$$

in inverse powers of a. The procedure employed is similar to the standard method: use of Watson's lemma and Laplace's method. A good starting point for the derivation is the Hermite integral representation for $\zeta(z,a)$

$$\zeta(z,a) = \frac{1}{2}a^{-z} + \frac{a^{1-z}}{z-1} + 2\int_0^\infty (a^2+t^2)^{-z/2}\sin\left[z\tan^{-1}\left(\frac{t}{a}\right)\right]\frac{dt}{e^{2\pi t}-1}. \tag{3}$$

The final, rigorous result turns out to be the same that one would obtain by naive derivation term by term of the asymptotic series (*tauberian th.*).

$$\zeta(z+1,a) = \frac{1}{z}a^{-z} + \frac{1}{2}a^{-z-1} + \frac{1}{z}\Sigma_0(z,a), \tag{4}$$

$$\Sigma_0(z,a) \equiv \sum_{k=2}^{\infty} \frac{B_k}{k!}(z)_k\, a^{-z-k}, \qquad (z)_k \equiv z(z+1)\cdots(z+k-1) = \frac{\Gamma(z+k)}{\Gamma(z)};$$

$(z)_k$ is Pochhammer's symbol (rising factorial) and B_k the Bernoulli numbers. Thus, the asymptotic series corresponding to $\zeta'(z+1,a)$ turn out to be[1,2]

$$\zeta'(z+1,a) = -\left(\frac{1}{z}+\ln a\right)\zeta(z+1,a) + \frac{1}{2z}a^{-z-1} + \frac{1}{z}\Sigma_1(z,a),$$

$$\Sigma_1(z,a) \equiv \sum_{k=2}^{\infty} B_k \sum_{j=0}^{k-1} \frac{(z)_j}{j!(k-j)}\, a^{-z-k}, \tag{5}$$

for large $|a|$ and $|Arg\ a| < \pi$. In particular, $\zeta(1-n,a) = -\frac{1}{n}B_n(a)$, (with $B_n(a)$ Bernoulli polynomial of degree n), and[1,2]

$$\zeta'(1-n,a) = \frac{1}{n}\left(\ln a - \frac{1}{n}\right)B_n(a) - \frac{1}{2n}a^{n-1} - \frac{1}{n}\sum_{k=2}^{n} B_k \sum_{j=0}^{k-1} \binom{n}{j}\frac{(-1)^j}{k-j}\, a^{n-k}$$

$$+ (-1)^{n-1}(n-1)! \sum_{k=n+1}^{\infty} \frac{B_k}{k(k-1)\cdots(k-n)}\, a^{n-k}. \tag{6}$$

1.2. Higher derivatives

Repeating for ζ'' the same procedure developed for ζ', i.e. replacement of $\tan^{-1}(t/a)$ and $\ln(a^2 + t^2)$ by their power series near $t = 0$,

$$\tan^{-1}\left(\frac{t}{a}\right) = \sum_{k=0}^{\infty} \frac{(-1)^k}{2k+1}\left(\frac{t}{a}\right)^{2k+1}, \qquad \ln(a^2+t^2) = 2\ln a + \sum_{k=1}^{\infty} \frac{(-1)^{k+1}}{k}\left(\frac{t}{a}\right)^{2k}, \tag{7}$$

we get

$$\zeta''(z+1,a) = -2\left(\frac{1}{z}+\ln a\right)\zeta'(z+1,a) - \left(2\frac{\ln a}{z}+\ln^2 a\right)\zeta(z+1,a) \tag{8}$$

$$+ \frac{1}{z}\Sigma_2(z,a), \qquad \Sigma_2(z,a) \equiv \sum_{k=2}^{\infty} B_k \sum_{j=0}^{k-1} \frac{1}{k-j}\sum_{h=0}^{j-1} \frac{(z)_h}{h!(j-h)}\, a^{-z-k}.$$

An alternative (standard) procedure is partial integration (it exhibits the asymptotic character of the series). We get the operational recurrence[3]

$$\frac{\partial^m}{\partial z^m}\zeta(z+1,a) = -\left[\left(\frac{\partial}{\partial z}+\ln a\right)^m - \left(\frac{\partial}{\partial z}\right)^m + \frac{m}{z}\left(\frac{\partial}{\partial z}+\ln a\right)^{m-1}\right]$$

$$\zeta(z+1,a) + \frac{1}{z}\Sigma_m(z,a), \tag{9}$$

being

$$\Sigma_m(z,a) \equiv \sum_{k=2}^{\infty} B_k \sum_{j_1=0}^{k-1} \frac{1}{k-j_1} \sum_{j_2=0}^{j_1-1} \frac{1}{j_1-j_2} \cdots \sum_{j_m=0}^{j_{m-1}-1} \frac{(z)_{j_m}}{j_m!\,(j_{m-1}-j_m)}\, a^{-z-k}, \qquad (10)$$

for large $|a|$ and $|Arg\ a| < \pi$. Using (9), the following general recurrence is obtained[3]

$$\zeta^{(m)}(z+1,a) = -\sum_{j=1}^{m} \binom{m}{j} \left(\frac{j}{z}+\ln a\right) \ln^{j-1} a\ \zeta^{(m-j)}(z+1,a) + \frac{1}{z}\Sigma_m(z,a), \quad (11)$$

In particular, for $z = -n$, $n \in \mathbf{N}$,

$$\zeta^{(m)}(1-n,a) = \sum_{j=1}^{m} \binom{m}{j} \left(\frac{j}{n}-\ln a\right) \ln^{j-1} a\ \zeta^{(m-j)}(1-n,a) - \frac{1}{n}\Sigma_m(-n,a), \quad (12)$$

where

$$\Sigma_m(-n,a) = \sum_{k=2}^{\infty} B_k \sum_{j_1=0}^{k-1} \frac{1}{k-j_1} \sum_{j_2=0}^{j_1-1} \frac{1}{j_1-j_2} \cdots \sum_{j_m=0}^{\mu_m} \binom{n}{j_m} \frac{(-1)^j}{j_{m-1}-j_m}\, a^{n-k}, \qquad (13)$$

being $\mu_m = \min(n, j_{m-1}-1)$.

This expression is very appropriate both for numerical and for analytical explicit calculations through the software packets commonly available. To prove this statement, we obtain immediately the particular cases[3]

$$\begin{aligned}
\zeta''(1-n,a) &= \left(-\frac{2}{n^2}+\frac{2\ln a}{n}-\ln^2 a\right)\frac{B_n(a)}{n} - \left(\frac{1}{n}-\ln a\right)\frac{a^{n-1}}{n} \\
&\quad - \frac{1}{n}\Sigma_2(-n,a) - \frac{2}{n}\left(\frac{1}{n}-\ln a\right)\Sigma_1(-n,a), \qquad (14)
\end{aligned}$$

$$\begin{aligned}
\Sigma_2(-n,a) &= \sum_{k=2}^{n} B_k \sum_{j=0}^{k-1} \frac{1}{k-j} \sum_{h=0}^{j-1} \binom{n}{h} \frac{(-1)^h}{j-h}\, a^{n-k} \\
&\quad + \sum_{k=n+1}^{\infty} B_k \sum_{j=0}^{n} \frac{1}{k-j} \sum_{h=0}^{j-1} \binom{n}{h} \frac{(-1)^h}{j-h}\, a^{n-k} \\
&\quad + (-1)^n n! \sum_{k=n+1}^{\infty} B_k \sum_{j=n+1}^{k-1} \frac{1}{(k-j)j(j-1)\cdots(j-n)}\, a^{n-k}
\end{aligned} \qquad (15)$$

and

$$\Sigma_1(-n,a) = \sum_{k=2}^{n} B_k \sum_{j=0}^{k-1} \binom{n}{j} \frac{(-1)^j}{k-j}\, a^{n-k} + (-1)^n n! \sum_{k=n+1}^{\infty} \frac{B_k}{k(k-1)\cdots(k-n)}\, a^{n-k}. \qquad (16)$$

With *Mathematica* in a *Sun Workstation* one gets in few seconds:

$$\zeta''(0,a) = -a\left(\ln^2 a - 2\ln a + 1\right) + \frac{1}{2}\ln^2 a - \frac{a^{-1}}{6}\ln a$$
$$+ \frac{a^{-3}}{60}\left(\frac{\ln a}{3} - \frac{1}{2}\right) - a^{-5}\left(\frac{\ln a}{630} - \frac{4}{1209}\right) + \cdots,$$

$$\zeta''(-1,a) = -\frac{a^2}{4}\left(2\ln^2 a - 2\ln a + 1\right) + \frac{a}{2}\ln^2 a + \frac{1}{12}\left(\ln^2 a - 2\ln a\right)$$
$$- \frac{a^{-2}}{360}\ln a + \frac{a^{-4}}{15120}\left(6\ln a - 5\right) + \cdots,$$

$$\zeta''(-2,a) = -\frac{a^3}{27}\left(9\ln^2 a - 6\ln a + 2\right) + \frac{a^2}{2}\ln^2 a - \frac{a}{6}\left(\ln^2 a - \ln a\right)$$
$$+ \frac{a^{-1}}{60}\left(\frac{\ln a}{3} + \frac{1}{2}\right) - \frac{a^{-3}}{3780}\ln a + \frac{a^{-5}}{1800}\left(\frac{\ln a}{7} - \frac{1}{12}\right) + \cdots,$$

$$\zeta''(-3,a) = -\frac{a^4}{32}\left(8\ln^2 a - 4\ln a + 1\right) + \frac{a^3}{2}\ln^2 a + \frac{a^2}{12}\left(3\ln^2 a + 2\ln a\right)$$
$$+ \frac{1}{60}\left(\frac{\ln^2 a}{2} + \frac{11\ln a}{6} + 1\right) + \frac{a^{-2}}{504}\left(\frac{\ln a}{5} + \frac{1}{6}\right) - \frac{a^{-4}\ln a}{16800} + \cdots.$$

Actually, one reaches values of n as high as $n = 20$ very quickly.

2. Uses of the zeta function in integration

We can obtain a general formula for the integral:

$$\int_a^b dt \, \ln \Gamma(t), \qquad a,b \in \mathbf{R}, \quad a,b > 0, \tag{17}$$

in terms of Hurwitz's ζ.

Starting from Binet's second expression for $\ln \Gamma(t)$:

$$\ln \Gamma(t) = \left(t - \frac{1}{2}\right)\ln t - t + \frac{1}{2}\ln 2\pi + 2\int_0^\infty du \, \frac{\tan^{-1}(u/t)}{e^{2\pi u} - 1}, \tag{18}$$

and using Fubini's theorem and my previous results on asymptotic expansion of these integrals, we get

$$\int_a^b dt \, \ln \Gamma(t) = \left[\frac{1}{2}\left(t^2 + t + \frac{1}{t}\right)\ln t - \frac{t}{4}(3t - 2) + \frac{t}{2}\ln 2\pi \right.$$
$$\left. - \sum_{h=1}^{\infty} \frac{B_{2h+2}}{(2h+2)(2h+1)2h} t^{-2h}\right]_{t=a}^{t=b}. \tag{19}$$

These expansions add up to the final expression[4]

$$\int_a^b dt \, \ln \Gamma(t) = \left[\zeta(-1,t) + \zeta'(-1,t) + \frac{t}{2} \ln 2\pi \right]_a^b . \tag{20}$$

For a consistency test, we reproduce the well-known expression (Erdélyi et al.)

$$\int_x^{x+n} dt \, \ln \Gamma(t) = \sum_{k=0}^{n-1} (x+k) \ln(x+k) - nx - \frac{n(n-1)}{2} + \frac{n}{2} \ln 2\pi, \tag{21}$$

for any $x \in \mathbf{R}$, $n \in \mathbf{N}$, as a particular case. Also interesting is the expression[4,5]

$$\int_m^{m+n+1/2} dt \, \ln \Gamma(t) = - \sum_{k=1}^{m-1} k \ln k + \frac{1}{2} \sum_{k=1}^{m+n-1} (2k+1) \ln(2k+1) \tag{22}$$

$$- \frac{1}{2}(m+n)^2(1 + \ln 2) + \frac{m(m-1)}{2} + \frac{1}{24}(3 - \ln 2) + \frac{1}{2}\left(n + \frac{1}{2}\right) \ln 2\pi - \frac{3}{2}\zeta'(-1).$$

We thus get an expansion for $\zeta'(-1)$, alternative to $\zeta'(-1,q)|_{q=1}$:

$$\zeta'(-1) = -\frac{1}{4} - \frac{7}{36} \ln 2 + \frac{1}{6} \ln \pi + \frac{\gamma}{12} - \frac{2}{3} \sum_{n=2}^{\infty} \frac{(-1)^n}{2^{n+1}n(n+1)} \zeta(n). \tag{23}$$

The numerical result is[4,6]

$$\zeta'(-1) = -0.16542115\ldots \tag{24}$$

We obtain *six* exact digits with just 11 terms.

3. The zeta-function regularization theorem

It provides a method for the computation of expressions of the form —and more involved ones (see later)

$$\sum_{n_1,\ldots,n_N=0}^{\infty} [a_1(n_1 + c_1)^{\alpha_1} + \cdots + a_N(n_N + c_N)^{\alpha_N}]^{-s}, \quad a_j, \alpha_j > 0, \tag{25}$$

for $\mathcal{R}(s)$ big enough, and their analytic (usually meromorphic) continuation to other values of s. In the zeta-function method this is provided by the ordinary Riemann and Hurwitz zeta-functions.

The simplest case corresponds to the Hamiltonian zeta-function[7] $\zeta(s) \equiv \sum_i E_i^{-s}$, with E_i eigenvalues of H. For a system of N non-interacting harmonic oscillators: $\alpha_j = 1$, $j = 1, 2, ..., N$, and a_j are the eigenfrequencies ω_j. Another case is partial toroidal compactification (spacetime $T^p \times \mathbf{R}^{q+1}$). Then $\alpha_j = 2$ and, usually, $c_j = 0, \pm 1/2$. We come to the Epstein zeta-functions[8]

$$Z_N(s) = \sum_{n_1,\ldots,n_N=-\infty}^{\infty} (n_1^2 + \cdots + n_N^2)^{-s}, \tag{26}$$

$$Y_N(s) = {\sum_{n_1,\ldots,n_N=-\infty}^{\infty}}' \left[\left(n_1 + \frac{1}{2}\right)^2 + \cdots + \left(n_N + \frac{1}{2}\right)^2 \right]^{-s}$$

(the prime means omission of the term $n_1 = n_2 = \ldots = n_N = 0$). Other powers α_j appear when one deals with the spherical compactification (spacetime $S^p \times \mathbf{R}^{q+1}$) and more involved ones. Hence the general expression (25). We shall denote the important particular case:

$$E_N(s; a_1, \ldots, a_N; c_1, \ldots, c_N) \equiv \sum_{n_1,\ldots,n_N=0}^{\infty} \left[\sum_{j=1}^{N} a_j (n_j + c_j)^2 \right]^{-s}, \tag{27}$$

where all $a_j > 0$ and not all the c_j are nonpositive integers.

The only precedents in the literature (to my knowledge) of this kind of evaluations have been restricted to few special cases other than $a_1 = a_2 = \ldots = a_N$ and $c_1 = c_2 = \ldots = c_N = 1$. A most famous expression is the celebrated result of Hardy[9] (a particular case of my final formula).

A basic point in the zeta-function regularization procedure is the interchange of order of summations of infinite series in expressions like

$$S_c^{(\alpha)}(s) \equiv \sum_{m=0}^{\infty} (m+c)^{-s-1} \sum_{a=0}^{\infty} \frac{(-1)^a}{a!} (m+c)^{\alpha a}. \tag{28}$$

For the case $c = 1$ and $\alpha < 2$ things are easy. The correct supplementary contributions for $\alpha \geq 2$ (always with $c = 1$) have been obtained by Elizalde and Romeo.[10] Write

$$S_c^{(\alpha)}(s) = \sum_{m=0}^{\infty} (m+c)^{-s-1} \oint_C \frac{da}{2\pi i} (m+c)^{-\alpha a} \Gamma(a), \tag{29}$$

with C a contour in the complex plane a, $C = L \cup K$, being L the straight line $\mathcal{R}(a) = a_0$, $0 < a_0 < 1$, and K a curved part, the semicircumference at infinity on the left of this line. For $\mathcal{R}(s)$ big enough,

$$S_c^{(\alpha)}(s) = \sum_{a=0}^{\infty} \frac{(-1)^a}{a!} \zeta(s+1-\alpha a, c) + \begin{cases} \frac{1}{\alpha} \Gamma(-\frac{s}{\alpha}) + \Delta_c^{(\alpha)}(s), & \frac{s}{\alpha} \notin \mathbf{N}, \\ (-1)^{s/\alpha} \left[\frac{\gamma}{\Gamma(\frac{s}{\alpha}+1)} - \frac{1}{\gamma \Gamma'(\frac{s}{\alpha}+1)} \right] + \Delta_c^{(\alpha)}(s), & \frac{s}{\alpha} \in \mathbf{N}, \end{cases} \tag{30}$$

where $\Delta_c^{(\alpha)}(s)$ is the following integral[10] over the curved part K of the contour C

$$\Delta_c^{(\alpha)}(s) \equiv \int_K \frac{da}{2\pi i} \zeta(s+1+\alpha a, c) \Gamma(a). \tag{31}$$

This is the main content of the zeta-function regularization theorem (for details see ref. 10 and papers mentioned therein).

The most interesting new case $(0 < c < 1)$: $S_c \equiv S_c^{(2)}(-1) = \sum_{m=0}^{\infty} e^{-(m+c)^2}$, can be expressed in terms of Hurwitz zeta-functions, as

$$S_c = \sum_{m=0}^{\infty} \frac{(-1)^m}{m!} \zeta(-2m, c) + \frac{\sqrt{\pi}}{2} + \sqrt{\pi} \cos(2\pi c) S(\pi^2), \tag{32}$$

where $S(t) \equiv \sum_{m=0}^{\infty} e^{-tm^2}$. This formula is *exact*[11] and holds for *any* value of c.

258

1. Particular case $c = 1$: we recover the known equality

$$S(1) = \frac{\sqrt{\pi} - 1}{2} + \sqrt{\pi} S(\pi^2). \tag{33}$$

2. For $c = 1/2$ we have $\zeta(-2m, 1/2) = 0$, $k = 0, 1, 2, ...$, and

$$\sum_{m=0}^{\infty} \exp\left[-\left(m + \frac{1}{2}\right)^2\right] = \frac{\sqrt{\pi}}{2} - \sqrt{\pi} \sum_{m=1}^{\infty} \exp(-m^2 \pi^2). \tag{34}$$

Permits us to obtain the value of the series with 10^{-10} accuracy, *with just two terms*

$$\sum_{m=0}^{\infty} \exp\left[-\left(m + \frac{1}{2}\right)^2\right] = \frac{\sqrt{\pi}}{2} - \sqrt{\pi} e^{-\pi^2} + \mathcal{O}(10^{-10}). \tag{35}$$

3. For $c = 0$

$$\sum_{m=0}^{\infty} e^{-m^2} = \frac{1}{2} + \frac{\sqrt{\pi}}{2} + \sqrt{\pi} S(\pi^2). \tag{36}$$

4. For $c = 1/4$,

$$\sum_{m=0}^{\infty} \exp\left[-\left(m + \frac{1}{4}\right)^2\right] = \frac{\sqrt{\pi}}{2} + \sum_{m=0}^{\infty} \frac{(-1)^m}{m!} \zeta\left(-2m, \frac{1}{4}\right). \tag{37}$$

The series is now asymptotic. It stabilizes between the 8th and the 12th summand and provides its best value (with $\simeq 10^{-7}$ accuracy) when we add the 10 first terms.

5. For $c = 1/3$ and $c = 1/6$:

$$\sum_{m=0}^{\infty} \exp\left[-\left(m + \frac{1}{3j}\right)^2\right] = \frac{\sqrt{\pi}}{2} + \sum_{m=0}^{\infty} \frac{(-1)^m}{m!} \zeta\left(-2m, \frac{1}{3j}\right)$$
$$+ (-1)^j \frac{\sqrt{\pi}}{2} \sum_{m=1}^{\infty} \exp(-m^2 \pi^2), \quad j = 1, 2. \tag{38}$$

6. Some more relations:

$$\sum_{m=-\infty}^{\infty} \exp\left[-(m + c)^2\right] = \sqrt{\pi} + 2\sqrt{\pi} \cos(2\pi c) S(\pi^2),$$
$$\sum_{m=0}^{\infty} m \exp\left[-(m + c)^2\right] = \frac{1}{2} + \sum_{m=0}^{\infty} \frac{(-1)^m}{m!} [\zeta(-2m - 1, c)$$
$$-c\zeta(-2m, c)] - \frac{\sqrt{\pi}}{2} c + \sqrt{\pi} [\pi \sin(2\pi c) - c \cos(2\pi c)] S(\pi^2),$$
$$\sum_{m=0}^{\infty} \exp[-a(m + c)^2] = \sum_{m=0}^{\infty} \frac{(-1)^m}{m!} a^m \zeta(-2m, c)$$
$$+\frac{1}{2}\sqrt{\frac{\pi}{a}} + \sqrt{\frac{\pi}{a}} \cos(2\pi c) S\left(\frac{\pi^2}{a^2}\right). \tag{39}$$

We get the general expresion (Epstein-Hurwitz type, for $N = 2$)[11]

$$E_2(s; a_1, a_2; c_1, c_2) = \sum_{n_1, n_2 = 0}^{\infty} \left[a_1(n_1 + c_1)^2 + a_2(n_2 + c_2)^2 \right]^{-s}$$

$$\times \quad \zeta(-2m, c_1)\zeta(2s + 2m, c_2) + \frac{a_2^{-s}}{2} \left(\frac{\pi a_1}{a_2} \right)^{1/2} \frac{\Gamma(s - \frac{1}{2})}{\Gamma(s)} \zeta(2s - 1, c_2)$$

$$+ \quad \frac{2\pi^s}{\Gamma(s)} \cos(2\pi c_1) a_1^{-\frac{s}{2} - \frac{1}{4}} a_2^{-\frac{s}{4} + \frac{1}{4}} \sum_{n_1 = 1}^{\infty} \sum_{n_2 = 0}^{\infty} \tag{40}$$

$$n_1^{s - \frac{1}{2}} (n_2 + c_2)^{-s + \frac{1}{2}} K_{s - \frac{1}{2}} \left[2\pi \sqrt{\frac{a_2}{a_1}} n_1 (n_2 + c_2) \right],$$

K_ν being the modified Bessel function of the second kind. It constitutes the general analytic continuation formula for two-dimensional series. In particular, for $s = 0$,

$$E_2(0; a_1, a_2; c_1, c_2) = \left(c_1 - \frac{1}{2} \right) \left(c_2 - \frac{1}{2} \right), \tag{41}$$

for $s = -1$,

$$E_2(-1; a_1, a_2; c_1, c_2) = a_2 \left(\frac{1}{2} - c_1 \right) \zeta(-2, c_2) + a_1 \left(\frac{1}{2} - c_2 \right) \zeta(-2, c_1) \tag{42}$$

$$= \frac{1}{3} \left(c_1 - \frac{1}{2} \right) \left(c_2 - \frac{1}{2} \right) \left[a_1 c_1 (1 - c_1) + a_2 c_2 (1 - c_2) \right],$$

and for $s = 2$,

$$E_2(2; a_1, a_2; c_1, c_2) = \frac{1}{a_2^2} \sum_{m=0}^{\infty} (-1)^m (m + 1) \left(\frac{a_1}{a_2} \right)^m$$

$$\times \quad \zeta(-2m, c_1)\zeta(2m + 4, c_2) + \frac{\pi}{4a_2} \frac{1}{\sqrt{a_1 a_2}} \zeta(3, c_2) \tag{43}$$

$$+ \quad \frac{\pi^2 \cos(2\pi c_1)}{a_1 a_2} \sum_{n=0}^{\infty} \left\{ (n + c_2)^{-2} \left[\exp\left(2\pi \sqrt{\frac{a_2}{a_1}} (n + c_2) \right) - 1 \right]^{-2} \right.$$

$$+ \quad \left. \left[(n + c_2)^{-2} + \sqrt{\frac{a_1}{a_2}} \frac{(n + c_2)^{-3}}{2\pi} \right] \left[\exp\left(2\pi \sqrt{\frac{a_2}{a_1}} (n + c_2) \right) - 1 \right]^{-1} \right\}.$$

The general expression[11] for arbitrary N turns out to be

$$E_N(s; a_1, \ldots, a_N; c_1, \ldots, c_N) = \frac{1}{\Gamma(s)} \sum_{m=0}^{\infty} \frac{(-1)^m}{m!} a_1^m \zeta(-2m, c_1)$$

$$\times \quad \Gamma(s + m) E_{N-1}(s + m; a_2, \ldots, a_N; c_2, \ldots, c_N) \tag{44}$$

$$+ \quad \frac{1}{2} \sqrt{\frac{\pi}{a_1}} \frac{\Gamma(s - \frac{1}{2})}{\Gamma(s)} E_{N-1}(s - \frac{1}{2}; a_2, \ldots, a_N; c_2, \ldots, c_N)$$

$$+ \quad \sqrt{\frac{\pi}{a_1}} \frac{\cos(2\pi c_1)}{\Gamma(s)} \sum_{n_1 = 1}^{\infty} \sum_{n_2, \ldots, n_N = 0}^{\infty} \int_0^{\infty} dt\, t^{s - 3/2} \exp\left[-\frac{\pi^2 n_1^2}{a_1 t} - t \sum_{j=2}^{N} a_j (n_j + c_j)^2 \right].$$

260

4. Multiple zeta-functions with arbitrary exponents: explicit expressions

Explicit expressions for the analytic continuation of the generalizations to arbitrary exponents of the inhomogeneous Epstein functions:

$$E_p^c(s) \equiv \sum_{n_1,\ldots,n_p=1}^{\infty} (n_1^2 + \cdots + n_p^2 + c^2)^{-s}, \tag{45}$$

(we allow for $c = 0$) are not easy to find for $p > 1$, the case $p = 1$ being the only one investigated in the literature.

Application of the zeta-function regularization theorem yields very general formulas for functions of the general type[11]

$$M_N^c(s; a_1, \ldots, a_N; \alpha_1, \ldots, \alpha_N) \equiv \sum_{n_1,\ldots,n_N=1}^{\infty} (a_1 n_1^{\alpha_1} + \cdots + a_N n_N^{\alpha_N} + c^2)^{-s}, \tag{46}$$

for any value of c, such as

$$
\begin{aligned}
M_N^c(s; a_1, \ldots, a_N; \alpha_1, \ldots, \alpha_N) &= \frac{1}{a_N^s \Gamma(s)} \sum_{p=0}^{N-1} \sum_{C_{N-1,p}} \prod_{r=1}^{p} \frac{b_{i_r}^{-1/\alpha_{i_r}}}{\alpha_{i_r}} \Gamma\left(\frac{1}{\alpha_{i_r}}\right) \\
&\times \sum_{k_{j_1},\ldots,k_{j_{N-p-1}}=0}^{\infty} \Gamma\left(s + \sum_{l=1}^{N-p-1} k_{j_l} - \sum_{r=1}^{p} \frac{1}{\alpha_{i_r}}\right) \\
&\times \prod_{l=1}^{N-p-1} \left[\frac{(-b_{j_l})^{k_{j_l}}}{k_{j_l}!} \zeta(-\alpha_{j_l} k_{j_l})\right] M_1^{c/\alpha_N}\left(\alpha_n\left(s + \sum_{l=1}^{N-p-1} k_{j_l} - \sum_{r=1}^{p} \frac{1}{\alpha_{i_r}}\right), 1, \alpha_N\right) + \Delta_{ER},
\end{aligned}
\tag{47}
$$

with $b_{i_r} \equiv a_{i_r}/a_N$ (notice the errata in eqs. (3.22) and (3.23) of my ref. J. Phys. A[11]) and $1 \leq i_1 < \cdots < i_p \leq N - 1$, $1 \leq j_1 < \cdots < j_{N-p-1} \leq N - 1$, being $i_1, \ldots, i_p, j_1, \ldots, j_{N-p-1}$ a permutation of $1, 2, \ldots, N - 1$. The sum on $C_{N-1,p}$ means sum over the $\binom{N-1}{p}$ choices of the indices $i_1, \ldots, i_p$ among the $1, 2, \ldots, N - 1$.

The most immediate case, $a_1 = \cdots = a_p = 1$ and $\alpha_1 = \cdots = \alpha_p = 2$, simplifies considerably. For $c = 0$, we get

$$E_p(s) = \frac{(-1)^{p-1}}{2^{p-1}} \frac{1}{\Gamma(s)} \sum_{j=0}^{p-1} (-1)^j \binom{p-1}{j} \Gamma(2s - j) \zeta\left(s - \frac{j}{2}\right) + \Delta_{ER}, \tag{48}$$

and, for $c \neq 0$,

$$E_p^c(s) = \frac{(-1)^{p-1}}{2^{p-1}} \frac{1}{\Gamma(s)} \sum_{j=0}^{p-1} (-1)^j \binom{p-1}{j} \Gamma\left(s - \frac{j}{2}\right) E_1^c\left(s - \frac{j}{2}\right) + \Delta_{ER}. \tag{49}$$

The poles of this last function arise from those of $E_1^c(s - j/2)$, which are obtained for the values of s such that $s - j/2 = 1/2, -1/2, -3/2, \ldots$. They are poles of order one at $s = p/2, (p-1)/2, p/2 - 1, \ldots$, except for $s = 0, -1, -2, \ldots$, then the function is finite (owing to the $\Gamma(s)$ in the denominator). These poles are removed by zeta-function regularization.[12]

Alternatively, a very useful and exact recurrent formula is[11]

$$
\begin{aligned}
E_N^c(s; a_1, \ldots, a_N) &\equiv \sum_{n_1, \ldots, n_N = 1}^{\infty} (a_1 n_1^2 + \cdots + a_N n_N^2 + c^2)^{-s} \\
&= -\frac{1}{2} E_{N-1}^c(s; a_2, \ldots, a_N) + \frac{1}{2} \sqrt{\frac{\pi}{a_1}} \frac{\Gamma(s - 1/2)}{\Gamma(s)} E_{N-1}^c(s - 1/2; a_2, \ldots, a_N) \\
&+ \frac{\pi^s}{\Gamma(s)} a_1^{-s/2} \sum_{k=0}^{\infty} \frac{a_1^{k/2}}{k!(16\pi)^k} \prod_{j=1}^{k} [(2s - 1)^2 - (2j - 1)^2] \sum_{n_1, \ldots, n_N = 1}^{\infty} n_1^{s-k-1} \\
&\times (a_2 n_2^2 + \cdots + a_N n_N^2 + c^2)^{-(s+k)/2} \exp\left[-\frac{2\pi}{\sqrt{a_1}} n_1 (a_2 n_2^2 + \cdots + a_N n_N^2 + c^2)^{1/2} \right].
\end{aligned}
\tag{50}
$$

The recurrence starts from expression

$$
E_1^c(s; 1) = -\frac{c^{-2s}}{2} + \frac{\sqrt{\pi}}{2} \frac{\Gamma(s - 1/2)}{\Gamma(s)} c^{-2s+1} + \frac{2\pi^s c^{-s+1/2}}{\Gamma(s)} \sum_{n=1}^{\infty} n^{s-1/2} K_{s-1/2}(2\pi n c).
\tag{51}
$$

We get for $c \neq 0$

$$
\begin{aligned}
E_2^c(s) &= -\frac{1}{2} E_1^c(s) + \frac{\sqrt{\pi}}{2} \frac{\Gamma\left(s - \frac{1}{2}\right)}{\Gamma(s)} E_1^c\left(s - \frac{1}{2}\right) + \Delta_{ER}, \\
E_3^c(s) &= \frac{1}{4} E_1^c(s) - \frac{\sqrt{\pi}}{2} \frac{\Gamma\left(s - \frac{1}{2}\right)}{\Gamma(s)} E_1^c\left(s - \frac{1}{2}\right) + \frac{\pi}{4(s-1)} E_1^c(s-1) + \Delta_{ER},
\end{aligned}
\tag{52}
$$

and similar expressions for $c = 0$. This case can be obtained from the former by analytically continuing in the parameter c.[12,13] Δ_{ER} is again the well-known term coming from (additional) series commutation.[10,11] Actually, for numerical evaluations we do not need to consider exponentially small terms in the asymptotic expansions above, which give a very good and quick approximation.

A couple of particularly useful expressions:

$$
\begin{aligned}
\sum_{n_1, n_2 = 1}^{\infty} \sqrt{\left(\frac{n_1}{a_1}\right)^2 + \left(\frac{n_2}{a_2}\right)^2} &= \frac{1}{24}\left(\frac{1}{a_1} + \frac{1}{a_2}\right) - \frac{\zeta(3)}{8\pi^2}\left(\frac{a_1}{a_2^2} + \frac{a_2}{a_1^2}\right) \\
&- \frac{\pi^{3/2}}{2\sqrt{a_1 a_2}}\left[\exp\left(-2\pi\frac{a_1}{a_2}\right)\left(1 + \mathcal{O}(10^{-3})\right)\right],
\end{aligned}
\tag{53}
$$

and (this one obtained after additional regularization)

$$
\sum_{n_1, n_2 = 1}^{\infty} \sqrt{\left(\frac{n_1}{a_1}\right)^2 + \left(\frac{n_2}{a_2}\right)^2 + c^2} = \frac{c}{4} - \frac{\pi}{6} a_1 a_2 c^3
$$

$$+ \left(\frac{1}{4\pi}\sqrt{\frac{c}{a_2}} - \frac{ca_1}{4\pi a_2}\right)\left[\exp\left(-2\pi ca_2\right)\left(1 + \mathcal{O}(10^{-3})\right)\right]. \tag{54}$$

In both cases we have assumed (this is, of course, no restriction) that $a_2 \leq a_1$.

5. Definition of the Casimir energy density and its relation with the vacuum energy

The Casimir energy is given by the expression[14,15]

$$E_{Cas} = \frac{1}{2}\hbar \sum_n \omega_n. \tag{55}$$

In ultrastatic space-time (4-dimensional) $g_4 = -(dx_0)^2 + g_3$. Correspondingly $D_4 = -d_0^2 + D_3$ and $\omega_n = c\sqrt{\lambda_n(D_3)}$. Then

$$E_{reg}(\epsilon) = \frac{1}{2}\hbar c\mu \sum_n \left(\frac{\lambda_n}{\mu^2}\right)^{\frac{1}{2}-\epsilon} = \frac{1}{2}\hbar c\mu \, \zeta_{D_3}\left(-\frac{1}{2} + \epsilon\right), \tag{56}$$

where

$$\zeta_{D_3}(s) = \frac{(4\pi)^{-d/2}}{\Gamma(s)}\left[\sum_{n=0}^{\infty}\frac{c_n}{s + n - d/2} + f(s)\right] \tag{57}$$

being $f(s)$ analytic and $c_n = \int_\Omega a_n + \int_{\partial\Omega} b_n$ (a volume and a surface integrals). $E_{reg}(\epsilon)$ is not analytic but meromorphic, with a pole (for $d = 3$) at $\epsilon = 0$ of residue: $-\hbar c\mu c_2(d = 3)/(32\pi^2)$. The regulator can be suppressed only if $c_2 = 0$, i.e. for flat space with thin boundaries and for massless particles. If boundaries are plates (of surface S and separation L):

$$E_{Cas}(L, S) = -\frac{\hbar cS}{L^3}\frac{\pi^2}{12}\zeta_R(-3), \tag{58}$$

this is the ordinary Riemann zeta function.

The pole can be absorbed into the bare action (term proportional to c_2), but there is no unique way to do this. The most economical one is taking the principal part.[15] Minimal subtraction yields

$$E_{Cas} \equiv \lim_{\epsilon \to 0}\frac{1}{2}\left[E_{reg}(+\epsilon) + E_{reg}(-\epsilon)\right] = \frac{1}{2}\hbar c\mu PP\zeta_{D_3}(-1/2), \tag{59}$$

and is valid for simple poles.[15] To be mentioned: (i) μ can introduce a second (finite) ambiguity; (ii) it can be proven that the difference between the Casimir energy and the effective action to one loop is finite, independent of μ and proportional to the (geometrical) term c_2.

The one-loop effective action is given by

$$S_{eff} = \frac{1}{2}\ln\det D_4, \tag{60}$$

where $D_4 = \partial_0^2 + D_3$ (Euclidean space),

$$\zeta_{D_4}(s) = \frac{\mu c T}{2\sqrt{\pi}} \frac{\Gamma(s-1/2)}{\Gamma(s)} \zeta_{D_3}\left(s - \frac{1}{2}\right), \tag{61}$$

with $T = \int dx/c$, age of the universe.

The following relation can be proven[15]

$$E_{eff} = E_{Cas} + \frac{1}{2}\hbar c \mu \left[\psi(1) - \psi\left(-\frac{1}{2}\right)\right] \frac{c_2}{(4\pi)^2}. \tag{62}$$

Here μc_2 is independent of μ and $\psi(s) = \Gamma'(s)/\Gamma(s)$ is the digamma function. Again,

if $c_2 = 0$: $E_{eff} = E_{Cas}$.

The vacuum energy

$$E_{vac} = \int < 0|T_{00}|0 > \tag{63}$$

is different (in general) both from E_{eff} and from[14] E_{Cas}. If $T_{00} = T_{00}^{free}$, then $E_{vac} = E_{Cas}$. Yet another definition of E_{vac} is the following, considering the total action (not just one-loop)

$$E'_{vac} = \frac{\Gamma_{eff}}{T}. \tag{64}$$

In all, there are up to four different definitions of the 'vacuum energy'.

6. The Casimir effect for flat spacetime and flat boundaries: zeta-regularization of the mode sum

This corresponds to the most simple situation. Some basic cases are.

6.1. Free massless scalar field in $S^1 \times \mathbf{R}^d$

The total (unregularized) energy density (55) is here

$$\mathcal{E}_0 = \frac{(2\pi)^{1-d}}{2L} \int_{\mathbf{R}^{d-1}} d\vec{k}_T \sum_{n=-\infty}^{+\infty} \omega_n. \tag{65}$$

L is the compactification length (it will also be the distance between the parallel plates).[16] The zeta-function regularization procedure consists in writing

$$\mathcal{E}_0 = \frac{(2\pi)^{1-d}}{2L} \frac{2\pi^{(d-1)/2}}{\Gamma((d-1/2))} \int_0^\infty dk\, k^{d-2} \sum_{n=-\infty}^{+\infty} \left[k^2 + \left(\frac{2\pi n}{L}\right)^2\right]^{-s/2}, \tag{66}$$

and in analytically continuing to $s = -1/2$. The outcome is[16]

$$\mathcal{E} = -\frac{\Gamma\left(\frac{d+1}{2}\right)\zeta(d+1)}{(\sqrt{\pi}L)^{d+1}}. \tag{67}$$

This is an *exact* result.

6.2. Free massless scalar field in $T^p \times \mathbf{R}^{d-p+1}$

We must use the recurrences for generalized Epstein-like series.[11] For instance:

$$\sum_{n_1,n_2=1}^{\infty} (a_1 n_1^2 + a_2 n_2^2)^{-s} = -\frac{1}{2} a_2^{-s} \zeta(2s) + \frac{1}{2} a_2^{-s} \left(\frac{\pi a_2}{a_1}\right)^{1/2} \frac{\Gamma(s-1/2)}{\Gamma(s)} \zeta(2s-1)$$

$$+ \frac{2\pi^s}{\Gamma(s)} a_1^{-s/2-1/4} a_2^{-s/2+1/4} \sum_{n_1,n_2=1}^{\infty} n_1^{s-1/2} n_2^{-s+1/2} K_{s-1/2}\left(2\pi\sqrt{\frac{a_2}{a_1}} n_1 n_2\right) \qquad (68)$$

very quickly convergent. In particular,[16] for $d = 3$ $p = 2$ (with all L's equal)

$$\mathcal{E} = -\frac{\beta(2)}{3L^4}. \qquad (69)$$

For interesting expressions with $a_1 \neq a_2$ see ref. 17.

6.3. Free massless scalar field, p pairs of walls, Dirichlet b.c.

Only slight modification of the same technique.[16] Example $d = 3$, $p = 2$, equal L's

$$\mathcal{E} = -\frac{1}{16L^4}\left[\frac{\zeta(3)}{\pi} - \frac{\beta(2)}{3}\right]. \qquad (70)$$

Again an *exact* expression.

6.4. Generalization

Further, the cases of free massless scalar field with p pairs of walls, of Neumann b.c., periodic b.c., and even the case of the electromagnetic field with p pairs of perfectly conducting plates (the original Casimir effect), are completely related to the former.[16,18] In particular

$$\mathcal{E}^{em}_{L_1 \cdots L_p}(d,p) = (d-1)\mathcal{E}^{D}_{L_1 \cdots L_p}(d,p) + \sum_{j=1}^{p} \frac{1}{L_j} \mathcal{E}^{D}_{L_1 \cdots L_{j-1} L_{j+1} \cdots L_p}(d-1,p-1). \qquad (71)$$

Finally, concerning the important question of the *sign* of the Casimir energy density see refs. 18, 19, 14. Nothing new can be said about this puzzling subject.

7. Experimental situation: history, new developments and related effects

The following references on applications and modern determination of the contribution of the Casimir effect to new wetting and nonwetting patterns of helium 4 adsorbed on alkali metals,[20] to critical fluctuations within very narrow fluid films confined by rigid walls (for different boundary conditions and temperatures),[21] and to related cavity effects in laser physics,[22] are highly interesting. A more classically minded, historical account can be found in ref. 19.

8. Acknowledgements

I thank the DGICYT, project PB90-0022, for finantial support.

9. References

1. E. Elizalde, *Math. Comput.* **47** (1986) 347.
2. S. Rudaz, *J. Math. Phys.* **31** (1990) 2832.
3. E. Elizalde, *A simple recurrence for the higher derivatives of the Hurwitz zeta function*, Univ. of Pennsylvania preprint.
4. E. Elizalde and A. Romeo, *Internat. J. Math.* **13** (1990) 453.
5. E. D'Hoker and D. H. Phong, *Commun. Math. Phys.* **B104** (1986) 537.
6. F. Steiner, *Phys. Lett.* **B188** (1987) 447; D. Fried, *Invent. Math.* **84** (1984) 523.
7. A. Actor, *J. Phys.* **A20** (1987) 927.
8. I. J. Zucker, *J. Phys.* **A7** (1974) 1568, and **A8** (1975) 1734; I. J. Zucker and M. M. Robertson, *J. Phys.* **A8** (1975) 874.
9. G. H. Hardy, *Mess. Math.* *49* (1919) 85.
10. E. Elizalde and A. Romeo, *Phys. Rev.* **D40** (1989) 436.
11. E. Elizalde, *J. Math. Phys.* **31** (1990) 170; ibid. *J. Phys.* **A22** (1989) 931.
12. E. Elizalde and A. Romeo, *Int. J. Mod. Phys. A*, to appear.
13. K. Kirsten, *J. Math. Phys.* **32** (1991) 3008.
14. G. Plunien, B. Müller and W. Greiner, *Phys. Rep.* **134** (1986) 87.
15. S. K. Blau, M. Visser and A. Wipf, *Nucl. Phys.* **B310** (1988) 163.
16. E. Elizalde, *Nuovo Cimento* **104B** (1989) 685.
17. E. Elizalde and S. D. Odintsov, *Effective potential and stability of the rigid membrane*, Univ. of Barcelona preprint UB-ECM-PF 92/11.
18. J. Ambjørn and S. Wolfram, *Ann. Phys. (N.Y.)* **147** (1983) 1.
19. E. Elizalde and A. Romeo, *Am. J. Phys.* **59** (1991) 711.
20. E. Cheng, M. W. Cole, W. F. Saam, and J. Treiner, *Phys. Rev. Lett.* **67** (1991) 1007.
21. M. Krech and S. Dietrich, *Phys. Rev. Lett.* **66** (1991) 345; **67** (1991) 1055; Wuppertal preprints WUB 92-03 and WUB 92-43 (1992).
22. F. De Martini and G. Jacobovitz, *Phys. Rev. Lett.* **60** (1988) 1711; F. De Martini et al., *Phys. Rev.* **A43** (1991) 2480.

WIGNER FUNCTION IN FIELD THEORY:
PHASE-SPACE DESCRIPTION OF THE DIRAC VACUUM

PAWEL GÓRNICKI[1]
*Max-Plank-Institut für Kernphysik, Postfach 10 39 80
D-6900 Heidelberg, Germany*

ABSTRACT

Formalism based on equal-time gauge invariant Wigner function for the Dirac vacuum is shortly presented. It allows us to study time evolution of the Dirac field. Some results obtained with this method are shortly indicated.

1. Introduction

According to one popular view the physical vacuum is a very complex object. Indeed, the description of the ground state of interacting quantum fields (QF) is barely possible.

The vacuum state of a *free* QF interacting with an external agent (*classical* field) forms a simpler problem. Some physical situations do not require the full quantum treatment of all fields involved allowing for the separation of classical and quantum fields. An useful example is particle creation by strong electromagnetic (or gluonic) fields. In this case the fermions require full quantum treatment while the electromagnetic (EM) field may be considered classically. It is sometimes necessary to treat these classical fields as fully dynamical objects rather than a fixed background. The study of back reaction of particles produced by some classical field upon this field is such a case.

To achieve our goals we had to simplify the notion of the vacuum. Within the Heisenberg picture the full description of a vacuum state at a given instant of time is provided by the set of all possible n-point equal time correlation functions of QF. The complexity of the vacuum is greatly simplified if one restricts himself to two point correlation functions only. Two point functions are sufficient to construct electric current. For this reason it is possible to describe the influence of the electron–positron vacuum on a classical EM field without going into further details of vacuum structure. In Ref. [1] we make an extensive use of the Fourier transformed equal-time two-point correlation functions for fermionic fields. This is a generalization of a Wigner function [2] known from nonrelativistic quantum mechanics.

---

[1]Permanent address: Centrum Fizyki Teoretycznej, Polish Academy of Sciences, Lotnikòw 32/46, 02-668 Warsaw, Poland.

The use of the Wigner functions to the description of the vacuum in QFT is not new [3, 4, 5]. The main difference is that we use equal-time correlation functions and do not perform the Fourier transform with respect to time variable. This seems to have great consequences of practical nature. This approach is somehow similar to that of Cooper and Mottola [6, 7, 8] which makes use of two-point Wightman functions (although these works do not state the full evolution equations). The similar treatment may be also found in [9] but these authors do not include the electromagnetic fields (they use Wigner function to study Nambu–Jona-Lasinio model).

We are going to concentrate on fermionic field but the method has broader applicability. The cases of scalar [10] and EM fields [11] were also studied.

2. Description of the Method

We are going to consider the Dirac vacuum evolving under influence of an external EM field. As already mentioned we will concentrate on two-point equal-time correlation functions for fermionic fields. They are summarized by the following matrix:

$$C_{\alpha\beta}^-(\vec{x}, \vec{y}, t) = \langle 0|[\Psi_\alpha(\vec{x}, t), \Psi_\beta(\vec{y}, t)]|0\rangle, \tag{1}$$

where α and β denote spinor indices.

Let us define:

$$W_{\alpha\beta}(\vec{x}, \vec{p}, t) = -\frac{1}{2} \int d^3s \exp\left[-i\vec{p}\cdot\vec{s} - ie\phi(\vec{x}, \vec{s}, \vec{A}, t)\right] C_{\alpha\beta}^-(\vec{x} + \vec{s}/2, \vec{y} - \vec{s}/2, t). \tag{2}$$

The above matrix valued function provides a phase-space description of the Dirac vacuum $|0\rangle$ and it is a close analog of the Wigner function. The phase:

$$\phi(\vec{x}, \vec{s}, \vec{A}, t) = \int_{-1/2}^{+1/2} d\lambda\, \vec{s}\cdot\vec{A}(\vec{x} + \lambda\vec{s}, t) \tag{3}$$

was introduced to make W gauge invariant. With this definition the argument $\vec{p}$ of the Wigner function may be interpreted as a kinetic momentum.

Our major goal is to study time evolution of the Wigner function W under an influence of electromagnetic fields. The appropriate equations are derived directly from the field equations. They read:

$$D_t W = -\frac{1}{2}\vec{D}\cdot\{\gamma^0\vec{\gamma}, W\} - i\vec{P}\cdot[\gamma^0\vec{\gamma}, W] - im[\gamma^0, W], \tag{4}$$

where γ are the Dirac matrices while $D_t, \vec{D}$ and $\vec{P}$ denote the following nonlocal operators:

$$D_t = \partial_t + e\int_{-1/2}^{+1/2} d\lambda \vec{E}(\vec{x} + i\lambda\vec{\partial}_p, t)\cdot\vec{\partial}_p, \tag{5}$$

$$\vec{D} = \vec{\nabla} + e\int_{-1/2}^{+1/2} d\lambda \vec{B}(\vec{x} + i\lambda\vec{\partial}_p, t)\times\vec{\partial}_p, \tag{6}$$

$$\vec{P} = \vec{p} - ie \int_{-1/2}^{+1/2} d\lambda \vec{B}(\vec{x} + i\lambda\vec{\partial}_p, t) \times \vec{\partial}_p. \tag{7}$$

For fields varying slowly with $\vec{x}$ these operators may be approximated by:

$$D_t = \partial_t + e\vec{E}(\vec{x}, t) \cdot \vec{\partial}_p, \tag{8}$$

$$\vec{D} = \vec{\nabla} + e\vec{B}(\vec{x}, t) \times \vec{\partial}_p, \tag{9}$$

$$\vec{P} = \vec{p}. \tag{10}$$

There is no assumption about time dependence of the fields.

The Wigner function for the free-field vacuum may be easily found:

$$W = -\gamma^0(m + \vec{\gamma} \cdot \vec{p})/\sqrt{\vec{p}^2 + m^2}. \tag{11}$$

It is also easy to check that the above function satisfies Eq. (4) (with $\vec{E} = \vec{B} = 0$).

Solution (11) may serve as a starting point in many physical situations. Using Eq. (4) we may trace its evolution when we switch on the fields. Of course these fields may produce some $e^+ e^-$ pairs so the vacuum will evolve into a complicated state. This is all contained in our basic equation (4)!

The Wigner function allows us to calculate a number of physical observables. Eg., density of electric current induced in the vacuum is given by:

$$\vec{j}_0^{ind}(\vec{x}, t) = e \int \frac{d^3 p}{(2\pi\hbar)^3} \, tr[\vec{\gamma}\gamma^0 W], \tag{12}$$

while induced charge density is:

$$\rho_0^{ind}(\vec{x}, t) = e \int \frac{d^3 p}{(2\pi\hbar)^3} \, tr \, W. \tag{13}$$

The densities given above are not renormalized and (in 3 dimensions) the dp^3 integrals are not convergent. The renormalization scheme begins with regularization of these integrals by introducing sharp cutoff Λ; we are going to denote the resulting values by $\vec{j}_0^{ind}(\Lambda)$ and $\rho_0^{ind}(\Lambda)$ respectively. In the next step one has to subtract the infinite parts:

$$\vec{j}^{ind} = \vec{j}_0^{ind}(\Lambda) - Z(\Lambda) \, [\partial_t \vec{E} - \vec{\nabla} \times \vec{B}], \tag{14}$$

$$\rho^{ind} = \rho_0^{ind}(\Lambda) - Z(\Lambda) \, \vec{\nabla} \cdot \vec{E} \tag{15}$$

and to take the limit $\Lambda \to \infty$. This procedure amounts to the redefinition of densities (12)(13). Such redefinition is physically meaningful since the subtracted parts are directly proportional to the sources of external fields and may be absorbed into these sources. The renormalization constant $Z(\Lambda)$ may be calculated directly from perturbation theory.

It may be shown that our evolution equation (4) leads to the usual continuity equation for the induced current.

More details may be found in the original work [1].

3. Some Results

Several simple results were obtained using the method described above. We were able to reproduce the old Schwinger [12] for the rate of pair creation in strong electric field [1]. The procedure involved numerical solution of Eq. (4) starting with the free-field vacuum and switching on the field. After the initial jump (related to the 'dynamical' effects) the production rate stabilized at the constant value exactly equal to the one obtained analytically by Schwinger. We were also able to obtain the rates for the fields varying in time – for weak fields the values agree with perturbation theory. Nevertheless our method is valid for strong fields as well.

The other interesting problem is the back reaction of the pairs upon the field that produced them. Pair production gives rise to the current that tends to damp the original field. This has been studied numerically [13] and the resulting oscillations of the field and current (plasma oscillations) were observed. The similar results were obtained by other group within different approach [14].

4. References

1. I. Bialynicki–Birula, P. Górnicki, J. Rafelski, *Phys. Rev.* **D 44**, 1825 (1991)

2. E. P. Wigner, *Phys. Rev.* **32**, 749 (1932).

3. J. M. Eisenberg, G. Kälberman, *Phys. Rev.* **D 37**, 1197 (1988).

4. D. Vasak, M. Guylassy, H.-Th. Elze, *Ann. Phys. (N.Y.)* **173**, 462 (1987).

5. H.-Th. Elze, U. Heinz, *Phys. Rep.* **183**, 81 (1989).

6. F. Cooper, E. Mottola, *Phys. Rev.* **D 36**, 3114, (1987).

7. F. Cooper, E. Mottola, *Phys. Rev.* **D 40**, 456, (1989)

8. F. Cooper, E. Mottola, B. Rogers, P. Anderson, in *Intermittence in High Energy Collisions*, ed. F. Cooper at al (World Scientific, Singapore, 1991), p.399.

9. J. da Providência, M. C. Ruivo, C. A. de Sousa, *Phys. Rev.* **D 36**, 1882 (1987).

10. Ch. Best, P. Górnicki, W. Greiner *to be published*

11. I. Bialynicki-Birula *private communication*

12. J. Schwinger, *Phys. Rev.* **82**, 664 (1951).

13. E.D. Davis, P. Górnicki *work in progress*

14. Y. Kluger, J.M. Eisenberg, B. Svetitsky, F. Cooper, E. Mottola, *Phys. Rev. Lett.* **67**, 2427 (1991); *Tel Aviv Univ. preprint* TAUP 1944-92 (1992).

THE EQUAL-TIME WIGNER TRANSFORM OF THE KLEIN-GORDON FIELD AND PAIR CREATION

CHRISTOPH BEST,* J. M. EISENBERG

*School of Physics and Astronomy, Raymond and Beverly Sackler Faculty
of Exact Sciences, Tel Aviv University, 69978 Tel Aviv, Israel,*

GERHARD SOFF

*Gesellschaft für Schwerionenforschung (GSI),
Postfach 110 552, 6100 Darmstadt, Germany,*

WALTER GREINER

*Institut für Theoretische Physik, Johann Wolfgang Goethe-Universität,
Postfach 11 19 32, 6000 Frankfurt am Main, Germany*

1 Introduction

Recently, the equal-time Wigner transform of the two-point Green function has been proposed as a tool to study the time evolution of the vacuum state[1]. Unlike the split-time Wigner transform, the equal-time Wigner function obeys an evolution equation *without* constraints and can thus be calculated from a given initial state, e.g., the vacuum. It is therefore a good candidate for the study of initial-value problems like pair creation in external fields (as in the formation of the quark-gluon plasma in heavy-ion collisions) and back reaction from the created pairs[2,3,4].

In the following we derive the equation that governs the Wigner function of spinless bosons in a classical electromagnetic field. It describes a relativistic gas of positively and negatively charged particles with quantum corrections due to interferences that cause pair creation. In homogeneous fields, the Schwinger term is recovered.

2 The equal-time Wigner function

2.1 Definition

The equal-time Wigner function[1,5] is defined from the symmetrized two-point function of the Klein-Gordon field:

$$P(q,p) = \int d^3y \, \langle 0|\Phi(q - \tfrac{1}{2}y, t)\Phi^+(q + \tfrac{1}{2}y, t)|0\rangle \exp\left(\frac{i}{\hbar}p \cdot y + \frac{ie}{\hbar}\int_{q-y/2}^{q+y/2} \mathcal{A}(x) \cdot dx\right),$$

$$(1)$$

---

*Permanent address: Institut für Theoretische Physik, Johann Wolfgang Goethe-Universität, Frankfurt, Germany. Email: cbest@sophia.th.physik.uni-frankfurt.de where $(\mathcal{A}^0, \mathcal{A})$ is a classical electromagnetic field, and the field Φ is in the Feshbach-Villars formalism, i.e.,

$$\Phi = \begin{pmatrix} \psi \\ \chi \end{pmatrix}, \quad \phi = \psi + \chi, \quad \left(i \frac{\partial}{\partial t} - e\mathcal{A}^0 \right) \phi = m(\psi - \chi), \tag{2}$$

with ϕ the ordinary Klein-Gordon field. The equation of motion for the Wigner function derives from the field equation

$$i \frac{\partial \Phi}{\partial t} = \left(\frac{(\hat{p} - e\mathcal{A})^2}{2m} a + m\,b + e\mathcal{A}^0 \mathbb{1} \right) \Phi, \tag{3}$$

where p designates the derivative operator $-i\,\partial/\partial q$, and a and b are the matrices

$$a = (\sigma_3 + i\sigma_2) = \begin{pmatrix} 1 & 1 \\ -1 & -1 \end{pmatrix}, \quad b = \sigma_3 = \begin{pmatrix} 1 & 0 \\ 0 & -1 \end{pmatrix}. \tag{4}$$

Using phase-space calculus[5], it can be cast in the form

$$\begin{aligned}
i &\left(\frac{\partial}{\partial t} + e\mathcal{E}(q) \cdot \frac{\partial}{\partial p} + \dots \right) P(q,p) \\
&= i \left[-\frac{p}{2m} \cdot \frac{\partial}{\partial q} + e\left(\mathcal{B}(q) \times \frac{p}{m} \right) \cdot \frac{\partial}{\partial p} + \dots \right] \left(a \cdot P(q,p) + P(q,p) \cdot a^+ \right) \\
&\quad + \frac{1}{m} \left(p^2 - \frac{1}{4} \frac{\partial^2}{\partial q^2} \right) \left(a \cdot P(q,p) - P(q,p) \cdot a^+ \right) \\
&\quad + m \left(b \cdot P(q,p) - P(q,p) \cdot b \right),
\end{aligned} \tag{5}$$

where the dots indicate quantum corrections involving higher derivatives of the electric field $\mathcal{E}(q)$ and the magnetic field $\mathcal{B}(q)$.

The phase-space function $P(q,p)$ is a two-by-two matrix, the meaning of whose components we shall see below. Eq. (5) contains flow terms as well as local couplings involving m and p^2. In particular, the equations have nonstationary solutions even in the case of a spatially homogeneous Wigner function, exhibiting oscillations in an internal degree of freedom related to particle-antiparticle conjugation.

2.2 Expansion in a diagonalizing basis

To analyze the local oscillations, we expand P into its free eigenmodes. Without an electromagnetic field, two of the eigenoscillations have the frequencies $\pm 2E_p$, while the other two have zero eigenfrequency and correspond to constant solutions of the equations of motion.

272

We expand $P(q,p) = \sum_{i=1}^{4} f_i \tilde{P}_i$ with the p-dependent basis,

$$
\tilde{P}_1 = \frac{1}{4} \begin{pmatrix} m/E_p + E_p/m & m/E_p - E_p/m \\ m/E_p - E_p/m & m/E_p + E_p/m \end{pmatrix} = \frac{1}{2} \left(u^{(+)} \otimes u^{(+)} + u^{(-)} \otimes u^{(-)} \right),
$$

$$
\tilde{P}_2 = \frac{1}{2} \begin{pmatrix} 1 & 0 \\ 0 & 1 \end{pmatrix} = \frac{1}{2} \left(u^{(+)} \otimes u^{(+)} - u^{(-)} \otimes u^{(-)} \right),
$$

$$
\tilde{P}_{3/4} = \frac{1}{4} \begin{pmatrix} m/E_p - E_p/m & m/E_p + E_p/m \mp 1/2 \\ m/E_p - E_p/m \pm 1/2 & m/E_p - E_p/m \end{pmatrix} = u^{(\mp)} \otimes u^{(\pm)},
$$

$$\tag{6}$$

where the representation in terms of positive- and negative-energy eigensolutions $u^{(\pm)}$ has been given. The corresponding equations of motion for $\underline{f}$ are

$$
\begin{aligned}
\left(\frac{\partial}{\partial t} + e\mathcal{E}(q) \cdot \frac{\partial}{\partial p} \right) f_1 &= -\frac{p}{E_p} \cdot \frac{\partial}{\partial q} f_2 + e \frac{\mathcal{E}(q) \cdot p}{E_p^2} (f_3 + f_4), \\
\left(\frac{\partial}{\partial t} + e\mathcal{E}(q) \cdot \frac{\partial}{\partial p} \right) f_2 &= -\frac{p}{E_p} \cdot \frac{\partial}{\partial q} (f_1 + f_3 + f_4), \\
\left(\frac{\partial}{\partial t} + e\mathcal{E}(q) \cdot \frac{\partial}{\partial p} \right) f_3 &= -\frac{1}{2} \frac{p}{E_p} \cdot \frac{\partial}{\partial q} f_2 + e \frac{\mathcal{E}(q) \cdot p}{E_p^2} \frac{f_1}{2} + 2i E_p f_3, \\
\left(\frac{\partial}{\partial t} + e\mathcal{E}(q) \cdot \frac{\partial}{\partial p} \right) f_4 &= -\frac{1}{2} \frac{p}{E_p} \cdot \frac{\partial}{\partial q} f_2 + e \frac{\mathcal{E}(q) \cdot p}{E_p^2} \frac{f_1}{2} - 2i E_p f_4.
\end{aligned}
$$

$$\tag{7}$$

These equations hold whenever the local variations of the Wigner function are small as compared with the Compton wavelength. For spatially homogeneous problems, they are exact. The self-coupling terms that give rise to local oscillations are now isolated in the third and fourth equation (which are complex conjugate to each other), rendering f_3 and f_4 oscillatory even in the case of a homogeneous Wigner function. These oscillations are caused by interferences between particles and antiparticles.

A new self-coupling term with the coefficient

$$
\frac{e\mathcal{E}(q) \cdot p}{E_p^2} \tag{8}
$$

has appeared. This term is due to the fact that the basis (6) depends on the coordinate p, i.e., the local oscillations f_3 and f_4 have different "directions" in the $\underline{f}$-space at different momenta of the particle. When an external force changes the momentum of a particle, its local oscillations will be slightly out of phase and bring about an oscillation in the f_1-component, which can be interpreted as pair creation. Inspection of (6) reveals that the basis changes most at low momenta, and thus the coupling term is maximal at $p \ll E$. Physically this means that the pair creation happens at approximately zero momentum for the particles.

In the nonrelativistic limit, the oscillations of f_3 and f_4 are so fast that they do not influence f_1 and f_2. Then the equations decouple, and the linear combinations

$$f_+ = \tfrac{1}{2}(f_1 + f_2), \qquad f_- = \tfrac{1}{2}(f_1 - f_2), \tag{9}$$

satisfy the Vlasov equation of a relativistic gas. Note that f_- describes particles of momentum $-p$, in accordance with the hole interpretation.

The phase-space densities of observables are as follows:

charge:	energy:	current:	momentum:
ef_2	$E_p(f_1 + f_3 + f_4)$	$(pe/E_p)(f_1 + f_3 + f_4)$	$-pf_2 + ip(f_3 - f_4)$

We have therefore been able to identify all of the components of the Wigner function. The quantities f_3 and f_4 are interference contributions that vanish in the semiclassical limit. As we shall see, they are responsible for pair creation.

The equal-time Wigner function is, by definition, not Lorentz-invariant. This reflects the fact that the identification of particles and antiparticles itself is not Lorentz-invariant. On the other hand, the underlying physics is Lorentz-invariant, and our description applies to any frame of reference. This leads to the apparent paradox that pair creation happens at approximately zero momentum in *any* frame of reference. A similar paradox is discussed by Zel'dovich[6].

3 Pair creation

As an example for the application of the equal-time Wigner function, let us consider pair production in a homogeneous constant electric field. In this case, all local derivatives drop out. The force terms in the left-hand side of (7) can be removed by the method of characteristics. The resulting equations are a set of ordinary differential equations in time,

$$\begin{aligned}
\frac{\partial f_1}{\partial t} &= \frac{e\mathcal{E} \cdot p}{E_p^2}(f_3 + f_4), \\[2mm]
\frac{\partial f_3}{\partial t} &= \frac{e\mathcal{E} \cdot p}{E_p^2}\frac{f_1}{2} + 2iE_p f_3, \qquad \frac{\partial f_4}{\partial t} = \frac{e\mathcal{E} \cdot p}{E_p^2}\frac{f_1}{2} - 2iE_p f_4,
\end{aligned} \tag{10}$$

where $p = p(p_0, t)$ is now a time-dependent quantity. The quantity f_2 vanishes because of spatial homogeneity. These equations represent the internal degrees of freedom of a test particle, moving in phase space according to Newton's law. They must be solved for all initial momenta p_0 of the test particle where f initially has the vacuum value $f_1 = 1$, $f_3 = f_4 = 0$. Since an electric field is present, the components f_3 and f_4 couple to f_1 and raise its value, indicating pair creation.

The solution of the set of ordinary differential equations (10) is aided by noticing that it is equivalent to a single differential equation in a new variable $\xi(t)$ satisfying

$$\ddot{\xi} + E_p^2(t)\xi = 0, \tag{11}$$

through the substitution

$$f_1' = \frac{E_p}{2E_p^{(0)}} \left(|\xi|^2 + \frac{1}{E_p^2} |\dot{\xi}|^2 \right), \qquad f_3' + f_4' = \frac{E_p}{2E_p^{(0)}} \left(|\xi|^2 - \frac{1}{E_p^2} |\dot{\xi}|^2 \right),$$
$$i(f_3' - f_4') = \frac{1}{2E_p^{(0)}} (\xi^* \dot{\xi} + \dot{\xi}^* \xi) \tag{12}$$

where dots denote derivatives with respect to t, and E_p^0 is the test-particle energy at $t \to -\infty$. The relationship between pair creation in a homogeneous field and oscillators of time-dependent frequency has been noted much earlier from the wave-function approach to pair creation[7].

Asymptotically, eq. (11) has the solution

$$\xi(t) = \frac{C_1}{\sqrt{E_p(t)}} e^{-iE_p t} + \frac{C_2}{\sqrt{E_p(t)}} e^{iE_p t}. \tag{13}$$

Since the variable-frequency oscillator is equivalent to the stationary Schrödinger equation of a barrier penetration problem, it can be solved approximately by means of the WKB- or imaginary-time method[7] and yields the barrier-penetration coefficient

$$\rho_0 = \left| \frac{C_2}{C_1} \right|^2 \approx \exp\left(-\pi \frac{m_\perp^2}{e|\mathcal{E}|} \right), \qquad t \to \pm\infty, \tag{14}$$

if $|C_2/C_1|^2 = 0$ at $t \to -\infty$. The current is then given by

$$j(q,p) \approx \frac{pe}{E_p} f_1 = \frac{pe}{E_p} \left(1 + \frac{2\rho}{1-\rho} \right). \tag{15}$$

The vacuum value of this quantity is pe/E_p; it represents the charge of the Dirac sea and must be subtracted to yield a measurable quantity. Thus the number of pairs produced is given by

$$n = \frac{\rho}{1-\rho}, \tag{16}$$

in full correspondence with eq. (3.7) of Ref.[7].

Further reasoning along these lines[8] allows one to recast the equation of motion of the Wigner function as a phase-space density with a Schwinger-type source term and a Bose enhancement factor. This agrees well with the results of Kluger et al.[3,4], who showed that a calculation based on a field-theory formulation of the back-reaction phenomenon can be quite well approximated by a phenomenological Schwinger term if Bose enhancement (or Pauli blocking for fermions) is taken into account.

4 Conclusions

By studying the equal-time Wigner function, we have easily been led to the correct semiclassical interpretation of the Klein-Gordon field in terms of particles and antiparticles. It is the special advantage of not having introduced an energy coordinate conjugate to the time coordinate that allowed us to solve the pair creation example. In this way, the equal-time Wigner transform provides a convenient tool to exhibit the semiclassical (particle-interpretation) features of a field theory. This advantage is partly offset by the disadvantage of not having a relativistically invariant description.

Since the Wigner function is defined from the two-point Green function, all the formalism of perturbative quantum field theory, such as Feynman graphs, can be translated into Wigner function language. In particular, the Wigner function must be renormalized in interacting fields by a process similar to Debye screening[1].

It is an open question as to how much one can further exploit the equal-time Wigner function. We have here only studied the case of an interaction with a classical field. As in the split-time Wigner function, one can generalize the equal-time Wigner function to interacting fields by analyzing higher Green functions at equal times, thus obtaining n-particle phase-space distributions. The resulting phase-space equations are then similar to the BBKGY hierarchy in classical kinetic theory, thus emphasizing the semiclassical picture of the vacuum as a gas of interacting particles. The Φ^4-interaction, for example, gives a short-range two-particle force. Whether this provides any new insights remains to be seen.

Acknowledgements: C.B. wishes to thank the School of Physics and Astronomy of Tel Aviv University for its hospitality and the German Israel Foundation for its support.

References

1. I. Bialynicki-Birula, P. Górnicki, and J. Rafelski, Phys. Rev. **D44**, 1825 (1991).

2. F. Cooper and E. Mottola, Phys. Rev. **D40**, 456 (1989).

3. Y. Kluger, J. M. Eisenberg, B. Svetitsky, F. Cooper, and E. Mottola, Phys. Rev. Lett. **67**, 2427 (1991).

4. Y. Kluger, J. M. Eisenberg, B. Svetitsky, F. Cooper, and E. Mottola, Phys. Rev. **D45**, 4659 (1992).

5. C. Best, P. Górnicki, G.Soff, and W. Greiner, *Phase-space structure of the Klein-Gordon field*, to be published.

6. Ya. B. Zel'dovich, in *Magic without Magic: John Archibald Wheeler*, ed. J. R. Klauder (W. H. Freeman, San Francisco, 1972).

7. M. S. Marinov and V. S. Popov, Forts. d. Physik **25**, 373 (1977).

8. C. Best and J. M. Eisenberg, *Pair creation in transport equations*, to be published.

THE INFRARED STRUCTURE OF HOT QCD

Michel LE BELLAC

and

Patrick REYNAUD

Institut Non Linéaire de Nice, Université de Nice*
Parc Valrose, 06108 Nice Cedex 2, France

ABSTRACT

We review some recent results on the infrared behaviour of quantum chromodynamics at high temperature; we discuss applications of Braaten-Pisarski's resummation scheme to dynamical screening and a possible generalization of the Kinoshita-Lee-Nauenberg theorem at finite temperature.

1. Introduction

Lattice calculations have given clear evidence for a phase transition in quantum chromodynamics at a temperature T_c of a few hundreds MeV[1]. The situation appears to be entirely clarified in the case of a pure (i.e. without fermions) SU(3)-gauge theory; in that case the order parameter is given by the Polyakov loop, and a Z_3 symmetry is broken above the critical temperature. Numerical simulations carefully interpreted thanks to finite size scaling show that the transition is a weak first order one. The situation is still somewhat confused when one adds fermions. For massless quarks, one expects chiral symmetry restoration above a critical temperature, the order parameter being the quark condensate $<\overline{\psi}\psi>$. In the realistic case of light up and down quarks and a strange quark with a mass around 200 MeV (if the quark mass $m \gg T_c$, then the corresponding flavour plays a negligible role), one may detect a transition by looking at a sharp variation of one of the previously mentionned order parameters, or of some other quantity like the energy density. Numerical simulations indicate that the character of the transition is sensitive to the quark masses and the number of flavours. The general consensus is that there is, in the realistic case, a phase transition at a temperature $T_c \simeq 150$ MeV.

In addition to the transition temperature, lattice calculations have produced a wealth of information on various thermodynamical properties of matter at high temperature: energy density, pressure, equation of state, screening lengths, etc. However, in order to compute dynamical properties, such as cross-sections, lifetimes, energy losses..., one has still to resort to perturbation theory (PT)[2]. Perturbation theory is

---

* Unité Mixte de Recherche 129 du CNRS

in principle justified at very high temperature: because of asymptotic freedom, the coupling constant $g(T)$ decreases like $1/\ln T$, and matter at very high temperature is a gas of almost free quarks and gluons, usually called the quark-gluon plasma. The high temperature phase corresponds thus to deconfinement, and the phase transition is the deconfinement transition.

It is possible to compare the results of lattice calculations of thermodynamical properties with those of perturbation theory in the deconfined phase. One finds reasonable agreement above a temperature of order $2T_c$, and thus for $T \geq 2T_c$ one may hope that PT can be a good guide, at least qualitatively. Of course PT breaks down when one gets too close to the transition temperature. Furthermore one must be aware that the perturbative calculations which will be presented in what follows are still very far from realistic applications to the phenomenology of heavy ion collisions.

Even with a small coupling constant, there are problems with PT at finite temperature because of infrared (IR) divergences. Of course one also encounters IR singularities at zero temperature, but the situation tends to be worse at non-zero T, because the Bose-Einstein factor induces further IR divergences. Indeed, if m_0 is a typical zero-temperature mass scale, the Bose-Einstein factor gives when $m_0 \ll k \ll T \ (k \sim |\omega|)$:

$$n_B(\omega) \equiv \frac{1}{e^{|\omega|/T} - 1} \sim \frac{T}{k} \tag{1}$$

Thus a further power of k^{-1} is introduced into the loop integrals for $T \gg m_0$, and of course for any T in the case of massless particles. In order to get reliable results, one must sum classes of diagrams, even in non-gauge theories. For example in the case of a massless $g^2\varphi^4$ theory, which is not IR divergent at $T = 0$, the pressure is given in PT by:

$$P = \frac{(\pi T)^4}{90}\left[1 - \frac{15}{8}\left(\frac{g}{\pi}\right)^2 + \frac{15}{2}\left(\frac{g}{\pi}\right)^3 + ...\right] \tag{2}$$

The first term in Eq.(2) corresponds of course to black-body radiation (free massless particles); the second one, proportional to g^2, is an ordinary perturbative correction, corresponding to the diagram in Fig. 1.a. The term in g^3 comes from a summation of the so-called "ring-diagrams" (Fig. 1.b): such a summation is necessary because of the IR divergences of the perturbative series.

(a)   (b)

Fig. 1 : a/ First order perturbative correction to the partition function.

b/ The ring diagrams: each bubble represents a self-energy insertion.

In this talk, we shall cover two main topics:

(i) First we shall explain some features of the Braaten-Pisarski[3] resummation scheme, starting from the concept of "Hard Thermal Loop" (HTL)[4] and working out one important consequence: dynamical screening.

(ii) Then we shall discuss briefly a possible generalization of the Kinoshita-Lee-Nauenberg theorem at finite temperature: infrared safe results are obtained by summing over all degenerate initial and final states.

2. Hard Thermal Loops (HTL)

The conventional way of introducing HTL is through the computation of one-loop graphs in finite temperature field theory. However, in the case of the two-point function, there exists a very simple derivation from kinetic theory, which allows to display most of the relevant features[5]. Let us consider a collisionless plasma and call $f_0(\mathbf{p})$ the equilibrium distribution function. The plasma is assumed to interact with a weak electromagnetic field; the particle distribution is then modified

$$f(\mathbf{x}, \mathbf{p}, t) = f_0(\mathbf{p}) + \delta f(\mathbf{x}, \mathbf{p}, t) \tag{3}$$

The deviation δf with respect to equilibrium obeys the standard Vlasov equation

$$\frac{\partial}{\partial t} \delta f + \mathbf{v} \cdot \frac{\partial}{\partial \mathbf{x}} \delta f = e\, \mathbf{E} \cdot \frac{\partial f_0}{\partial \mathbf{p}} \tag{4}$$

This equation is solved in Fourier space, for an electric field $\widetilde{\mathbf{E}} \exp\left[i\left(\mathbf{k}.\mathbf{x} - \omega t\right)\right]$ and a deviation $\delta \widetilde{f} \exp\left[i\left(\mathbf{k}.\mathbf{x} - \omega t\right)\right]$:

$$\delta \widetilde{f} = \frac{e\, \widetilde{\mathbf{E}}}{i\left(\mathbf{k}.\mathbf{v} - \omega - i\eta\right)} \cdot \frac{\partial f_0}{\partial \mathbf{p}} \tag{5}$$

The prescription $\omega \to \omega + i\eta$ $(\eta \to 0^+)$ is dictated by causality, and was first established by Landau. Using Maxwell's equations, one can translate Eq.(5) into an expression for the dielectric tensor $\varepsilon_{ij}(\omega, k)$, which can be decomposed into a transverse (ε_t) and a longitudinal (ε_l) component $(\widehat{k}_i \equiv k_i/k)$:

$$\begin{aligned} \varepsilon_{ij} &= (\delta_{ij} - \widehat{k}_i \widehat{k}_j)\varepsilon_t + \widehat{k}_i \widehat{k}_j\, \varepsilon_l \\ &= \delta_{ij} - \frac{e^2}{\omega} \int \frac{\mathrm{d}^3 p}{(2\pi)^3} \frac{v_i}{\mathbf{k}.\mathbf{v} - \omega - i\eta} \frac{\partial f_0}{\partial p_j} \end{aligned} \tag{6}$$

Silin[6] was first to use this result in the case of an ultrarelativistic plasma: a remarkable simplification occurs, because $|\mathbf{v}| = c$ so that the momentum and angular integration decouple. Let us for example write explicitly ε_l:

$$\varepsilon_l(\omega, k) = 1 + \frac{1}{2\pi^2} \frac{e^2}{k} \int_0^\infty \mathrm{d}p\, p\, f_0(p) \int_{-1}^1 \mathrm{d}\cos\theta \frac{\cos\theta}{k\, c \cos\theta - \omega - i\eta} \tag{7}$$

We now specialize to the case of a QED plasma with zero chemical potential using for $f_0(p)$ a Fermi-Dirac distribution:

$$f_0(p) = n_F(p) \equiv \frac{1}{e^{p/T} + 1} \tag{8}$$

Taking into account the charge and spin degrees of freedom and setting as usual $c = 1$, we find

$$\varepsilon_l(x, k) = 1 + \frac{1}{3}\left(\frac{eT}{k}\right)^2 \left[1 + \frac{x}{2}\ln\frac{x+1}{x-1}\right] \tag{9}$$

It is convenient to introduce the following convention: upper case letters denote four-momenta while the corresponding lower case letters denote energies and three-momenta:

$$K_\mu = (\omega, \mathbf{k}) \quad \text{or} \quad (x\,k, \mathbf{k}) \quad \text{with} \quad x \equiv \omega/k \quad \text{and} \quad k \equiv |\mathbf{k}| \tag{10}$$

Many features of Eq.(8) and Eq.(9) are worth noticing.

(i) In field theory, one computes to the one-loop approximation the photon self-energy, which is directly related to the dielectric tensor. The present kinetic derivation gives only part of the field theoretic result: although the proportionality to e^2 shows that we found a perturbative correction, we have only obtained the leading term in T of the photon self-energy, which is proportional to T^2. One sees from Eq.(7) and Eq.(8) that the particles contributing to this term have momenta of order T:

$$\int_0^\infty \mathrm{d}p\, p\, n_F(p) = \frac{\pi^2}{12}\, T^2 \tag{11}$$

One can show that the leading power in T of any one-loop graph in a renormalizable field theory is at most T^2. Those one-loop graphs which do behave as T^2 are called Hard Thermal Loops (HTL). The reason for this terminology will be explained shortly.

(ii) Static screening ($x = 0$, k fixed $\neq 0$):

$$\varepsilon_l(x = 0, k) = 1 + \left(\frac{m_D}{k}\right)^2 \tag{12}$$

with the Debye mass

$$m_D = \frac{eT}{\sqrt{3}} \tag{13}$$

This means that the (static) Coulomb potential is screened with a Debye length $r_D = 1/m_D$. One sees that a mass scale eT is naturally introduced. It will be convenient to define the following thermal mas m:

$$m^2 = \frac{(eT)^2}{6} = \frac{1}{2}\, m_D^2 \tag{14}$$

In fact this mass is nothing but the (gauge invariant) photon mass induced by the temperature for transverse photons with momenta $k \gg eT$.[2]

(iii) From the $x \to x + i\eta$ prescription, one derives the imaginary part of the dielectric constant:

$$\mathrm{Im}\,\varepsilon_l(x,k) = \pi \left(\frac{m_D}{k}\right)^2 x\,\Theta(1 - x^2) \tag{15}$$

where $\Theta(1 - x^2) = 1$ if $x^2 \leq 1$, $= 0$ otherwise.

It is important to understand that $\mathrm{Im}\,\varepsilon_l(x,k) \neq 0$ for *space like* momenta $K^2 < 0$. This comes from the fact that the imaginary part of HTL is due to the scattering of electrons and positrons from the bath, with momenta of order T, on photons with momenta $k \sim eT$, while at $T = 0$ the imaginary part is related to the decay rate of heavy photons with time like momenta. This phenomenon is known as Landau damping.

(iv) Soft and hard scales: the loop integral in Eq.(7) is dominated by momenta $k \sim T$ which will be referred to as the scale of hard momenta. Looking at Eq.(9), one realizes that for the region $\omega, k \sim eT$, referred to as the region of soft momenta, the perturbative correction is as important as the zeroth-order term, and we see that naïve PT breaks down at this scale. This is the reason why one must use a resummation, in order for PT to make sense.

(v) Generalization of the kinetic derivation of HTL: Silin's derivation[6] of the photon self-energy has been generalized by Blaizot and Iancu[7]. Their work relies on the following observations:

a/ Hard particles are the essential ones for the calculation of polarization phenomena and for them radiative corrections are negligible. They do not interact directly but only through their coupling with the average photon and electron fields.

b/ The average photon and electron fields are weak, corresponding to small deviations from equilibrium.

c/ The average fields are slowly varying, their Fourier transforms carry soft momenta of order eT.

The validity of a/, b/ and c/ leads to a consistent approximation scheme and allows to recover all results from field theoretical calculations.

3. Properties of Hard Thermal Loops

As already mentionned, it is possible to show that in a renormalizable field theory, at one-loop order, the leading power in T of the one-particle irreducible N-point function $\Gamma^{(N)}$ is at most T^2. Of course not all $\Gamma^{(N)}$ behave as T^2, and those which do behave as T^2 are HTL. It has been shown by Frenkel and Taylor[4] and by Braaten and Pisarski[3] that in QCD, the only HTL are the one-loop graphs with N external gluons or $(N-2)$ external gluons and 2 quarks $(N \geq 2)$. In particular there are no HTL with external ghosts.

It is quite remarkable that HTL which are off shell Green's functions are gauge-fixing independent. Furthermore they obey Ward identities similar to those of tree graphs; denoting by $\delta\Pi_{\mu\nu}$ and $\delta\Gamma_{\mu\nu\rho}$ the HTL approximation to the gluon self-energy and the three-gluon vertex, we have, schematically (colour indices have been suppressed)

$$P^{\mu}\,\delta\Pi_{\mu\nu}(P) \;=\; 0 \tag{16}$$

$$P^{\mu}\,\delta\Gamma_{\mu\nu\rho}(P,Q,K) \;=\; \delta\Pi_{\nu\rho}(K) - \delta\Pi_{\nu\rho}(Q) \tag{17}$$

Using gauge invariance, one can write an effective action for HTL, which is essentially determined by the two-point functions. The extra term reads for gluons[8]:

$$\delta L_g \;=\; m^2\,\mathrm{Tr}\left[G^{\mu\alpha} \int \frac{d\Omega_k}{4\pi}\, \frac{\widehat{K}^{\alpha}\,\widehat{K}^{\beta}}{(\widehat{K}\cdot D)^2}\, G^{\mu\beta} \right] \tag{18}$$

where $G^{\mu\alpha}$ is the gluon field tensor, $\widehat{K}^{\alpha} = (1,\widehat{\mathbf{k}})$ and $D_{\mu} = \partial_{\mu} - ig[A_{\mu}, \ ]$ is the covariant derivative in the adjoint representation; $d\Omega_k$ is the element of angular integration and the thermal mass m^2 is given, for N_c colours and N_f flavours, by

$$m^2 \;=\; \frac{1}{6}\,(g\,T)^2\,[\,N_c + \frac{1}{2}\,N_f\,] \tag{19}$$

This expression simplifies in the case of QED, because the covariant derivative acting on the field strength tensor reduces to the ordinary derivative; a simple transformation of Eq.(18) gives

$$\delta L_{\gamma} \;=\; m^2\left[A_0^2 + \int \frac{d\Omega_k}{4\pi}\,(\widehat{K}\cdot A)\,\frac{\partial_0}{\widehat{K}\cdot\partial}\,(A\cdot\widehat{K}) \right] \tag{20}$$

the thermal mass m^2 being that of Eq.(14). This expression is interesting since it allows to isolate the static part leading to Debye screening: in the static limit A_0 acquires a mass $\sqrt{2}\,m$. Futhermore it shows that in QED all HTL with external photons only vanish for $N \geq 3$. The effective action for fermions is quite similar in QED and QCD:

$$\delta L_f = i\,m_f^2\,\overline{\psi}\,\gamma_{\mu}\left[\int \frac{d\Omega_k}{4\pi}\,\frac{\widehat{K}^{\mu}}{\widehat{K}\cdot D} \right]\psi \tag{21}$$

with

$$m_f^2 = C_f\,\frac{(g\,T)^2}{8} \quad (\,\frac{(e\,T)^2}{8}\ \text{ in QED }) \tag{22}$$

Now comes the crucial observation, already alluded to in the previous section: for soft external momenta HTL are of the same order as the zeroth order term of PT. Actually it is necessary that all external momenta be soft: for example if two external momenta of a four-point function are hard, then naïve PT holds. However, when all external momenta are soft, we must use, instead of bare propagators and vertices, effective ones, namely propagators and vertices corrected by taking into account the HTL. It may be worth noting that in some cases there exist effective vertices which do not have bare counterparts, such as the 2 quark-2 gluon vertex. This effective PT has been developed by Braaten and Pisarski[3], and is also called a resummation scheme. This method has allowed for example to compute in gauge invariant manner the plasmon damping rate[9]. In the next section, we want to examine another interesting property of the resummation scheme which can often cure infrared divergences, even if there is no screening mass in the static limit. This phenomenon is known as dynamical screening.

4. Dynamical screening

Let us recall the expression of the longitudinal component Π_l of the photon self-energy in the HTL approximation[2,3,4]:

$$\Pi_l(x) = 2\,m^2\,[\,1 + \frac{x}{2}\ln\frac{x+1}{x-1}\,] \tag{23}$$

In the static limit $x = 0$:

$$\Pi_l(x=0) = m_D^2 \tag{24}$$

The situation is completely different for the transverse component Π_t :

$$\Pi_t(x) = m^2\,[\,x^2 + (1-x^2)\frac{x}{2}\ln\frac{x+1}{x-1}\,] \tag{25}$$

The leading behaviour for fixed k and $x \to 0$ comes from the imaginary part of Π_t, which is proportional to $m^2\,x$. In any case:

$$\lim_{x\to 0}\Pi_t(x) = 0 \tag{26}$$

In other words, there is no static screening for transverse gluons (photons), or no static magnetic screening. Nevertheless one will observe dynamical magnetic screening. In order to illustrate this feature, we shall examine the energy loss of a heavy fermion propagating in a plasma. The study of such processes was initiated by Bjorken[10], who argued that the energy loss of a heavy quark in a quark-gluon plasma would be much less important than in conventional nuclear matter, and that this "jet quenching" could be a signal of plasma formation. We limit our discussion to the QED case to keep things simple; a heavy fermion ("muon") of mass M propagating in a QED plasma[11], the following conditions being realized: $m_e \ll T \ll M$. In QED the Compton effect is negligible, which makes the problem easier than in QCD; the reaction rate $\Gamma(E)$ and the energy loss dE/dx are controlled by photon exchange between the muon and the electrons and positrons of the plasma: see Fig. 2, which also defines the kinematics.

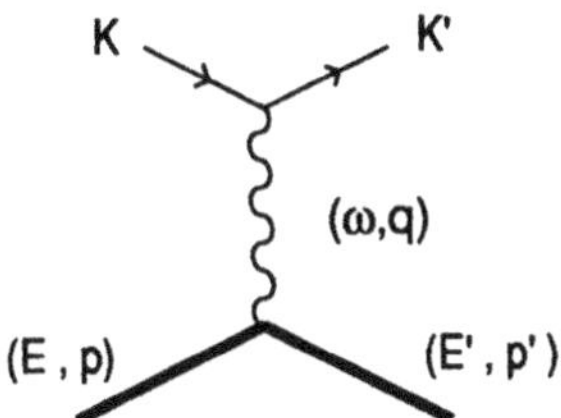

Fig. 2 : Coulomb scattering of the heavy fermion (double line) on electrons of the plasma (full line) through photon exchange (wavy line).

It is straightforward to write down $\Gamma(E)$ in bare PT:

$$\Gamma(E) = \frac{1}{2E} \int \frac{\mathrm{d}^3 p'}{(2\pi)^3} \frac{1}{2E'} \int \frac{\mathrm{d}^3 k}{(2\pi)^3} \frac{1}{2k} n_F(k) \int \frac{\mathrm{d}^3 k'}{(2\pi)^3} \frac{1}{2k'} [1 - n_F(k')]$$
$$\times (2\pi)^4 \, \delta^{(4)}(P + K - P' - K') |\mathcal{M}|^2 \qquad (27)$$

where $|\mathcal{M}|^2$ is the matrix element squared, summed over final and averaged over initial spins, while $\mathrm{d}E/\mathrm{d}x$ reads

$$\frac{\mathrm{d}E}{\mathrm{d}x} = \frac{1}{v} \int_M^\infty \mathrm{d}E' \, \omega \, \frac{\mathrm{d}\Gamma}{\mathrm{d}E'} \qquad (28)$$

where v is the muon velocity. Thus $\mathrm{d}E/\mathrm{d}x$ is given by an expression almost identical to Eq.(27), except that an extra factor $\omega \equiv E - E'$ is inserted in the integrand. $\Gamma(E)$ exhibits a Coulomb divergence $\mathrm{d}q/q^3$ $(Q \equiv P - P')$, while $\mathrm{d}E/\mathrm{d}x$ is only logarithmically divergent because of the ω-factor.

Now, there is an alternative expression for the reaction rate $\Gamma(E)$, which can be expressed in terms of the imaginary part of the heavy fermion self-energy Σ (that such a relation may hold is suggested by the drawing of Fig. 3).

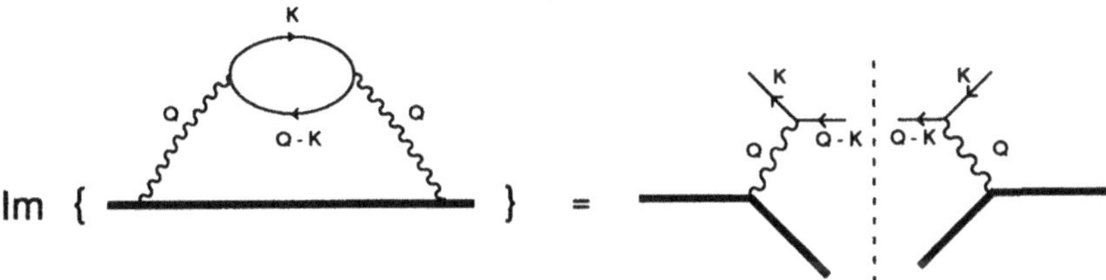

Fig. 3 : Relation between the fermion self-energy and the scattering cross-section.

It has been shown by Weldon[12] that:

$$\Gamma(E) = -\frac{1}{2E}\left[1 - n_F(E)\right]\mathrm{Tr}\left[(\gamma\cdot P + M)\,\mathrm{Im}\,\Sigma\,(E + i\eta, \mathbf{p})\right] \tag{29}$$

The imaginary part of Σ may be computed by using the resummed photon propagator; note that the fermion line is hard, since $M \gg T$, so that the fermion propagator need not be resummed. It is convenient to introduce the longitudinal and transverse spectral functions ρ_l and ρ_t of the photon propagator:

$$\rho_l \equiv \frac{1}{\pi}\,\mathrm{Im}\,\Delta_l = \frac{1}{\pi}\,\frac{\mathrm{Im}\,\Pi_l}{(Q^2 - \mathrm{Re}\,\Pi_l)^2 + (\mathrm{Im}\,\Pi_l)^2} \tag{30}$$

$$\rho_t \equiv \frac{1}{\pi}\,\mathrm{Im}\,\Delta_t = \frac{1}{\pi}\,\frac{\mathrm{Im}\,\Pi_t}{(Q^2 - \mathrm{Re}\,\Pi_t)^2 + (\mathrm{Im}\,\Pi_t)^2} \tag{31}$$

When Q is hard, $\mathrm{Re}\,\Pi$ and $\mathrm{Im}\,\Pi$ can be neglected in Eq.(30) and Eq.(31); one then recovers bare PT. This observation suggests that the calculation should be organized in the following way. Introduce a scale $q^\star$: $eT \ll q^\star \ll T$; for $q > q^\star$, use bare PT with exact kinematics, Eq.(27); this will give the hard component $(\mathrm{d}E/\mathrm{d}x)_H$ to the energy loss. For $q < q^\star$, use resummed PT, with suitable kinematic approximations. Since $|\mathcal{M}|^2$ is gauge independent, this procedure is perfectly gauge invariant. Of course the arbitrary scale $q^\star$ should disappear from the final result. We now concentrate on the soft piece, which is in fact the interesting one, since it makes use of resummation. After some kinematic approximations which are valid in the region $q < q^\star$, one finds, starting from Eq.(29), a simple expression for $\Gamma_S(E)$:

$$\Gamma_S(E) = \frac{1}{2\pi}\frac{e^2}{v}\int_0^{q^\star}\mathrm{d}q\,q^2\int_{-v}^{v}\mathrm{d}x\,[1+N_B(q\,x)]\left(\rho_l(x,q)+(v^2-x^2)\rho_t(x,q)\right) \tag{32}$$

where

$$N_B(q\,x) \equiv \frac{1}{e^{q\,x/T} - 1} \tag{33}$$

Notice that the Bose-Einstein factor defined in Eq.(33) is not that of Eq.(1) and that $q\,x$ may be negative. The thermal structure of this expression may look puzzling at first sight, since the Fermi-Dirac factors present in Eq.(27) have disappeared. However, from detailed balance, one has the following identity:

$$n_F(E)\left[1 - n_F(E+\omega)\right] = \left[1 + N_B(\omega)\right]\left[n_F(E) - n_F(E+\omega)\right] \tag{34}$$

and the combination $[n_F(E) - n_F(E+\omega)]$ appears precisely in the calculation of the photon self-energy[17]. In order to find the IR behaviour of Eq.(32), we only need ρ_l and ρ_t in the region $x \ll 1$. From Eq.(23) and Eq.(25) we get:

$$\rho_l(x,q) \simeq m^2\,\frac{x}{(q^2 + 2m^2)^2} \tag{35}$$

$$\rho_t(x,q) \simeq \frac{m^2}{2} \frac{x}{(q^2 + 2\,m^2\,x^2)^2 + (\pi\,m^2\,x/2)^2} \tag{36}$$

In the small x region:

$$1 + N_B(q\,x) \simeq \frac{T}{q}\frac{1}{x} + \frac{1}{2} + O(x) \tag{37}$$

and the dominant behaviour comes from the first term. It is easy to check that ρ_l gives a convergent contribution to Eq.(32), while ρ_t gives a logarithmic divergence:

$$\int_0^{q^*} dq\, q \int_0^v dx\, \frac{(v^2 - x^2)}{x}\, \rho_t(x,q) \sim \int_0^{q^*} \frac{dq}{q} \tag{38}$$

The Coulomb singularity (dq/q^3) of bare PT has been weakened into a logarithmic one, entirely due to magnetic scattering. This suggests that there is another relevant scale of order $g^2 T$, the magnetic scale; at this scale, Braaten-Pisarski's resummation is not sufficient to screen all singularities, and a magnetic mass $\sim g^2 T$ may be necessary in order to get finite results. Incidentally, it is very easy to recover the bare singularities by neglecting m^2 in the denominators of Eq.(35) and Eq.(36). Let us now turn to the soft contribution $(dE/dx)_S$ to the energy loss, which is given by:

$$\left(\frac{dE}{dx}\right)_S = \frac{1}{2\,\pi}\frac{e^2}{v} \int_0^{q^*} dq\, q^3 \int_{-v}^v dx\, x\, [1 + N_B(q\,x)] \left(\rho_l(x,q) + (v^2 - x^2)\,\rho_t(x,q) \right) \tag{39}$$

Because of parity, the first term in Eq.(37) gives a vanishing contribution and the dominant one comes from the $1/2$ term: $(dE/dx)_S$ differs from $\Gamma_S(E)$ by a factor $(qx)^2/2vT$ in the integrand

$$\left(\frac{dE}{dx}\right)_S = \frac{1}{4\,\pi} \left(\frac{e}{v}\right)^2 \int_0^{q^*} dq\, q^3 \int_0^v dx\, x\, \left(\rho_l(x,q) + (v^2 - x^2)\,\rho_t(x,q) \right) \tag{40}$$

In order to display the general features of the result, let us examine the integral:

$$I = \int_0^{q^*} dq\, q^3 \int_0^v dx\, x\, \rho_t(x,q) \tag{41}$$

Using Eq.(36) we find:

$$I \simeq \frac{m^2}{2} \int_0^1 dx\, x^2 \int_0^{q^*} dq\, \frac{q^3}{q^4 + (\pi\,m^2\,x/2)^2} \tag{42}$$

It is clear that the term $(\pi\,m^2\,x/2)^2$ acts as a (frequency dependent) IR cut-off: the potential logarithmic divergence in the q-integration is cut-off at $q_c = m\,\sqrt{\pi\,x/2}$, leading to a soft part proportional to $m^2 \ln(q^*/q_c)$. This is precisely the phenomenon of dynamical screening, which is able to turn the logarithmic divergence of the bare calculation into a finite result. The total energy loss dE/dx is obtained by adding the hard and soft pieces:

$$\left(\frac{dE}{dx}\right)_H = (e^2\,T)^2\,g(v)\left[\ln\left(\frac{E\,T}{M\,q^\star}\right) + f_H(v)\right] \qquad (43)$$

$$\left(\frac{dE}{dx}\right)_S = (e^2\,T)^2\,g(v)\left[\ln\left(\frac{q^\star}{e\,T}\right) + f_S(v)\right] \qquad (44)$$

where $g(v)$, $f_H(v)$ and $f_S(v)$ are functions of v which have been explicitly computed. The arbitrary scale q^* disappears from the final result, leaving a $\ln(1/e)$ factor which is typical of dynamical screening.

There are other interesting examples where dynamical screening leads to finite rsults, while bare PT was logarithmically divergent.

(i) The production of real photons in a quark-gluon plasma has been computed independently by Baier et al.[13] and by Kapusta et al.[14]. The relevant graph in the soft approximation is drawn in Fig. 4. The calculation is relatively simple because it does not involve effective vertices: at least one of the quarks lines must be hard, since the photon is hard.

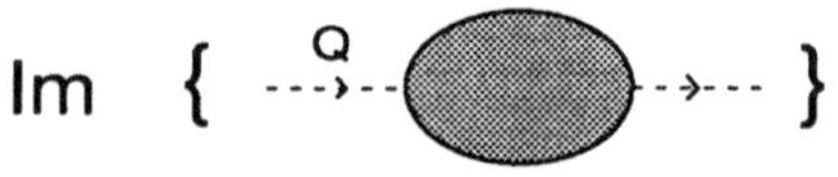

<u>Fig. 4</u> : The photon self-energy.

(ii) The Primakoff production of axions[15].

(iii) The production of massive photon pairs in a quark-gluon plasma[16].

(iv) The calculation of transport coefficients in a quark-gluon plasma, which involves quark-quark, quark-gluon and gluon-gluon scattering [17]. All corresponding cross-section exhibit a Coulomb divergence ($d\sigma/d\Omega \sim \theta^{-4}$), which is turned into a logarithmic one by dynamical screening. Fortunately transport coefficients depend on the so-called transport cross-section

$$\sigma_{tr} = \int d\Omega\,(1 - \cos\theta)\,\frac{d\sigma}{d\Omega} \sim \int \frac{d\theta}{\theta} \qquad (45)$$

because small angle scattering is inefficient for the transfer of momentum. Dynamical screening cures the logarithmic divergence and again a finite result is obtained, depending on a typical $\ln(1/\alpha_S)$, where $\alpha_S \equiv g^2/4\,\pi$ is the strong coupling constant. One finds for example in the case of viscosity:

$$\eta \sim \frac{T^3}{\alpha_S^3 \ln(1/\alpha_S)} \qquad (46)$$

All these results are very nice from a theoretical point of view; in practice they can only be used as a qualitative guide, because the coupling constant g is not really small, and the hierarchy of scales not so well defined. Slightly above T_c, one can estimate that $\alpha_S \simeq 0.3$, so that $g \simeq 2$!

5. The Kinoshita-Lee-Nauenberg theorem at finite temperature

Our last topic will be the $T \neq 0$ generalization of the Kinoshita-Lee-Nauenberg (KLN) theorem. Recall that at $T = 0$ the KLN theorem allows for example to state that the total decay rate of a heavy particle remains finite order by order in perturbation theory, even if there are massless particles in the final state. A standard example is the total annihilation cross section $e^+ - e^- \to$ hadrons (which can of course be viewed as the total decay rate of a heavy virtual photon into hadrons): in perturbative QCD, one computes the decay rate into quarks and gluons, and one finds a finite result despite the fact that gluons are massless. In fact one can even consider the case of massless quarks and still get a finite result. On the contrary the total cross-section for other processes, for instance the Drell-Yan process (lepton pair production in hadronic collisions) is not a priori finite in perturbative QCD: the singularities which occur in the perturbative calculation have to be absorbed via the factorization theorem into hadronic structure functions. One may wonder what happens at non-zero T if one tries for example to compute the total rate for muon-pair production in a quark-gluon plasma (one of the possible signals of this plasma). It is reasonable to assume that the weakly interacting muons do not have time to interact with the plasma once they have been produced, and they are not thermalized. Thus there is a second scale in the problem, in addition to the temperature, namely the invariant mass Q of the muon-pair. Of course this mass cannot be much larger than T: otherwise the production rate would be negligible, since muon pairs arise from the annihilation of quarks and antiquarks in the thermal bath. At finite T one may hope cross-sections to be finite because of screening. However if screening is responsible for the finiteness of cross-sections, one will find logarithms like $\ln^p(Q/gT)$ or even powers like $(Q/gT)^p$ or $(Q/g^2T)^p$. Then the cross-section is presumably not computable in perturbation theory. On the contrary if there are no singularities in (bare) perturbation theory, then such terms are absent and everything is computable in perturbation theory. A proof of the absence of singularities in the calculation of rates would constitute the required generalization of the KLN theorem.

The rate $\Gamma(Q)$ for lepton-production in a quark-gluon plasma can be obtained by computing the imaginary part of the photon self-energy (Fig. 5) which has been evaluated at two loop order by several groups[18], in the case where the virtual photon is at rest in the thermal bath (the general case has been worked out recently by Cleymans and Dadic[19]):

$$Q = (q, 0) \qquad \text{and} \qquad q \gg gT \qquad (47)$$

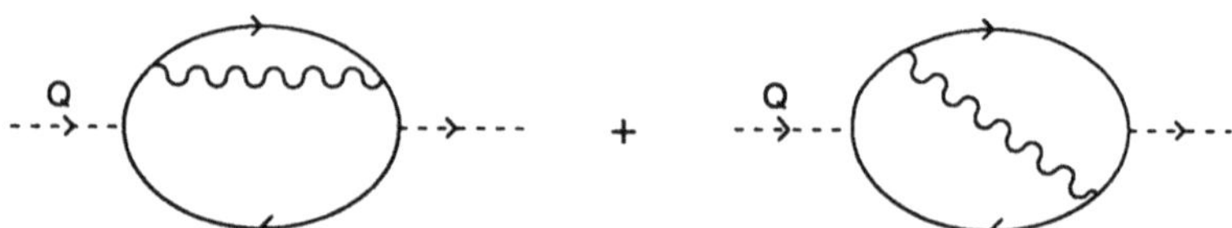

<u>Fig. 5</u> : The photon self-energy at two loop (wavy line = gluons, dotted lines = photons and full lines = quarks).

The condition $q \gg g\,T$ ensures that the external photon is not soft, so that bare PT is correct; the situation $q \sim g\,T$ has been treated by Braaten and Pisarski, using resummation[20]. For massless particles, the final formula is:

$$\Gamma = \Gamma_B \left[1 + \frac{\alpha_S C_F}{\pi} \left(\frac{3}{4} + F(\frac{q}{T}) \right) \right] \tag{48}$$

The factor of 3/4 corresponds to the standard result obtained at $T = 0$ in $e^+ - e^-$ annihilation. The function $F(q/T)$ is finite, at least to this order of PT, and one finds a finite-T generalization of the KLN theorem.

It would be interesting to go to higher orders. However the computations in QCD are very complicated, and no general argument has been found. A three loop calculation has been performed in the framework of a renormalizable toy model: massless $g\,\phi^3$ in space-time dimension $D = 6$, for a particular class of graphs[21] (Fig. 6). This graph is interesting because there appears a new IR singularity, not present at $T = 0$ ($g\,\phi_6^3$ is not IR divergent at $T = 0$!), when the momentum k goes to zero. This new IR singularity shows up only at three loop order and higher, and this is the reason why we undertook the calculation. This strong IR singularity however disappears when one sums over the three possible cuts of the diagram drawn in Fig. 6; similarly all collinear singularities cancel out and one is left with a finite result. Thus it is reasonable to hope that there exists a finite temperature generalization of the KLN theorem.

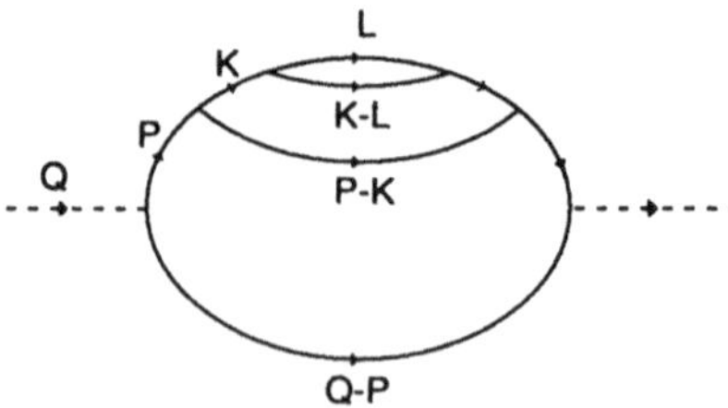

<u>Fig. 6</u> : A three-loop graph.

6. Conclusion

Perturbation theory at finite-T is strongly IR divergent. Resummation methods have been developed, which allow to control the singularities in some, but not all, cases. Processes controlled by the scale gT appear to be well understood: in particular the phenomenon of dynamical screening is able to convert logarithmic singularities of bare perturbation theory into finite results. For stronger (Coulomb like) singularities, it is likely that the magnetic scale g^2T plays a crucial role, but this scale is at present poorly understood. Finally it seems that there exists infrared safe processes, when one sums over all degenerate initial and final state, which generalizes the familiar $T = 0$ KLN theorem.

7. References

1. For a recent review see e.g. A. Ukawa, *Plenary talk at the 25th International Conference on High Energy Physics*, Singapore, preprint UTHE-P213 (1990).
2. For reviews of finite temperature field theory see e.g.: J. Kapusta, *"Finite temperature field theory"*, Cambridge University Press (1989); N. Landsman and Ch. van Weert, *Phys. Rep.* **145** (1987) 142; M. Le Bellac, *Lectures notes in Physics* **396** (1991) 275.
3. E. Braaten and R. Pisarski, *Nucl. Phys.* **B339** (1990) 310.
4. J. Frenkel and J.C. Taylor, *Nucl. Phys.* **B334** (1990) 199.
5. E. Lifchitz and L. Pitayevski, *"Physical Kinetics"* Pergamon Press (1980).
6. V. Silin, *Sov. Phys. JETP* **11** (1960) 1136.
7. J.P. Blaizot and E. Iancu, *Saclay Preprint* SPhT/92-071 (1992).
8. E. Braaten and R. Pisarski, *Phys. Rev.* **D45** (1992) 1827.
9. E. Braaten and R. Pisarski, *Phys. Rev.* **D42** (1990) 2156.
10. J. Bjorken, *Fermilab preprint* PUB-82/59-THY (1982) unplublished.
11. E. Braaten and M. Thoma, *Phys. Rev.* **D44** (1991) 1298.
12. H. A. Weldon, *Phys. Rev.* **D28** (1983) 2007.
13. R. Baier, H. Nakkagawa, A. Niégawa and K. Redlich, *Zeit. Phys.* **C53** (1992) 433.
14. J. Kapusta, P. Lichard and D. Seibert, *Phys. Rev.* **D44** (1991) 2774.
15. E. Braaten and T. Yuan, *Phys. Rev. Lett.* **66** (1991) 2183.
16. R. Baier, H. Nakkagawa, A. Niégawa and K. Redlich, *Phys. Rev* **D45** (1992) 4323.
17. G. Baym, H. Monien, C. Pethick and D. Ravenhall, *Phys. Rev. Lett.* **64** (1990) 1867.
18. R. Baier, B. Pire and D. Schiff, *Phys. Rev.* **D38** (1988) 2814.
    T. Altherr, P.Aurenche and T. Becherrawy, *Nucl. Phys.* **B315** (1989) 436.
    Y. Gabellini, T. Grandou and D. Poizat, *Ann. Phys. (N.Y.)* **202** (1990) 436.
19. J. Cleymans and I. Dadic, *Bielefeld preprint* 135-TP 92-17 (1992).

20. E. Braaten, R. Pisarski and T. Yuan, *Phys. Rev. Lett.* **64** (1990) 2242.
21. T. Grandou, M. Le Bellac and D. Poizat, *Nucl. Phys.* **B358** (1991) 408.
    M. Le Bellac and P. Reynaud, *Nucl. Phys.* **B380** (1992) 423.

TRANSPORT COEFFICIENTS IN THERMAL QCD

H. Heiselberg,[a,b] G. Baym,[a] C. J. Pethick,[a,c] J. Popp,[a] and E. F. Staubo[a,c]

[a] *University of Illinois at Urbana-Champaign,*
1110 W. Green St., Urbana, IL 61801, USA
[b] *Niels Bohr Institute, Blegdamsvej 17, DK2100 Copenhagen Ø., Denmark*
[c] *Nordita, Blegdamsvej 17, DK2100 Copenhagen Ø., Denmark*

ABSTRACT

We summarize here calculations of transport coefficients for quark-gluon plasmas to leading order in the interaction strength, including rates of momentum and thermal relaxation, electric conductivity, and viscosities of quark-gluon plasmas, as well as energy loss and damping of quarks and gluons in quark-qluon plasmas. We discuss consequences for ultra-relativistic heavy ion collisions and neutron stars.

1. Introduction

The inclusion of Landau damping, or dynamic screening, in the calculation of the properties of relativistic plasmas solves [1, 2] the Rutherford divergence problem, i.e., the strong divergence of cross-sections for elementary scattering processes between quarks and gluons or electromagnetic charges at small angles, $d\sigma/d\Omega \sim (1 - \cos\theta)^{-2}$. Although Coulomb-like interactions, arising from longitudinal photon or gluon fields, are screened in plasmas at distances of order the Debye screening length, λ_D, magnetic-like interactions, from exchange of transverse bosons, are not similarly screened in the static limit. However, finite-frequency (ω) magnetic-like interactions in a finite temperature plasma are dynamically screened at the length scale $\sim \omega^{-1/3}\lambda_D^{2/3}$ in the weak-coupling limit [1]. Here we summarize recent calculations [2] of transport coefficients of relativistic QED and QCD plasmas, found by solving the kinetic equation, including dynamic screening of the transverse interactions. In addition we discuss recent calculations of energy loss and damping of quarks and gluons in quark-gluon plasmas.

2. Dynamic Screening

The bare longitudinal and transverse QCD or QED interactions, arising from exchange of a gluon or photon in a plasma is modified in a medium by the boson self-energies, Π_L and Π_T:

$$D_L^{-1}(q,\omega) = q^2 + \Pi_L; \qquad D_T^{-1}(q,\omega) = \omega^2 - q^2 - \Pi_T, \tag{1}$$

where q is the 3-momentum transfer and ω the energy transfer. In the RPA, $\Pi_{L,T}$ are given by the simple bubble diagram describing coupling to particle-hole and particle-antiparticle intermediate states, which for small momentum transfers reduces to

$$\Pi_L = q_D^2 \left(1 - \frac{1}{2}\mu \log \frac{\mu+1}{\mu-1}\right) \; ; \quad \Pi_T = q_D^2 \left(\frac{1}{2}\mu^2 + \frac{1}{4}\mu(1-\mu^2)\log\frac{\mu+1}{\mu-1}\right) , \qquad (2)$$

where $\mu = \omega/q$ and $q_D = 1/\lambda_D$ is the Debye wavenumber. In a weakly-interacting hot quark-gluon plasma $q_D^2 = (1 + N_f/6)g^2T^2$, where N_f is the number of quark flavors with masses less than the temperature, T.

3. Transport coefficients for hot quark-gluon plasmas

In this section we give a number of transport coefficients for high temperature plasmas. We work in the kinematic limit of massless particles, $m \ll T$, and zero chemical potential, $|\mu| \ll T$.

3.1. Momentum and thermal relaxation times

The characteristic timescales, τ, describing the rate at which a plasma tends towards equilibrium if it is initially produced out of equilibrium, as in a scattering process, or if driven by an external field, are determined by solving the kinetic equation. For a system with well defined quasiparticles we have

$$\left(\frac{\partial}{\partial t} + \mathbf{v_p}\cdot\nabla_\mathbf{r} + \mathbf{F}\cdot\nabla_\mathbf{p}\right)n_\mathbf{p} = \left(\frac{\partial n_\mathbf{p}}{\partial t}\right)_{\text{coll}} , \qquad (3)$$

where $\mathbf{p}$ is the quasiparticle momentum, $n_\mathbf{p}$ the quasiparticle distribution function, $\mathbf{F}$ the driving force on the quasiparticles, and $(\partial n_\mathbf{p}\partial t)_{\text{coll}}$ the collision integral.

Momentum relaxation times are computed by considering two interpenetrating, spatially uniform plasmas of different species of particles at a common temperature T, but with relative fluid flow. [By taking different species we avoid complications of the initial conditions for identical fermions.] A relativistic plasma with flow velocity $\mathbf{u}$ is described by a local equilibrium distribution function $n_\mathbf{p} = (\exp[(p - \mathbf{u}\cdot\mathbf{p})/T] \mp 1)^{-1}$ for bosons and fermions respectively. The momentum density of each plasma species is

$$\mathbf{g} = \nu \sum_\mathbf{p} \mathbf{p}n_\mathbf{p} \equiv w\mathbf{u} , \qquad (4)$$

where w is the rest-frame enthalpy density, $= \nu(2\pi^2/45)T^4$ for relativistic bosons, and $\nu(7\pi^2/180)T^4$ for relativistic fermions; ν is 16 for gluons, $6N_f$ for quarks, 2 for photons, and $2N_l$ for leptons.

To be specific we take the first species to be fermions with flow velocity $\mathbf{u}_1$ and the second to be bosons with flow velocity $\mathbf{u}_2$. We assume, in the sense of the variational method for calculating transport coefficients [3], that each component remains in equilibrium, neglecting distortions of the individual distribution functions caused by collisions between the two plasmas. By scattering individual fermions from initial state 1 with bosons from initial state 2 to final states 3 and 4, respectively, the first component transfers a net momentum to the second at a rate given in terms of the collision integral in the kinetic equation by

$$
\begin{aligned}
\frac{d\mathbf{g}_1}{dt} &= -\frac{d\mathbf{g}_2}{dt} = \nu_1 \sum_{\mathbf{p}_1} \mathbf{p}_1 \left(\frac{\partial n_{\mathbf{p}}}{\partial t} \right)_{\text{coll}} \\
&= -2\pi \nu_1 \nu_2 \sum_{1234} \mathbf{p}_1 |\mathcal{M}_{12 \to 34}|^2 \cdot \delta_{\mathbf{p}_1+\mathbf{p}_2,\mathbf{p}_3+\mathbf{p}_4} \delta(\varepsilon_1 + \varepsilon_2 - \varepsilon_3 - \varepsilon_4) \\
&\qquad \times [n_{\mathbf{p}_1} n_{\mathbf{p}_2}(1 - n_{\mathbf{p}_3})(1 + n_{\mathbf{p}_4}) - n_{\mathbf{p}_3} n_{\mathbf{p}_4}(1 - n_{\mathbf{p}_1})(1 + n_{\mathbf{p}_2})] \\
&\equiv -\frac{(\mathbf{u}_1 - \mathbf{u}_2)}{\tau_{\text{mom}}} \frac{w_1 w_2}{w_1 + w_2}.
\end{aligned}
\tag{5}
$$

Equation (5) defines the momentum relaxation time, τ_{mom}, for two interpenetrating plasmas. In the center-of-mass frame, $d\mathbf{g}_i/dt = -\mathbf{g}_i/\tau_{\text{mom}}$, and the momentum density decreases exponentially on a timescale τ_{mom}.

The vacuum Lorentz invariant matrix element for quark-gluon scattering is dominated by a $t^2 = (\omega^2 - q^2)^2$ singularity, $|\mathcal{M}_{qg \to qg}|^2 \simeq g^4(u^2 + s^2)/t^2$, and must be screened dynamically, as in Eq. (1). To calculate transport coefficients to leading logarithmic order in α_s only small q need be taken into account. We refer to Ref. [2] for details concerning the evaluation of the collision term of Eq. (5). The essential ingredient is $\int |D_{L,T}|^2 q^3 dq \sim \log(q_{max}^2/\Pi_{L,T}(\mu))$, where the maximum momentum transfer is $q_{max} \sim T$. This expression is then integrated over $\mu = \omega/q$, $(|\mu| \leq 1)$, resulting in a $\log(T^2/q_D^2)$ term and a μ-integral. The latter is finite because the energy transfer is generally non-zero. For small μ the leading term in $\Pi_T(\mu)$, which comes from Landau damping, is $\simeq -i\pi\mu q_D^2/4$. The resulting momentum relaxation rate τ_{mom}^{-1} for gluons colliding with quarks and antiquarks ($\nu_2 = 12N_f$), exact to leading logarithmic order in $\alpha_s = g^2/4\pi$, is

$$
1/\tau_{\text{mom}} = \frac{20\pi}{7}\left(1 + \frac{21}{32}N_f\right)\alpha_s^2 |\log \alpha_s| T .
\tag{6}
$$

Momentum relaxation times for two plasmas with different quark flavors, spins, or colors, or for different gluon colors or spins have the same form, $\sim \alpha_s^2 |\log \alpha_s| T$.

Thermal conduction is not a hydrodynamic mode in relativistic plasmas with zero chemical potential. However, thermal relaxation times may be computed by considering two plasmas at rest with respect to each other but at slightly different temperatures. The thermal relaxation time is simply $\tau_{\text{therm}} = 3\tau_{\text{mom}}$ [2].

3.2. Viscosities

In Ref. [1], the first viscosity of a quark-gluon plasma was calculated to logarithmic order in the QCD coupling strength by solving the kinetic equation. The viscosity is a sum of the gluon and quark viscosities, $\eta = \eta_g + \eta_q$; writing each η_i $(i = q, g)$ in terms of the viscous relaxation time, $\tau_{\eta i}$, as $\eta_i = w_i \tau_{\eta i}/5$, one has for gluons: $\tau_{\eta g}^{-1} = 4.11T(1 + N_f/6)\alpha_s^2 |\ln \alpha_s|$, and for quarks: $\tau_{\eta q}^{-1} = 0.39\tau_{\eta g}^{-1}$. These rates are very similar to τ_{mom}^{-1}. The second viscosity ζ of a gas of massless relativistic particles is zero [4].

3.3 Electric conductivity

Another transport coefficient of interest is the electric conductivity, σ_{el}, of the early universe. The principal conduction process is flow of leptons as limited by electromagnetic scattering on other charged particles; strongly interacting particles have much shorter mean-free paths. The infrared singularity of the transverse interaction in QED is treated as in QCD, only now $q_D^2 = N_l e^2 T^2/3$, where N_l is the number of charged lepton species present at temperature T. To calculate σ_{el} we consider the current of charged leptons (1) and antileptons (2) in a static electric field, $\mathbf{E}$. Taking the components to be thermally distributed with opposite fluid velocities, $\mathbf{u_1} = -\mathbf{u_2}$, the total electric current is $\mathbf{j}_{\ell\bar{\ell}} = -en_{\ell\bar{\ell}}\mathbf{u_1}$, where the density of electric charge carriers is $n_{\ell\bar{\ell}} = 3\zeta(3)N_\ell T^3/\pi^2$. Solving the kinetic equation (3) we find the electrical conductivity

$$\sigma_{\ell\bar{\ell}} = j_{\ell\bar{\ell}}/E = (3\zeta(3)/\log(2))(\alpha|\log\alpha|)^{-1}T\,. \tag{7}$$

Although quarks (of charge Q_q) contribute negligibly to the electrical current, their presence leads to additional stopping of the leptons and thus smaller conductivity. Adding contributions from ℓq and $\ell\bar{q}$ collisions we obtain the total electrical conductivity $\sigma_{el} = \sigma_{\ell\bar{\ell}}/(1 + 3\sum_{q=1}^{N_f} Q_q^2)$.

3.5 Damping of quarks and gluons

The damping rate of quasiparticles in a quark-gluon plasma has been a source of controversy since it was first calculated to be negative (i.e., unstable) and furthermore gauge dependent. We discuss here the quasiparticle lifetime in the framework of the kinetic equation. The quasiparticle decay rate for a gluon of momentum $\mathbf{p_1}$ scattering on other gluons of momentum $\mathbf{p_2}$ is

$$1/\tau_{p_1}^{gg} = 2\pi\nu_2 \sum_{\mathbf{q},\mathbf{P_2}} \frac{n_2(1 + n_3)(1 + n_4)}{1 + n_1}|M_{gg \to gg}|^2\delta(\varepsilon_1 + \varepsilon_2 - \varepsilon_3 - \varepsilon_4)\,. \tag{8}$$

The leading contribution comes from small momentum transfers $q \sim q_D \sim gT$ and so assuming $p_1 \sim T$, Eq. (8) reduces to

$$1/\tau_{p_1}^{gg} \simeq 2\pi\nu_2 \sum_{\mathbf{q},\mathbf{P_2}} n_2(1 + n_2)\frac{9}{8}g^4 \left|\frac{1}{q^2 + \Pi_L} - \frac{(1 - \mu^2)\cos(\phi)}{q^2 - \omega^2 + \Pi_T}\right|^2 \delta(\varepsilon_1 + \varepsilon_2 - \varepsilon_3 - \varepsilon_4)\,. \tag{9}$$

By introducing an auxiliary integral over the energy transfer, i.e., introducing the variable $\mu = \omega/q$, the angular integrals are easily performed (see [2] for details). The transverse contribution actually diverges for small momentum and energy transfers even when integrating over energy transfers, i.e., dynamic screening is insufficient to obtain a non-zero quasiparticle lifetime. Concentrating on the transverse interactions for small $\mu = \omega/q$, where $\Pi_T \simeq i(\pi/4)q_D^2\mu$, we find to leading logarithmic order

$$1/\tau_{p_1}^{gg} = \frac{3}{8\pi}g^4 T^3 \int_\lambda^{\sim T} q\,dq \int_{-1}^{1} d\mu \frac{1}{q^4 + (\pi/4)^2 q_D^4 \mu^2} = \frac{3}{2\pi}g^4 \frac{T^3}{q_D^2}\log(\frac{q_D}{\lambda})\,, \tag{10}$$

where we have introduced an infrared cutoff, λ. Adding gluon scattering on quarks just adds a factor $(1 + N_f/6)$. The infrared cutoff is on the order of the inherent scale of the "magnetic mass" $\lambda \sim g^2 T$ and so we finally obtain

$$1/\tau_{p_1}^g \;=\; 3\alpha_s |\log \alpha_s| T. \tag{11}$$

The quasiparticle decay rate for quarks can be calculated analogously and is just $4/9$ of that for the gluon, the factor coming from the matrix elements at small momentum transfer for quark scattering on quarks and gluons compared with those for gluon scattering.

Pisarski et al. [5] have developed a resummation technique of soft thermal loops which provides screening so that the damping, γ, is positive and gauge independent. Recently, Burgess, and Marini and Rebhan [6] have obtained the leading logarithmic order for quarks and gluons with momenta $p_1 \gg gT$. They evaluate the gluon self-energy, given by the gluon bubble, to leading order by including screening in the propagator of the soft gluon in the bubble and introducing the same cutoff. Their result for the imaginary part of the selfenergy, which is one half the quasiparticle decay rate, agrees with ours, Eq. (11), because exchange contributions (vertex corrections), which are automatically included in the kinetic equation, do not contribute to leading order.

The relaxation rates in transport processes [2] are of order $\sim \alpha_s^2 T$, i.e., suppressed by a factor α_s with respect to the damping rates. This is because in transport processes one has an extra factor q^2 in integrals like (9) which suppresses the rate by a factor q_D^2/T^2.

4. Transport coefficients in cold quark matter

Momentum transfer processes in degenerate quark matter, $T \ll \mu$ (chemical potential), as for example in neutron stars, are characterized by the rate of momentum relaxation for strong interactions, τ_{mom}^{-1}. For u, d, or s quarks with the same chemical potentials we find, neglecting the strange quark mass, that [7]

$$1/\tau_{\mathrm{mom}} = \frac{8N_f}{3\pi} T\alpha_s^2 \times \left\{ \begin{array}{ll} (3/2)\log(T/q_D) + 2.72, & T \gg q_D \\ a(T/q_D)^{2/3} + (\pi^3/12)(T/q_D), & T \ll q_D \end{array} \right\}, \tag{12}$$

where $a = (2\pi)^{2/3}\Gamma(8/3)\zeta(5/3)/6 \simeq 1.81$. For $T \gg q_D$ this is the standard result [7], with $q_D^2 = 6\alpha_s\mu^2/\pi$. For $T \ll q_D$ the result is qualitatively different due to Landau damping of modes with $q \lesssim (\omega q_D^2)^{1/3}$, where $\omega \sim T$. Applications to burning of nuclear matter into strange quark matter in the interior of a neutron star are described in Ref. [7]. Transport coefficients may, however, be significantly different in a complex mixed phase of quark and nucleon matter in cores of neutron stars [8].

5. Stopping in ultrarelativistic heavy ion collisions

For the situation considered above the characteristic relaxation times that emerge are $\tau_{\mathrm{mom}} \sim (T\alpha_s^2 |\log \alpha_s|)^{-1}$. Taking the perturbative running coupling constant

$\alpha_s(T) \simeq 6\pi/[(33 - 2N_f)\log(T/\Lambda)]$, with $\Lambda \sim 150\text{MeV}$, we would estimate $\tau_{\text{mom}} \sim$ 3 fm and 1 fm for $T = 400$ MeV and 1 GeV respectively. These results are derived only in the weak-coupling limit and are therefore inapplicable for temperatures near the QGP phase transition. Non-perturbative effects are expected to lead to stronger interactions, and plasmas formed in heavy ion collisions are likely to experience considerably faster relaxation. In addition, the Debye screening length, $\lambda_D \sim (gT)^{-1}$, for $T \sim 200$ MeV is $\sim$ a few fm which is comparable to the size of the plasma formed in a heavy ion collision; finite size effects can also contribute to the screening of infrared divergences. Furthermore the creation of excitations and dissociation of the confined quarks and gluons in the nucleons lead to an initial plasma whose components are not in local equilibrium.

Since the relative flow velocities of the colliding nuclei are close to the velocity of light the calculation of the relaxation times needs to be generalized. The stopping or energy loss due to elastic collisions of a fast ($E \gg T$) light quark or gluon penetrating a quark-gluon plasma was estimated by Bjorken [9]; for light quarks

$$\frac{dE^q}{dx} = -\frac{4\pi}{3}(1 + \frac{N_f}{6})\alpha_s^2 T^2 \log\left(\frac{E}{\alpha_s T}\right), \tag{13}$$

to leading logarithmic order. More detailed calculations are given in Refs. [10] and [11]. The energy loss of a fast gluon differs only by a prefactor 9/4, again the ratio of gluon to quark interaction strengths for small momentum transfers. The corresponding stopping rate, $\tau_E^{-1} \equiv (-dE/dx)/E \sim \alpha_s^2 \log(E/\alpha_s T)T^2/E$, decreases almost linearly with collision energy.

Recently Shuryak [12] estimated the equilibration times of quark and gluon plasmas at CERN and RHIC energies. For a typical relaxation rate he applies a *total* collision rate (not a transport rate) $\tau_{\text{coll}}^{-1} \sim \alpha_s T$, which is derived by assuming screening of the longitudinal as well as transverse interactions for $q \lesssim q_D$. This collision rate is larger than the typical transport and momentum relaxation rates, $\tau_\eta^{-1} \sim \tau_{\text{mom}}^{-1} \sim \alpha_s^2 |\log \alpha_s| T$. Secondly, he mainly applies the quark and gluon scattering rates at large angles and finds that the gluons stop faster than quarks by almost an order of magnitude. However transport and momentum stopping are dominated by forward scattering processes in weak coupling, for which the gluon scattering rate is 9/4 times that for quarks at the same temperature. Since gluons interact more strongly they equilibrate earlier, at a higher temperature,[1] and so the equilibration time for quarks is $(9/4)^{3/2} \simeq 3.4$ times that for gluons; gluons stop faster than quarks although not as fast as in [12]. Gyulassy et al. [13] have, however, estimated the energy loss by gluon radiation and find an expression around five times that of Eq. (13) for both quarks and gluons, indicating that radiative energy loss is dominant; since it is the same for quarks and gluons the stopping times will be the same.

---

[1] The equilibration temperature is determined as in [12] by matching the entropy density of a quark-gluon plasma in thermal equilibrium, $S \simeq 7.0T^3$, to that in the Bjorken scenario, $S \sim (dN/dy)/(\pi R_A^2 \tau)$, at the time τ equal to the quark or gluon collision or relaxation time $\propto T^{-1}$.

This work was supported in part by U. S. National Science Foundation Grants PHY89-21025, PHY91-00283, DMR91-22385, and NASA Grant NAGW-1583.

8. References

1. G. Baym, H. Monien, C. J. Pethick, and D. G. Ravenhall, *Phys. Rev. Letters* **64**, 1867 (1990); G. Baym, H. Monien, and C. J. Pethick, *Proc. XVI Int. Workshop on Gross Properties of Nuclei and Nuclear Excitations, Hirschegg* (ed. H. Feldmeier, GSI and Institut für Kernphysik, Darmstadt, 1988), p. 128; Nucl. Phys. **A498** (1989) 313c; Nucl. Phys. **A525** (1991) 415c; G. Baym, H. Heiselberg, H. Monien, C. J. Pethick, and J. Popp, *Nucl. Phys.* **A**544 (1992) 569c.

2. G. Baym, H. Heiselberg, H. Monien, C. J. Pethick, and J. Popp, to be published.

3. G. Baym and C. J. Pethick, *Landau Fermi-liquid theory: concepts and applications* (J. Wiley and Sons, New York, 1991).

4. S. Gavin, *Nucl. Phys.* **B435**, 826 (1984).

5. E. Braaten and R. D. Pisarski, *Phys. Rev.* **D42** (1990) 2156.

6. C. P. Burgess and A. L. Marini, *Phys. Rev.* **D45** (1992) R17; A. Rebhan, *Phys. Rev.* **D48** (1992) 482.

7. H. Heiselberg, G. Baym, and C. J. Pethick, *Nucl. Phys. B* (Proc. Suppl.) **24B** (1991) 144; H. Heiselberg and C. J. Pethick, *Transport coefficients of degenerate plasmas*, to be published.

8. H. Heiselberg, C. J. Pethick, and E. Staubo, Nordita preprint 92/39 A, submitted to *Phys. Rev. Letters*.

9. J. D. Bjorken, Fermilab Report No. Pub-82/59/thy, 1982, unpublished.

10. Y. Koike and T. Matsui, *Phys. Rev.* **D45** (1992) 3237.

11. E. Braaten and M. H. Thoma, *Phys. Rev.* **D44** (1991) R2625; M. H. Thoma, *Phys. Letters* **B273** (1991) 128.

12. E. Shuryak, *Phys. Rev. Letters* **68** (1992) 3270.

13. M. Gyulassy, M. Plümer, M. Thoma, and X. N. Wang, *Nucl. Phys.* **A538** (1992) 37c.

MEDIUM EFFECTS ON HADRON STRUCTURE

M. Asakawa[1], C. M. Ko[1] and P. Lévai[1,2]
[1]*Cyclotron Institute, Texas A & M University*
College Station, TX 77843, USA

[2]*KFKI Institute for Particle and Nuclear Research*
H-1525 Budapest 114. POB 49., Hungary

ABSTRACT

Theories based on quark degrees of freedom predict decreasing hadron masses with increasing temperature and density as a general in-medium effect on hadronic structure. We review the results of our investigations where hadronic degrees of freedom were used and the influence of high baryon density was calculated. Using Vector Dominance Model we could obtain the density dependent spectral function for heavy mesons such as ρ and Φ. It was found decreasing mass for ρ-meson and approximately constant mass for Φ-meson. These results are consistent with QCD Sum Rule calculations. The width of both particles increased drastically.

1. Introduction

In relativistic heavy-ion collisions the nuclear matter can be compressed to densities which are many times that in normal nuclei. This has recently generated great interests in theoretical studies of hadron properties under extreme conditions[1]. In studies based on quark degrees of freedom both Nambu-Jona-Lasinio model[2] and QCD Sum Rules[1,3] result that hadron masses in general decrease with increasing density and temperature as a result of the partial restoration of chiral symmetry. This phenomena can yield interesting consequencies at lower density, namely it can cause observable change in different particle production rate as a basic result of in-medium effects. However the main consequency connects to the predicted phase transition at high energy density where deconfinement and chiral restoration would lead to the appearance of quark-gluon plasma. If hadrons can loose their mass close to the chiral restoration then the hadronic system does not need to pass through a phase transition drived by deconfinement because chiral symmetry is restored already and the hadronic degrees of freedom with zero mass can reproduce in general important behaviours predicted by lattice-QCD calculation. In such a way the chiral restoration with decreasing mass hadrons is a real alternative to describe hadronic matter at high energy density.

However, there is an interesting self-contradiction: models based on quark degrees of freedom predict the unneeded role of these quarks to understand the high energy density behaviour of hadronic matter. Our purpose is to use models with hadronic degrees of freedom for investigating in-medium effects and to find agreement or disagreement with the results of Nambu-Jona-Lasinio model and QCD Sum Rules.

For nucleons and nucleon resonances one of the most popular model at hadronic level is the Walecka-type mean field theory with an attractive scalar and a repulsive vector field[4]. Beyond the succesfull reconstruction of ground state properties of nuclear matter the model has introduced the effective mass for baryons coming from the presence of the scalar field. As the scalar field increases with increasing temperature and density the mass will decrease, even at normal baryon density the nucleon mass is less with 20-25 % as the bare mass. On the other hand, using QCD Sum Rule calculations, recent theoretical investigations[5] reconstruct this scalar and vector field with quark degrees of freedom and yield the same result as the mean field theory. This indicates that one can find good agreement between models based on hadrons or quarks to determine in-medium effects for baryons.

To describe heavy mesons such as ρ and Φ-meson in hot and dense hadronic matter the best candidate is the Vector Dominance Model (VDM)[6]. In VDM one can calculate the in-medium propagation of these heavy composite particles via the medium effects on light mesons: $\rho \to \pi\pi \to \rho$; $\Phi \to K\overline{K} \to \Phi$. Early calculations have shown already that the ρ-mass depends on temperature very weakly[7]. There is a special difficulty with the density dependent in-medium effects because the calculations yield a complicated spectral function which can not be described as a simple particle. One will see the importance of $\Delta - N - hole$ excitation coming from π absorption at finite nucleon density. Using zero width for Δ-particle an approximately constant mass and an increasing width was obtained for ρ-meson[8]. However, taking into account the finite (constant) width for the Δ-particle, the result of this more realistic calculation can be interpreted with a decreasing effective mass for the ρ-meson determined from the spectral function[9]. This result is consistent with the QCD Sum Rule.

For Φ-meson previous works applied unchanged mass and obtained large increase of the width in hot mesonic matter[10] and in dense hadronic matter[11]. Applying the Vector Dominance Model one can calculate the mass and the width at the same time taking into account the density-dependent kaon effective mass as given by the linear chiral perturbation theory[12,13]. The mass of Φ-meson depends on density very weakly but the width increases drastically[14]. This result is also consistent with the prediction of QCD Sum Rule where one can obtain a weak density dependence coming from the mentioned weak coupling between light and strange quarks.

Thus we have found a good agreement between the predictions of QCD Sum Rule method and the results of hadronic descriptions for the mass of baryons and heavy mesons. However these calculations contain more ambiguities and further work is necessary to prove the complete equivalence between models with quark and with hadronic degrees of freedom. For further development we will summarize briefly the recent results, concentrating mostly on heavy composite mesons.

In Chapter 2. we consider the density and temperature dependence of in-medium hadron mass coming from QCD Sum Rule method. In Chapter 3. we review our results in details for ρ and Φ-meson coming from VDM model at finite baryon density.

2. In-medium hadronic mass with quark degrees of freedom

Considering the models with quark degrees of freedom such as the QCD Sum Rule, the quark condensate $\langle \bar{q}q \rangle$ has basic role to describe in-medium effects. This condensate has a finite value at low density/temperature, violating the chiral symmetry. However, at high density/temperature this finite value disappears and the chiral symmetry is restored by the value of zero for quark condensate. Detailed calculations in chiral perturbation theory show the following expression for temperature dependence of the quark condensate[15]

$$\frac{\langle 0|\bar{q}q|0\rangle_T}{\langle 0|\bar{q}q|0\rangle_{T=0}} \cong \left[1 - \left(\frac{T}{T_c}\right)^2\right]^{1/2} \tag{1}$$

where $T_c \simeq 2f_\pi$ with f_π the pion decay constant in the chiral limit.

Using a linear approximation in the density, the following result was obtained for the density dependence of the quark condensate[3,16.17]

$$\langle \bar{q}q \rangle_\rho = \langle \bar{q}q \rangle_0 + \frac{\Sigma_{\pi N}}{2m_q}\rho \tag{2}$$

where $\Sigma_{\pi N} \equiv m_q \langle \bar{u}u + \bar{d}d \rangle_N = (45 \pm 7) \; MeV$ is the $\pi - N$ sigma term in vacuum with $m_q = (7 \pm 2) \; MeV$ being the averaged value of the light current quark mass.

The strange quark condensate has a similar temperature dependence, because at high temperature many strange meson can appear. However the density dependence is different. Since strange quarks couple very weakly to light quarks, strange quarks are not effected by dense u and d quarks. Of course, even the proton has a small strangeness content, thus one can expect a weak baryon density dependence[3]

$$\langle \bar{s}s \rangle_\rho = \langle \bar{s}s \rangle_0 + y \cdot \frac{\Sigma_{\pi N}}{2m_q}\rho \tag{3}$$

Here $y \equiv 2\langle s\bar{s} \rangle_N / (\langle u\bar{u} \rangle_N + \langle d\bar{d} \rangle_N)$ is the strangeness content of the proton, $y \approx 0.12$.

These expressions suggest that at normal nuclear matter density ($\rho_0 = 0.17 \; fm^{-3}$) absolute value of the light and strange quark condensate decreases by 20-30 % and 4-8 %. The mass shift for ρ is much faster at finite density then that at finite temperature because one is linear and the other is quadratic. Φ-meson mass changes slightly.

Now, as we know the density/temperature dependence of the condensates, one needs to find a method to determine the hadronic mass. Let us consider the results of QCD Sum Rule which describes the static hadronic properties very well.

Early calculations in vacuum gave a good approximation for the nucleon mass[18]

$$m_N = \left[-2 \cdot (2\pi)^2 \cdot \langle 0|\bar{q}q|0\rangle\right]^{1/3} \tag{4}$$

and similarly for the ρ-meson mass[19]

$$m_\rho = 2.8 \cdot [\sqrt{\alpha_s}\langle 0|\bar{q}q|0\rangle]^{1/3} \tag{5}$$

where α_s is the QCD running coupling constant. Beyond the leading term $\langle \bar{q}q \rangle$ more detailed calculations result the contributions of terms $\langle m_q \bar{q}q \rangle$, $\langle \bar{q}q \rangle^2$, $\langle \bar{s}s \rangle$ and $\langle m_s \bar{q}q \rangle$[18]. The last two terms are important to obtain the mass of Φ-meson[20].

Now, if we simply substitute the temperature and density dependence of the quark condensate given above into the expression derived by QCD Sum Rule, then we can obtain an estimation for the in-medium mass of hadrons. However, at finite baryon density new terms arise, such as the vector-type condensate $\langle (q^+q)^2 \rangle$ which has zero contribution in vacuum and other higher derivatives e.g. $\langle \bar{q}\gamma_\mu \partial_\mu q \rangle$ [3,21]. Detailed calculations give the following linear density dependence for ρ and Φ-meson mass[21]:

$$\frac{m^*}{m_0} \simeq 1 - C \left(\frac{\rho}{\rho_0} \right) \tag{6}$$

where $C \simeq 0.18$ for ρ-meson and $C \simeq 0.15 \cdot y$ for Φ-meson.

3. ρ and Φ-meson in dense hadronic matter

The medium effects on heavy composite mesons will be calculated by means of Vector Dominance Model[6] using the dressed graphs of Fig. 1. We will consider both ρ and Φ-meson at rest in the nuclear matter. Then the strong tensor coupling of the ρ-meson to the nucleon can be ignored because it is proportional to the ρ-meson momentum. Also, the vector coupling of the ρ-meson to the nucleon can be neglected as the non-relativistic nucleon particle-hole polarization vanishes at zero momentum. The self-energy of a ρ-meson in the nuclear medium is thus determined by its coupling to the pions, which are modified by the $\Delta - hole$ polarization in the nuclear matter.

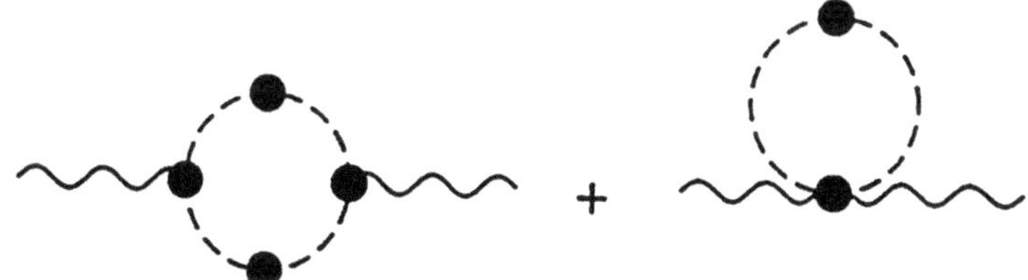

Fig.1. The ρ and Φ-meson self-energy diagrams in Vector Dominance Model.

For a ρ-meson with four momentum k its self-energy can be expressed as

$$\Sigma^{\mu\nu}(k) = ig_\rho^2 \int \frac{d^4q}{(2\pi)^4} \left[\frac{\Gamma^\mu_{\rho\pi\pi}(k,q)\Gamma^\nu_{\rho\pi\pi}(k,q)}{[q^2 - m_\pi^2 - \Pi(q)][(q-k)^2 - m_\pi^2 - \Pi(q-k)]} - \frac{2\Gamma^{\mu\nu}_{\rho\rho\pi\pi}(k,q)}{q^2 - m_\pi^2 - \Pi(q)} \right] \tag{7}$$

In the above, g_ρ is the $\rho\pi\pi$ coupling constant and has a value $g_\rho^2/(4\pi) \approx 2.94$ determined from the ρ-meson width in free space; the pion mass is denoted by m_π.

302

The $\rho\pi\pi$ vertex $\Gamma^\mu_{\rho\pi\pi}$ and $\rho\rho\pi\pi$ vertex $\Gamma^{\mu\nu}_{\rho\rho\pi\pi}$ are given by

$$\Gamma^\mu_{\rho\pi\pi}(k,q) \approx (2q-k)^\mu + (q-k)_\nu \Pi^{\nu\mu}(q-k) + q_\nu \Pi^{\nu\mu}(q) \tag{8}$$

$$\Gamma^{\mu\nu}_{\rho\rho\pi\pi}(k,q) \approx g^{\mu\nu} + \Pi^{\mu\nu}(k-q) \tag{9}$$

In the above Γ-s the first terms are the bare $\rho\pi\pi$ and $\rho\rho\pi\pi$ vertices and the other terms connect to the pion self-energy $\Pi(q)$ due to the $\Delta - hole$ polarization, i.e. $\Pi(q) = k_\mu \Pi^{\mu\nu}(q) k_\nu$. The approximation means that we considered only the dominant graphs. To prove the complete satisfaction of Ward-Takahashi identity one needs to consider all of the graphs and full vertex corrections[8,9].

The ρ-meson propagator in the medium D is related to the free propagator D_0 by $(D^{-1})^{\mu\nu} = (D_0^{-1})^{\mu\nu} - \Sigma^{\mu\nu}$. The imaginary part of the ρ-meson self-energy is obviously finite. But the real part of the self-energy is divergent and needs to be renormalized. This is done by writing it as $\Re e\, \Sigma^{\mu\nu} = (\Re e\, \Sigma^{\mu\nu} - \Re e\, \Sigma_0^{\mu\nu}) + \Re e\, \Sigma_0^{\mu\nu}$. The difference in the bracket is finite as the pion self-energy in the nuclear matter vanishes at large momenta. The remained divergent self-energy contribution in free space $\Re e\, \Sigma_0^{\mu\nu}$ is then replaced by the square of the measured mass, $m_\rho = 770\ MeV$.

Since we can satisfy the Ward-Takahashi identity, the gauge invariance is kept in the calculation. Thus the renormalized self-energy tensor $\tilde\Sigma^{\mu\nu}$ has only three nonvanishing components of the same magnitude $\tilde\Sigma^{11} = \tilde\Sigma^{22} = \tilde\Sigma^{33} = \tilde\Sigma$. Using the above mentioned approximation for the vertices in eqs. (8) (9) the gauge invariance is violated but only slightly, which can be seen from the relatively small value of $\tilde\Sigma^{00}$.

The property of ρ-meson in the medium can be expressed by its spectral function which is defined on the following way:

$$S(M) \equiv 2\pi \cdot \Im m\, D = -\frac{2 \cdot \Im m\, \tilde\Sigma(M)}{[M^2 - m_\rho^2 - \Re e\, \tilde\Sigma(M)]^2 + [\Im m\, \tilde\Sigma(M)]^2} \tag{10}$$

The calculated spectral function at different baryon densities is shown on Fig. 2. One can see the shift of the original ρ-peak into higher invariant mass region and its width becomes larger and larger. The spectral function becomes fragmented and another low mass peak arises around $3m_\pi$ coming from the presence of $\Delta - hole$ excitation. The complex dispersion relation of this excitation is very simple: $\omega_R = \sqrt{\mathbf{q}^2 + m_\Delta^2} - m_N - i\Gamma_\Delta/2$. Since we chose constant nucleon mass, Δ mass and Δ width, this lower peak is shifting down only very slightly.

If we would like to see the effect of the density dependent mass in the original effective Lagrangian we can adopt the scaling low of Ref.[22] and use the empirical density-dependent nucleon effective mass, i.e.

$$\frac{m_N}{m_N^{(0)}} \approx \frac{m_\Delta}{m_\Delta^{(0)}} \approx \frac{\Gamma_\Delta}{\Gamma_\Delta^{(0)}} \approx \frac{1}{1 + 0.25\rho/\rho_0} \tag{11}$$

Repeating the full calculation with in-medium baryons the results are shown in Fig. 3. It is seen that the ρ-peak in the spectral function moves to a smaller invariant mass with diminishing strength when the density becomes higher. The low mass peak also shifts down with increasing density and it becomes more pronounced.

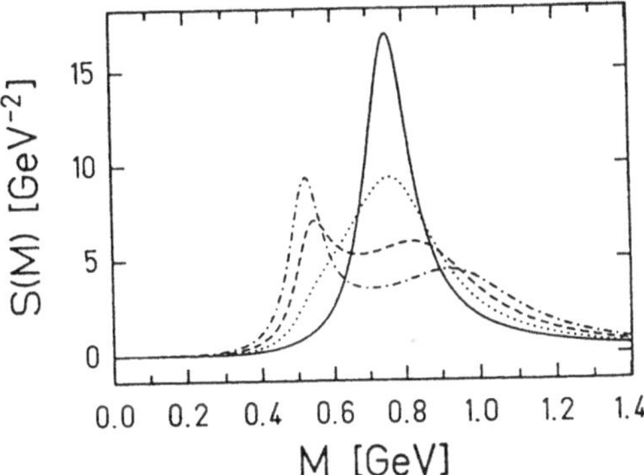

Fig.2. The spectral function of ρ-meson in vacuum and at densities ρ_0, $2\rho_0$, $3\rho_0$.

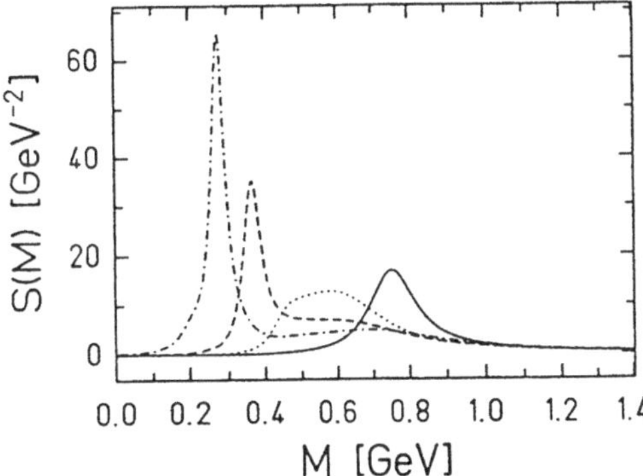

Fig.3. Same as Fig. 2 with density-dependent hadron masses.

In the case of Φ-meson the eq.(7) for the self-energy becomes simpler, because the K-meson has a scalar type interaction with nuclear matter and one can introduce an effective mass m_K^* to describe in medium effects on kaon and anti-kaon. As it was shown in linear perturbation theory[12,13], in dense matter this effective mass decreases with increasing nuclear density:

$$m_K^* = m_K \left[1 - \frac{\rho_B}{\rho_c} \right]^{1/2} \tag{12}$$

with critical density $\rho_c = f_K^2 m_K^2 / \Sigma^{KN}$ where m_K is the kaon mass in free space and

$f_K \sim 93 \; MeV$ is the kaon decay constant. The KN sigma term Σ^{KN} is defined by

$$\Sigma^{KN} = \frac{1}{2}(m_u + m_s)\langle N|\overline{u}u + \overline{s}s|N\rangle \tag{13}$$

There is a considerable uncertainty in the strangeness content of the nucleon and one can obtain the limits $2.5\rho_0 < \rho_c < 5\rho_0$. We will use the upper limits, $\rho_c \approx 5\rho_0$, corresponding to zero strangeness content for the nucleon.

Applying the Vector Dominance Model, the one-loop self-energy of the Φ-meson is shown by the diagrams in Fig.1 is given by

$$\Sigma^{\mu\nu}(k) = ig_\Phi^2 \int \frac{d^4q}{(2\pi)^4}\left[\frac{(2q-k)^\mu(2q-k)^\nu}{[q^2 - m_K^{*2} + i\varepsilon][(q-k)^2 - m_K^{*2} + i\varepsilon]} - \frac{2g^{\mu\nu}}{q^2 - m_K^{*2} + i\varepsilon}\right] \tag{14}$$

Because of the momentum independent effective mass for K-meson this self-energy $\Sigma^{\mu\nu}$ can be evaluate easily, using general regularization methods, such as the dimension regularization[7]. The detailed calculation[14] results the mass and width of Φ-meson in the matter as they are shown in Fig. 4. One can see that the Φ-meson mass decreases slightly with increasing density, except near the critical density ρ_c where the Φ-meson mass increases again with the density but remains below the free mass. At the same time the width increases with the density and becomes an order of magnitude larger than its value in the free space. This density effect on the width is much larger than the temperature effect studied in Ref.[10]. According to this calculation the mass and width remain unchanged beyond the critical density.

This slightly decreasing mass of the Φ-meson is consistent with the basic estimation of the QCD Sum Rule method at lower density, see eq. (6). At higher density one needs to consider further contributions in the QCD Sum Rule method also, which would modify the recent low density linear dependence.

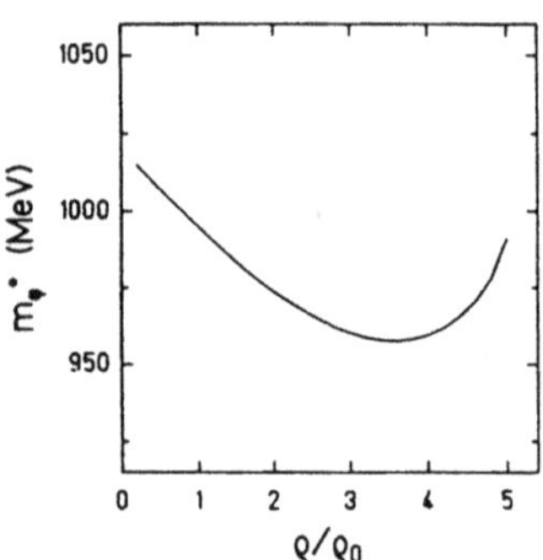
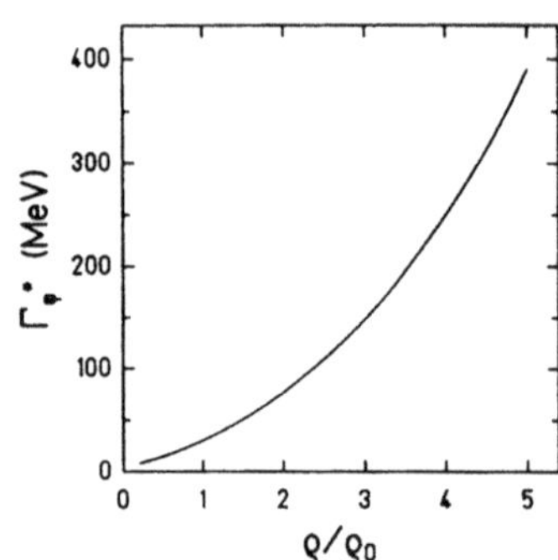

Fig.4. The density dependence of Φ-meson mass and width.

5. Acknowledgements

This work was supported in part by the National Science Foundation under Grant No. PHY-8907986, the Welch Foundation under Grant No. A-1110 and the Hungarian National Science Foundation (OTKA) under Grant No. 2973.

6. References

1. G.E. Brown, Nucl. Phys. **A522** (1991) 397c. and references therein.
2. V. Bernard, U.G. Meissner and I. Zahed, Phys. Rev. Lett. **59** (1987) 966.; V. Bernard and U.G. Meissner, Phys. Lett. **B227** (1989) 465.;
3. T. Hatsuda and S.H. Lee, YSTP-91-10.
4. J. D. Walecka, Annals of Physics **83** (1974) 491.; B. D. Serot, J. D. Walecka: *The Relativistic Nuclear Many-body Problem*, Advances in Nuclear Physics **16**, Ed. J. W. Negele and E. Vogt, Plenum Press 1986
5. T. D. Cohen, R. J. Furnstahl, D. K. Griegel, Phys. Rev. Lett. **67** (1991) 961.
6. J. J. Sakurai, *Currents and Mesons*, University of Chicago Press, Chicago, 1969.
7. C. Gale, J. Kapusta, Phys. Rev. **D43** (1991) 3080.
8. M. Herrmann, PhD. Thesis, GSI-92-10 report, 1992.
9. M. Asakawa, C. M. Ko, P. Lévai, X. J. Qiu, Phys. Rev. **C** in press.
10. D. Lissauer, E. V. Shuryak, Phys. Lett. **B253** (1991) 15.
11. P. Z. Bi, J. Rafelski, Phys. Lett. **B262** (1991) 485.
12. D. B. Kaplan, A. E. Nelson, Phys. Lett. **B175** (1986) 57.
13. G. E. Brown, K. Kubodera, M. Rho, Phys. Lett. **B192** (1987) 273.
14. C. M. Ko, P. Lévai, X. J. Qiu, C. T. Li, Phys. Rev. **C45** (1992) 1400.
15. P. Gerber, H. Leutwyler, Nucl. Phys. **B321** (1989) 387.
16. C. Adami, T.Hatsuda, I. Zahed, Phys. Rev. **C43** (1991) 921.
17. M. A. Shifman, A. I. Vainshtein, V. I. Zakharov, Nucl. Phys. **B147** (1979) 385.
18. L. J. Reinders, H. Rubinstein, S. Yazaki, Phys. Rep. **127** (1985) 1.
19. N. V. Krasnikov, A. A. Pivovarov, N. N. Tavkhelidze, Z. Phys. **C19** (1983) 301.
20. L. J. Reinders, H. R. Rubinstein, Phys. Lett. **B145** (1984) 108.
21. T. Hatsuda, H. Høgaasen, M. Prakash, Phys. Rev. Lett. **66** (1991) 2851.
22. G. E. Brown, M. Rho, Phys. Rev. Lett. **66** (1991) 2720.

SURFACE TENSION IN HOT LATTICE QCD

S. HUANG

Department of Physics, University of Washington, FM-15
Seattle, WA 98195, USA

J. POTVIN

Department of Science and Mathematics, Parks College of Saint Louis University
Cahokia, IL 62206, USA

and

C. REBBI

Department of Physics, Boston University
Boston, MA 02215, USA

ABSTRACT

We review the results of recent calculations of the surface tension in lattice QCD
for flat and spherical interfaces separating coexisting phases of hadronic matter and
quark-gluon plasma. We compare these results with the the Bag model calculation of Mardor and Svetitsky and discuss their implications for the 'Swiss Cheese'
instability. Presented by J. Potvin.

1. Introduction

Recent lattice simulations of quenched and full QCD suggest that the phase transition between hadronic matter and quark-gluon plasma may be weakly 1st order (and
possibly 2nd order in 2 flavor QCD).[1] First order phase transitions are particularly
interesting since they feature metastable states which decay by nucleating droplets of
stable matter. In strongly first order phase transitions the lifetime of such a decay is
calculated from classical nucleation theory[2,3] and depends on the value of the surface
tension on a flat interface separating macroscopic domains of hadronic matter and
quark-gluon plasma. As reviewed in the following section, several lattice calculations
of the surface tension in quenched QCD have given a value of order $\alpha = 50 MeV/fm^2$
or less.[4-11] Compared to the scale of the strong interactions such a value is very small,
confirming further the weak character of the phase transition.

Weak first order phase transitions are outside the range of validity of classical nucleation theory. Studying droplet decay will therefore require a more detailed knowledge of the droplets' free energy at various droplet sizes.[3] One standard parametrization of the radius-dependence of the free energy is the following:

$$\Delta F(R) = -\Delta f\, V + \alpha_{oo} A - 8\pi C R + \dots\dots \tag{1}$$

were V, A and R are the droplet's volume, surface and radius respectively; f is the bulk free energy difference between the metastable and the stable states and the factor $8\pi C$ is the so-called curvature coefficient. Computing the value of the curvature coefficient on gluonic and hadronic droplets has motivated some of the most recent lattice work in finite temperature QCD.[12,13] These will be discussed in Section 3 below.

The droplet surface free energy being size-dependent may imply a phase kinetics (i.e. droplet time-evolution) very different from the kinetics characteristic of classical nucleation. An example is the 'Swiss cheese' instability suggested by Mardor and Svetitsky[14], in which droplets of sizes smaller than the size of the 'saddle-point' configuration in the classical theory may exist as metastable states (these sub-critical droplets are unstable in the classical theory). Because scenarios like the Swiss Cheese instability depend on the precise size-dependence of the free energy, lattice calculations are particularly useful in determining the likely forms of droplet evolution as we will see in Section 4.

2. Surface Tension of Flat Interfaces

Most of the studies done so far have focused on studying the surface free energy in quenched QCD.[4–6,8–11] Work in QCD with 4 flavors of dynamical (heavy) fermions was done by Hackel et al.[7]

The surface tension of a flat interface separating hadronic and quark-gluon phases is calculated from the free energy needed to create a macroscopic interface of surface area $A = 2\,L^2$ on a $L * L * 2L$ spatial lattice with periodic boundary conditions (see figure 1). It is sufficient to compare F_{12}, the free energy associated with a mixed phase system containing equal volumes of phase 1 (hadronic matter) and phase 2 (gluonic plasma), with the free energy (F_1, F_2) associated with each bulk phase separately. By parametrizing ($F_{12} - F_1$) and ($F_{12} - F_2$) as a superposition of a volume term and a surface term, the contributions from the bulk free energy can be subtracted out, giving:

$$\Delta F_s = \frac{[V_1(F_{12} - F_1) - V_2(F_2 - F_{12})]}{V_1 + V_2}. \tag{2}$$

V_1 and V_2 correspond to the volume occupied by phase 1 and 2 respectively. A similar formula can be used in the case of droplets of phase 1 immersed in bulk matter of phase 2 (or vice versa), using instead V_{in} and V_{out}, the volumes inside and outside the droplet respectively. Eq. (2) is equivalent to an experiment in which the system is initially in phase 1 and is allowed to 'grow' the other phase in one half of the lattice, leading to a mixed phase. The interface is then allowed to "melt" when the rest of the system is transformed to the second phase.

In practice, the free energy is obtained from a suitable derivative of the partition function Z, since $F = -\ln Z$. (Here, $F/k_B T$ has been rescaled to F). Some investigators have used a derivative with respect to a change of the interface's surface.[6] Others have used a derivative with respect to the gauge coupling β ($\beta = 6/g^2$)[4] or to

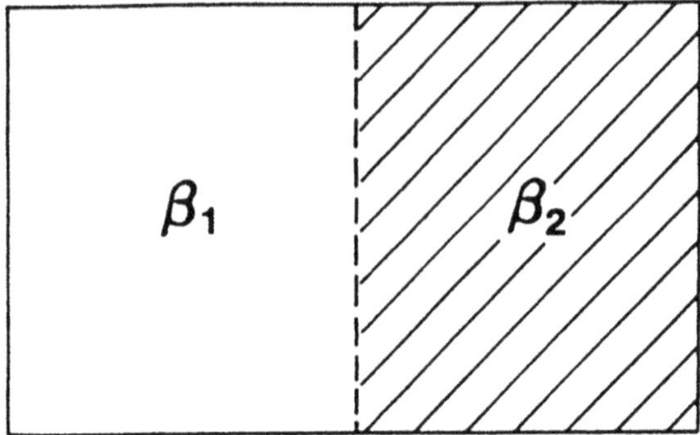

Fig. 1. A mixed phase configuration on a lattice, created by a small gauge coupling (temperature) difference across the interface (ref. [4]).

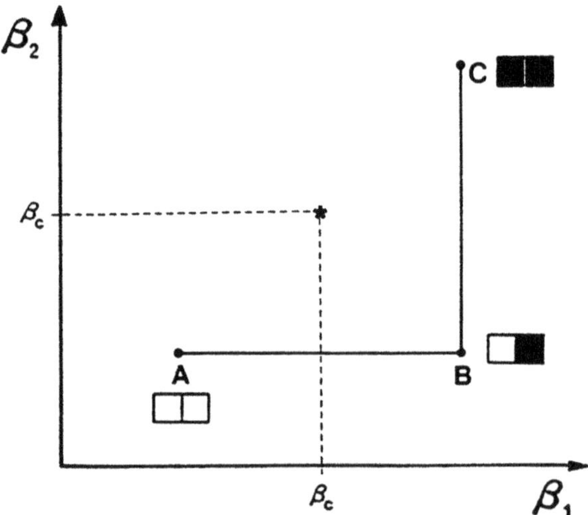

Fig. 2. Path in coupling space for the calculation of the surface tension.

external Polyakov line couplings[5] in order to simulate an actual creation/destruction of phases through small changes of temperature or external fields. Others have explicitely calculated the probability to generate a two-phase system.[8-10]

Derivatives of the partition function are usually obtained from Monte Carlo simula An example is the average action $\langle S \rangle$ calculated from the derivative of $\ln Z$ with respect to β. The free energy is then simply obtained from the following integral:

$$F(\beta'') - F(\beta') = \int_{\beta'}^{\beta''} d\beta \, \langle S \rangle_\beta \qquad (3)$$

The calculation of the hadronic and quark-gluon droplets free energy is obtained from the integral of $\langle S \rangle$ along paths shown in figure 2. Each path is defined by tuning the gauge coupling β inside V_1 (β_1) and inside V_2 (β_2) in succession (changing the coupling is equivalent to increasing and/or decreasing the temperature). In order to have a stable interface the integration starts at temperatures slightly smaller than T_c, that is $\beta = \beta_c - \delta\beta$ and ends at temperature slighlty higher, or $\beta = \beta_c + \delta\beta$.[4] To obtain the physical value of the surface free energy, the calculation is repeated for smaller values of the temperature difference $\delta\beta$, and an extrapolation to zero-$\delta\beta$ is performed in the end.

Lattices of several volumes have been used for surface tension studies, namely lattices with 2 and 4 times slices (or $N_t = 2, 4$) with spatial volumes ranging from $6*6*12$ up to $16*16*32$,[4,5] from $8*8*16$ to $8*8*40$,[6,7] and finally from $12*12*6$ up to $12*12*12$ (on a $N_t = 2$ lattice),[8] and $12*12*12$ on $N_t = 2$ as well as $20*20*20$, $24*24*24$ on $N_t = 4$.[10]

The computation of the surface tension has a recent history. The world data can be listed as follows:

First, on $N_t = 2$ lattices, $\alpha/T^3 = 0.12(2)$ [ref.4] , 0.08(1) [ref. 6 (approximate error)] , 0.071(8) [ref. 8], 0.140(6) [ref. 9] and, 0.078(8) [ref. 10].

On $N_t = 4$ lattices, on the other hand: $\alpha/T^3 = 0.024(4)$ [ref. 5] and 0.007(2) [ref. 10].

We see that all studies are consistent with the statement that the surface tension is indeed small in QCD. There are some discrepencies between the different methods of computation however. Because the results within each group of methodologies do not agree, it is hard at this time to identify clearly each source of systematic error. Here it seems that errors due to small lattice volume, Monte Carlo sampling, ignorance of the so-called Karsch coefficients at intermediate coupling, etc. may be affecting the results in an uneven manner. More work is clearly needed here to determine how large a lattice or how long a Monte Carlo run need to be in order to see agreeement between these different methods.

It is important to point out that all the recent computations of the surface tension in the 3-dimensional Ising model agree nicely over a wide range of temperatures. [15] On the other hand, a major disagreement do exist in the 2-dimensional 7-state Potts model between the 'integral' method[16] and the probability distribution method.[8] The source of the discrepency is not understood at the moment.

It seems clear that the surface tension on lattices with 2 and 4 time slices differ substancially, indicating lack of scaling. This is consistent with the scaling properties of other physical parameters . In the case of the transition temperature T_c for example, studies have shown that scaling appears only on lattices with more than 10 time slices $(N_t = 10)$.[17]

A computation of the surface tension in 4-flavor QCD has been performed by Hackel et al.[7] These authors have used the derivative with respect to a change of the interface area to compute the free energy.[6] They have obtained a very small (negative) value, also consistent with zero within errors. It is worth mentioning that Hackel et al. have found that the value of the surface tension results from a delicate balance between the (positive) contribution of the gluonic degrees of freedom and the (negative) contribution of the fermionic degrees of freedom. Much remains to be done, however, regarding the estinate of the systematic errors present in this calculation, since approximate algorithms were used for both the computation of the surface tension and the Monte Carlo update of the gauge configurations.

3. Surface Tension of Spherical Interfaces

Unlike the surface tension of a flat interface (or wall), the definition of the surface tension on a droplet is more ambiguous. The reason rests with the fact that at $T = T_c$, walls are macroscopic objects which are stable in the infinite volume limit. Droplets, on the other hand, are unstable objects which either shrink or grow during the phase transition. Of course, stable droplets do exist in macroscopic sizes in the real world, when a pressure difference between the droplet's interior and exterior balances out the surface tension. In principle, such an environment could be created in supercooled and superheated matter.

Another ambiguity in the definition of a droplet's free energy has to do with the location of the droplet's surface when the interface has a finite thickness w. This problem can also be solved for macroscopic droplets supporting a pressure difference across their interface by using Laplace's equation $\Delta p = 2\alpha/R$ as a constraint.[3,18,19]

The radius dependence of a stable droplet surface tension was first computed by Gibbs.[3,18,19] Following general thermodynamics and using Laplace's equation, one can derive this radius dependence in terms of the wall surface tension α_{oo} and of the surface thickness w (assumed to be thin), thus obtaining the Tolman-Gibbs equation:

$$\alpha = \frac{\alpha_{oo}}{1 + 2w/R} \; . \tag{4}$$

As discussed in reference [19], it can be argued that the Tolman-Gibbs formula holds for slowly growing non-equilibrium macroscopic droplets as well.

Practically speaking, it is very difficult to adequately simulate metastable states or slowly growing droplets in lattice QCD simulations because of limitations on the system's volume and on the actual identification of metastable states (not to mention the definition of real time in Eucledian field theories). One alternative is to artificially create a droplet of a given size at $T = T_c$ using some external fields or by applying a (small) temperature difference across the interface. The problem is of course whether this is at all meaningful, particularly in the case of small (microscopic) droplets. However this procedure may be justified for large droplets which feature flat walls on a microscopic scale. These artificially created droplets would correspond to droplets which in the real world are in equilibrium or grow at a very slow rate. These assumptions can be checked indirectly by comparing the surface free energies obtained from the use of different types of external fields and/or by comparing with the large R-limit of the Tolman-Gibbs equation (Eq. (4)) as will be done below.

Two series of calculations in quenched $N_t = 2$ lattice QCD have been performed recently,[12,13] in wich droplets were created by tuning the gauge coupling (and hence the temperature) inside the droplet to artificially stabilize its surface. Huang et al.[12] have measured the derivative of the partition function with respect to the gauge coupling β and integrated with respect to β in order to obtain the free energy. Kajantie et al.[13] have computed the derivative with respect to the droplet's radius R and integrated with respect to R to compute F. Both studies have used Eq. (1) to extract the value of the curvature coefficient; in both cases also, an extrapolation to a zero value of the stabilizing temperature difference had to be performed.

The values of the surface tension obtained in reference [12] are shown in figure 3: $\alpha/T_c^3 = -0.060(7), -0.042(5), +0.055(14)$ for hadronic droplets with radii $R = 4, 5, 8$ respectively, and $\alpha/T_c^3 = 0.000, -0.076(5), +0.061(14)$ for gluonic droplets with radii $R = 4, 5, 8$. These values were obtained directly from the integration of the action, without using equation (1). The results are compared with the Tolman-Gibbs formula (Eq. 4), using a value for the interface width of $w = 4$,[6] and a wall surface tension of $\alpha_{oo}/T_c^3 = 0.12(2)$.[4] The agreement with the Tolman-Gibbs formula for $R > 5$ is rather surprising given the fact that no free parameters were available. Perhaps the size of $R = 8$ droplets could be considered as macroscopic; given the droplet width $w = 4$,[6] droplets with radii smaller than 5 are clearly microscopic.

It is important to notice that having $w = 4$ implies that the definition of a droplet's radius is ambiguous at T_c (Laplace's equation cannot be used at T_c).[19] However, one can estimate the error due to this ambiguity by using calculus of variations with respect to the transformation $R \to R + \delta$ on the surface term of the droplet's total free energy. One finds an error of $\frac{R}{(R+w/2)} - 1$, or, of about 20 percent for $R = 8$ droplets and up to 60 percent for $R = 4, 5$. Qualitatively, the ambiguity due to interface thickness does not change the picture shown in figure 3.

Kajantie, Karkkainen and Rummukainen[13] have expressed their computed free energy using equation (1) as an ansatz. The values of the curvature coefficients thus obtained were $C = -0.076(25)$ and $C = +0.076(120)$ (in units of T^3) for hadronic and gluonic droplets respectively. Their data on hadronic droplets have smaller errors than the data on gluonic droplets. These numbers can be compared with the Bag model calculation of Mardor and Svetitsky[14] (see also Section 4. below) which predicts $C = -0.0707$ (hadronic) and $C = +0.0707$ (gluonic), thus showing a nice agreement particularly in the case of the hadronic droplets.

A consistency check can be made by comparing the coefficient of the surface-dependent term which should be identical to the flat wall surface tension: Kajantie

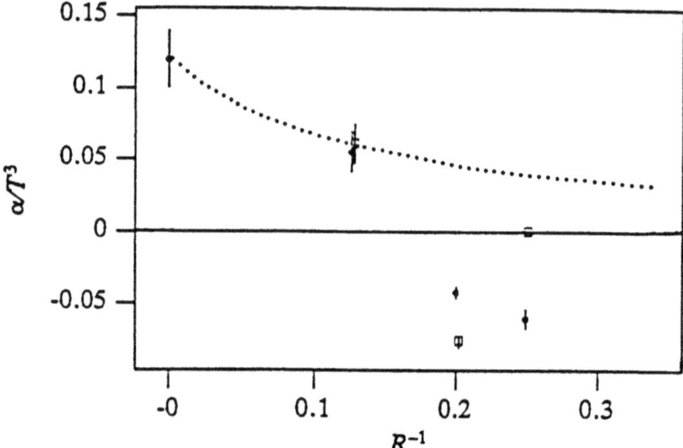

Fig. 3. Radius dependence of the surface tension on hadronic (filled circles) and gluonic (squares) droplets. The data point at $R^{-1}=0.$ corresponds to the wall surface tension (Ref. 4). The dashed curve represents the Tolman-Gibbs equation.

et al. have obtained 0.10(14) and $-0.37(80)$ (in units of T^3) for hadronic and gluonic droplets, values which are consistent with the flat interface studies[4-10] but also consistent with zero surface tension.

Huang et al.[12] have also computed the values of the curvature coefficient, by fitting their data to equation (1). In units of T_c^3 the fit gave a the following values for the curvature term C: $-0.051(9)$ and $-0.077(13)$ for hadronic and gluonic droplets respectively. There is (some) quantitative agreement between these results and those of reference [13] for the hadronic droplets but *not* for the gluonic droplets. One reason for the discrepency may be in the different extrapolation schemes used, particularly in the computation of the free energy itself.

We note that given the large values of the ratio w/R used so far, one should be suspicious of truncated series such as equation (1), especially when terms in powers of $1/R$ may be important. That the fits may be incorrect can be seen by looking at the fitted value of the wall surface tension (α^{fitted}), which in reference [12] is *twice* the value of α_{oo}, [4-10] instead of being the same.

4. Implications for the Swiss Cheese Instability

Mardor and Svetitsky have studied the surface tension in the Bag Model, [14] using equation (1) as an ansatz. They found a value of the curvature coefficient given by $C = \pm \left(\frac{4}{9}\right) T^2$ ($-$ for hadronic droplets and $+$ for gluonic droplets). Moreover, the flat wall surface tension α_{oo} in QCD with NO strange quarks turned out to be zero.

At least for hadronic droplets, the lattice results are consistent with the Bag model calculations at least qualitatively, in that the value of the wall surface tension is small on the scale of the strong interactions, and that the curvature coefficient is negative. Comparisons in the case of gluonic droplets may be premature at this time, given the discrepencies found in different lattice studies regarding the sign of the curvature coefficient (negative in reference [12], positive in reference [13]).

An interesting consequence of Mardor and Svetitsky's results is that the negative value of the curvature term leads to a secondary extremum at small R in the free energy thus generating the so-called "Swiss cheeze" instability. In this scenario, sub-critical hadronic droplets of a certain characteristic size are not destroyed by fluctuations as they would be in classical nucleation theory.[14] At temperatures smaller than T_c, the metastable vacuum would not consist of bulk quark-gluon plasma, but instead of plasma with 'droplets' of hadronic matter.[14] At temperatures larger than T_c, on the other hand, no such 'dirty' metastable hadronic state could exist since the curvature coefficient is positive. Some uncertainties concerning the viability of this scenario remain, however, especially regarding the contribution of $1/R$-terms to the free energy,[19] as well as the inherent (in)stability of these droplets.[20]

5. Acknowledgements

The authors would like to thank W. Janke, K. Kajantie, L. Karkkainen, F. Karsch, G. Lana, M. Ogilvie, K. Rummukainen, B. Svetitsky, T. Trappenberg and L. Viehland for stimulating discussions. The calculations performed by the authors were partly carried out at the Pittsburgh Supercomputing Center and at Boston University. J. P. thanks the Beaumont Professional Development Fund and Saint Louis University for financial support. C. R. acknowledges support from the U.S. Department of Energy (contracts DE-AC02-89ER40509 and DE-FC05-85ER250000).

1. "Lattice 89": Proceedings of the International Symposium on Lattice Gauge Theory, Capri, Italy, September 1989, N. Cabbibo *et al.* eds., *Nucl. Phys. B (Proc. Suppl.)* **17**, 223 (1990). "Lattice 90": Proceedings of the International Symposium on Lattice Gauge Theory, Florida State University, Tallahassee, October 1990, U. M. Heller *et al.* eds., *Nucl. Phys. B (Proc. Suppl.)* **20**, (1990); "Lattice 91": Proceedings of the International Symposium on Lattice Gauge Theory, September 1991, Iwasaki et al *et al.* eds., *Nucl. Phys. B (Proc. Suppl.)* in press.

2. L. D. Landau and E. M. Lifschitz, *Statistical Physics*, Pergamon Press (1980).

3. F.D. Abraham, *Homogeneous Nucleation Theory; The Pretransition Theory of Vapor Condensation*, Academic Press (New York, 1974)

4. S. Huang, J. Potvin, C. Rebbi, and S. Sanielevici, *Phys. Rev. D* **42**, 2864 (1990); *ibid* **43**, 2056 (1991) (Erratum).

5. R. Brower, S. Huang, J. Potvin and C. Rebbi; *Phys. Rev. D*, to appear.

6. K. Kajantie, L. Kärkkäinen, and K. Rummukainen, *Nucl. Phys. B* **333**, 100 (1990) and **357**,693 (1991).

7. M. Hackel, M. Faber, H. Markum, and M. Müeller, in "Nuclei in the Cosmos", Proceedings of the International Symposium on Nuclear Astrophysics, Munich, Germany, 1990, H. Oberhummer and W. Hillebrandt eds.; MPA/MP4, 44, Max-Plank-Institut (Munich, 1990);and Phy. Rev. D, to appear.

8. W. Janke, B. A. Berg and M. Katoot, *Nucl. Phys. B*, to appear.

9. B. Grossman, B. Laursen, T. Trappenberg and U. Wiese; private communication and: Proceedings of the "HLRZ Workshop on First Order Phase Transitions"; Juelich, June 1-3 1992, World Scientific, to appear.

10. N. A. Alves, Sao Paulo preprint 1992.

11. Z. Frei and A. Patkos, *Phys. Lett.*B **222**, 469 (1989).

12. S. Huang, J. Potvin and C. Rebbi; Proceedings of the "HLRZ Workshop on First Order Phase Transitions"; Juelich, June 1-3 1992, World Scientific, to appear.

13. K. Kajantie, L. Kärkkäinen, and K. Rummukainen, *Phys. Lett. B*; to appear.

14. I. Mardor and B. Svetitsky, *Phys. Rev. D* **44**, 878 (1991).

15. H. Gausterer, J. Potvin, C. Rebbi and S. Sanielevici, *Physica* A, to appear, and references therein.

16. J. Potvin and C. Rebbi, *Phys. Rev. Lett.* **62**, 3062 (1989).

17. S. Aoki et al. *Int. J. Mod. Phys. C* **2**, 829 (1991).

18. J. S. Rowlinson and B. Widom, *Molecular Theory of Capillarity*, Revised Edition, Clarendon Press, (Oxford, 1989).

19. K. Kajantie, J. Potvin and K. Rummukainen, Institute for Theoretical Physics/Santa Barbara, April 1992, NSF-ITP-92-55.

20. G. Lana and B. Svetitsky; Tel Aviv preprint, March 1992; see also these proceedings.

INSTABILITY OF BUBBLES
NEAR THE HADRON QUARK-GLUON-PLASMA PHASE TRANSITION

GIDEON LANA

and

BENJAMIN SVETITSKY

School of Physics and astronomy
Raymond and Beverly Sackler Faculty of Exact Sciences
Tel Aviv University ,69978 Tel Aviv, Israel

ABSTRACT

The small surface tension of the interface between hadronic and quark-gluon-plasma domains, along with a negative curvature tension, implies that the uniform plasma is unstable against spontaneous formation of hadronic bubbles. We furthermore show that spherical bubbles are not stable against small perturbations.

1. Introduction

In what follows, we shall adopt the common assumption that the quark-gluon-plasma phase turns into the hadron phase via a first order transition. In other words, that at some temperatures the two phases may coexist.

First order phase transition usually proceeds by nucleation of bubbles. Consider for example, boiling water at atmospheric pressure near the transition temperature $T_0 = 373K$. The free energy of a bubble of vapor in the water has a free energy which depends on its radius:

$$F(R) = \Delta P \,\frac{4}{3}\pi R^3 + \sigma\, 4\pi R^2 + \dots .\tag{1}$$

ΔP is the difference between the pressures of the two phases (zero at the transition temperature $T = T_0$, and positive above); σ is the excess of free energy per unit area for a planar interface.

$F(R)$ is shown schematically in figure 1. For $T < T_0$, both the volume and surface terms act to suppress bubble formation; any bubble formed by a fluctuation collapses to $R = 0$. For $T > T_0$, however, the volume term $\Delta P\, 4\pi R^3/3$ encourages growth of bubbles while the surface term $\sigma\, 4\pi R^2$ creates a barrier. The maximum of $F(R)$ occurs at some $R = R_c$. A bubble formed in the superheated liquid with $R < R_c$ will

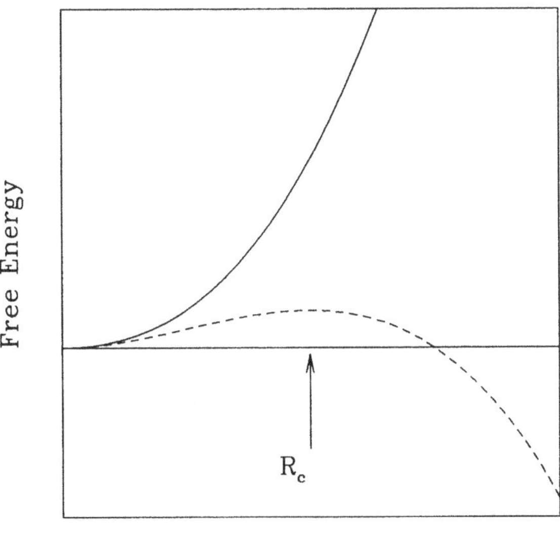

Figure 1: Schematic representation of bubble free energy in boiling water, retaining only volume and area terms, for $T < T_0$ (solid) and $T > T_0$ (dashed).

shrink away, but a bubble formed with $R > R_c$ will grow until the entire system is vapor.

In Equation 1 we have indicated that there might be more terms beyond the area term, having in mind an expansion in powers of $1/R$. In condensed-matter contexts, a 'curvature' term, $\alpha \cdot 8\pi R$ is clearly negligible; by dimensional analysis, $\alpha/\sigma \sim \lambda$, where λ is a microscopic scale, and $(\alpha R)/(\sigma R^2) \sim \lambda/R$, an exceedingly small number since the typical nucleating bubble is much larger than either the inter-molecular spacing or the correlation length which are usually of the order of angstroms. The picture in Figure 1, based on volume and area terms, is then the whole story.

Note that our picture will be substantially the same for liquid droplets in the vapor as for vapor bubbles in the liquid, since σ is the same in the two cases. This is because the surface tension may be measured in planar interfaces, and it doesn't matter which phase is on which side.

2. The QCD Transition

In QCD (at least in the limit of massless quarks) there is only one scale. Masses, radii, the transition temperature, the latent heat, etc. are all on the order of 200 MeV or 1 fm. This is then the only "microscopic" size, and 1 fm should also be the typical

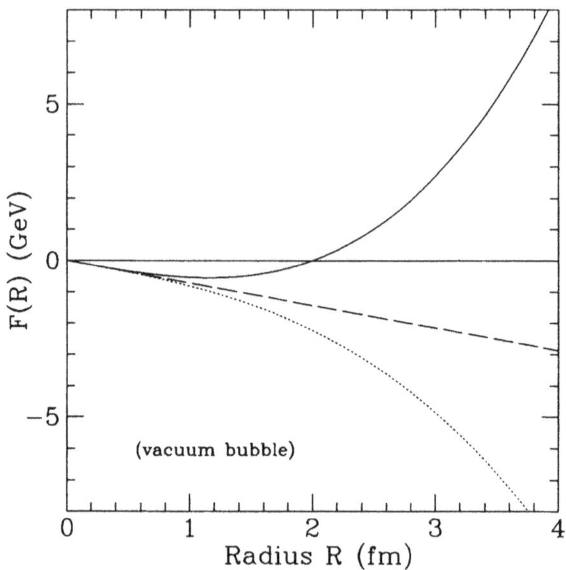

Figure 2: Free energy of a hadron gas bubble in the plasma for temperatures near the transition temperature, fixed at $T_0 = 150\ MeV$. $T = 155\ MeV$ (solid), $T = T_0 = 150\ MeV$ (dashed) and $T = 147\ MeV$ (dotted). The figures are taken from a bag model calculation of ref. 3.

size of a bubble, and one cannot ignore additional terms in (1). Indeed, Monte Carlo measurements [2,5,7], and MIT bag model calculations[1,3], found that for QCD σ/T_0^3 turn out to be a small number, and one is forced to include additional terms,

$$F(R) = \Delta P \frac{4}{3}\pi R^3 + \sigma\, 4\pi R^2 + \alpha\, 8\pi R + \dots .\tag{2}$$

Furthermore, it was found [3] that α is sizeable and negative. Fig. 2. depicts $F(R)$ for a hadron bubble within the quark gluon plasma. It is qualitatively different than that shown in fig. 1. Indeed, the minimum in the free energy for $T > T_0$ implies that vacuum bubbles are stable. The minimum of $F(R)$ is not a mere artefact of using the expansion (2); but an exact bag model calculation [3] exhibited a very similar behavior.

It was further suggested 3, 4 that this result might transcend the bag model: All that is required is a thin interface, the energy of which is due primarily to the reaction of the fields on either side. A lattice calculation of α in the pure glue theory [5] apparently supports this picture.

The fact that one can lower the free energy of the plasma phase by creating a vacuum bubble seems at first glance to imply that the equilibrium plasma is a Swiss cheese of bubbles, the density of which is determined by the (unknown) bubble–bubble interaction. This would only be true, however, if (1) the bubbles do not affect each other, and (2) each bubble is stable in itself. A simple calculation will teach us, however, that spherical bubbles are unstable against deformation, and that growth of long, narrow structures is favored. We do not address the problem of determining the actual equilibrium configuration of the high-temperature phase, a problem which will require much more sophisticated analysis.

In order to study the stability of the spherical bubble against small perturbations, we expand the free energy around the equilibrium spherical bubble, of radius R_0. Minimizing (2), we find that R_0 satisfies

$$4\pi\Delta P\,R_0^2 + 8\pi\sigma R_0 + 8\pi\alpha = 0 \ . \tag{3}$$

The generalization of (2) to non-spherical shapes is [6]

$$F = \Delta P\,V + \sigma\int dS + \alpha\int dS\left(\frac{1}{R_1} + \frac{1}{R_2}\right) + \cdots \ , \tag{4}$$

where the integrand in the third term is the local extrinsic curvature, with R_1 and R_2 representing the principal radii of curvature. Let the bubble surface be specified by

$$R(\theta,\phi) = R_0 + \delta R(\theta,\phi) \ . \tag{5}$$

The total $O(\delta R)$ contribution to F vanishes identically on the equilibrium radius $R = R_0$. The correction to the free energy is then second order and may be written as

$$\delta F = \int \sin\theta\, d\theta\, d\phi\, \delta R\left(A + BL^2\right)\delta R \ . \tag{6}$$

$-L^2$ is just the angular part of the Laplacian operator, and

$$A = \Delta P\,R_0 + \sigma \ , \qquad B = \frac{\sigma}{2} + \frac{\alpha}{R_0} \ . \tag{7}$$

Recall that we presume $\alpha < 0$. When R_0 is determined by (3), it is easy to see that A is positive and B negative.

Expanding δR in a multipole expansion,

$$\delta R(\theta,\phi) = \sum_{l,m} c_l^m Y_{lm}(\theta,\phi) \ , \tag{8}$$

we have

$$\delta F = \sum_{l,m}[A + Bl(l+1)]\,|c_l^m|^2 \ . \tag{9}$$

Evidently $\delta F > 0$ for a monopole perturbation, since (2) has been minimized; for a dipole perturbation we obtain $\delta F = 0$, consistent with the fact that $l = 1$ corresponds to a translation of the bubble. For quadrupole and higher deformations, however, δF is always negative, showing instability against arbitrary changes of shape.

3. Discussion

We have here considered only static properties of the quark–hadron interface. Dynamical processes in the course of the phase transition will destabilize the interface even more. For example, the necessity of carrying heat and baryon number away from a growing vacuum bubble will lead to a negative effective surface tension and hence to complex phenomena of pattern formation. On the other hand, interaction of bubbles, which was completely ignored in all the above treatments, may play a crucial role in stabilizing the bubbles' surface.

A word of caution: the above analysis relies on a bag model calculation[3] . And although Monte Carlo measurements confirmed the smalleness of σ as well as the sign and magnitude of α of the bag model calculation, it is still far from actually confirming even qualitatively the appearance of a minimum in $F(R)$ for a hadron bubble with finite radius surrounded by the quark gluon plasma, as depicted in fig. 2. As a matter of fact, the lattice results show that the interface is about 2fm thick, and it is clearly not negligible compared to the temperature or to R_0 , whereas the stability analysis is based on a thin wall approximation.

4. Acknowledgements

This work was supported by a Wolfson Research Award administered by the Israel Academy of Sciences and Humanities.

5. References

1. E. Farhi and R. L. Jaffe, Phys. Rev. D **30** (1984) 2379.

2. K. Kajantie and L. Kärkkäinen, Phys. Lett. B **214** (1988) 595; K. Kajantie, L. Kärkkäinen, and K. Rummukainen, Nucl. Phys. **B333** (1990) 100.
   S. Huang, J. Potvin, C. Rebbi, and S. Sanielevici, Phys. Rev. D **42** (1990) 2864; **43** (1991) 2056(E).
   R. Brower, S. Huang, J. Potvin, and C. Rebbi, Boston University preprint BUHEP-92-3 (unpublished, 1992).

3. I. Mardor and B. Svetitsky, Phys. Rev. D **44** (1991) 878.

4. B. Svetitsky, lecture given at the Workshop on QCD at Finite Temperature and Density, Brookhaven National Laboratory, August 1991, Tel Aviv preprint TAUP 1909–91 (1991), to appear in proceedings.

5. K. Kajantie, L. Kärkkäinen, and K. Rummukainen, Helsinki preprint HU-TFT-92-1 (unpublished, 1992).

6. R. Balian and C. Bloch, Ann. Phys. (NY) **60** (1970) 401.

7. J. Potvin, these Proceedings.

IDEAL GAS OF STRINGS AND QCD AT HADRONIC SCALES

CHUNG-I TAN

Physics Department, Brown University, Providence, RI 02912, USA

ABSTRACT

By using lessons learned from modern string studies, we show how interesting non-perturbative features of QCD can be learned from future heavy ion collisions even if the deconfinement density is not reached.

1. Introduction

It is the hope of many people that experimental studies of ultra-relativistic nuclear collisions in the near future could reach sufficiently energy densities so as to reveal the physics of deconfined quarks and gluons. In this talk, I would like to suggest, based on lessons learned from modern string studies,[1-3] that interesting non-perturbative features of QCD can already be learned even if the deconfinement energy density, $\mathcal{E}_d$, is not reached.

It is well understood that the character of QCD changes depending on the nature of available probes. At short distances, the basic degrees of freedom are quarks and gluons. As one moves to larger distance scales, the QCD coupling increases and one enters the non-perturbative regime. Short of resulting to lattice Monte Carlo studies, the most promising tool for a non-perturbative treatment of QCD which builds in naturally quark-gluon confinement remains the large-N expansion. In this scheme, although the vacuum of QCD at hadronic scales is complicated, model studies suggest that the effective degrees of freedom of QCD there can most profitably be expressed in terms of "extended objects". Indeed, low-lying hadron spectrum suggests that they can be understood as "string excitations". In high-energy soft hadronic collisions[4] where the interactions are mostly peripheral, it is possible to "see" the dominant excitations in terms of the exchanges of Regge trajectories.

We would like to suggest that, in heavy ion collisions, it is possible to probe the structure of string excitations inaccessible to other types of high energy hadronic collisions. This can be accomplished even if heavy ion collision experiments in the near future fail to reach past the deconfinement scale $\mathcal{E}_d$.

Assuming thermo-equilibrium can be achieved in hadronic multi-particle production, one expects that the energy density $\mathcal{E}$ can be parametrized monotonically in terms of a temperature, T. At low temperatures, $\mathcal{E}(T)$ can be given effectively in terms of pion gas. At high temperature and after deconfinement, one has Stefan-Boltzmann law, (appropriate for gluons and light quarks), which, for $SU(2)$ flavor, leads to $\mathcal{E} \simeq 12T^4$. Indeed, this expectation has been substantiated by lattice

322

calculations,[5] as depicted below. These "numerical experiments" suggest that the deconfinement density is of the order $\mathcal{E}_d \simeq 12T_H^4$, where $T_H \sim 200 MeV$. We would like to stress that they also indicate the existence of another density scale, $\mathcal{E}_H \simeq cT_H^4$, where $c = 0(1)$. We believe that, for $\mathcal{E}_H < \mathcal{E} < \mathcal{E}_d$ where the notion of temperature loses much of its usefulness, the effective degrees of freedom for QCD are "string-like". Probing this kinematical regime should reveal interesting non-perturbative features of confined QCD at hadronic scales.

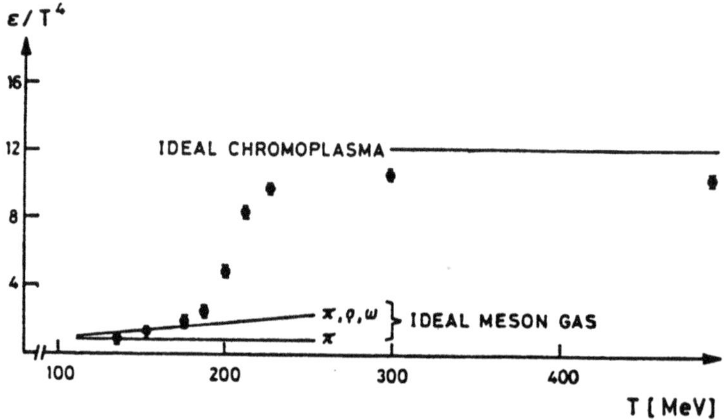

Fig. 1. The energy density in finite temperature QCD with Wilson fermions, taken from Ref. 5.

2. Counting Effective Degrees of Freedom at Hadronic Scales

The basic degrees of freedom of a string theory are string excitations. Each excitation can be given a particle attribute, *i.e.*, a mass m, with its center of mass energy p_0 and spatial momentum vector $\vec{p}$, $p_0^2 = m^2 + \vec{p}^2$. The mass is due to internal string oscillations; as such, it can take on increasing values, with a corresponding increase in degeneracies. For instance, the original dual model leads to an operator expression for the mass squared, $\alpha'm^2 = -1 + \sum_{i=1}^{3} \sum_{n=1}^{\infty} n\hat{N}_n^i$, where α' is the "Regge slope" and $\hat{N}_n^i$ is an ocillator number operator taking on eigenvalues $0, 1, 2. \cdots$. Well-known features of this model include: (i) equal-spacing rule for the mass spectrum, and (ii) exponentially increasing mass degeneracies, [*i.e.*, for $m^2 = \alpha'N$, the degeneracy $d(m) = \sum_{N_n^i=0}^{\infty} \delta(N + 1 - \sum_{i=1}^{3} \sum_{n=1}^{\infty} nN_n^i)$ increases exponentially with $\sqrt{N}$.]

Formulating a consistent *effective string theory* for QCD has been one of the major challenges for string theorists. Since we are still far from accomplishing this task, any insight into the problem, either theoretical or experimental, could prove to be useful. A common feature of all string-like theories is the rapid increase of mass degeneracies. We shall assume that the desired effective QCD string theory has an asymptotic exponential mass degeneracy, which charaterizes the growth of its effective degrees of freedom. Alternatively, one finds that the single-particle density behaves similarly at high energies

$$f(E, V) = V \sum_i d(m_i) \int \frac{d^3k}{h^3} \delta(E - \sqrt{k^2 + m_i}) \propto E^{-(\gamma+1)} e^{\beta_H E}. \tag{1}$$

That is, we shall characterize a string theory minimally by two critical exponents, γ and β_H, (the latter is often referred to as the inverse Hagedorn temperature.)

It can be demonstrated that, for systems which can be charaterized by Eq. (1), there always exists an energy density scale, $\mathcal{E}_H$, above which the conventional canonical description becomes inapplicable.[1-3] A microcanonical analysis can nevertheless be carried out which is valid both above and below $\mathcal{E}_H$. On the other hand, under an ideal gas approximation, the treatment is incapable of yielding the "deconfinement transition scale", $\mathcal{E}_d$, which could come about only when interactions are included. Since Monte Carlo experiments seem to indicate the existence of a sizable density interval, $[\mathcal{E}_H, \mathcal{E}_d]$, in which a large number of string states are excited, our subsequent analysis should be meaningful there.

3. Ideal Gas of Strings at High Energy Densities

For statistical systems with a finite number of fundamental degrees of freedom, it is well-known that a microcanonical and a canonical descriptions are equivalent. For strings, this equivalence breaks down at high energy densities. However, it turns out that the canonical partition function remains useful when considered as an analytic function of the inverse temperature, $\beta \equiv 1/T$.

The fundamental quantity in a canonical approach is the partition function: $Z(\beta, V) \equiv Tr e^{-\beta \hat{H}} = \sum_\alpha e^{-\beta E_\alpha}$, where the sum is over all possibe multiparticle states of the system. For a microcanonical approach, one works with a density function, which counts the number of microstates, $\Omega(E, V)dE \equiv \sum_\alpha \delta(E - E_\alpha) \, dE$. Statistical mechanics based on a microcanonical ensemble is more general, even though it is often more convenient to work with $Z(\beta, V)$, *e.g.*, when interactions must be included.

Representing the δ-function by an integral along an imaginary axis, we find that $\Omega(E, V) = \sum_\alpha \int_{-i\infty}^{+i\infty} \frac{d\beta}{2\pi i} e^{\beta(E - E_\alpha)}$. Note that the integral is in the form of an inverse Laplace transform over the complex-β plane. If one can deform the contour into a region where interchanging the order of sum and integral is allowed, one obtains

$$\Omega(E, V) = \int_{\beta_0 - i\infty}^{\beta_0 + i\infty} \frac{d\beta}{2\pi i} Z(\beta, V) \, e^{\beta E}. \tag{2}$$

The allowed region is labelled by the interception of the contour with the real axis,

β_0. One can then recover $Z(\beta, V)$ from $\Omega(E, V)$ via a Laplace transform, $Z(\beta, V) = \int_0^\infty dE\, \Omega(E, V)\, e^{-\beta E}$, which provides an alternative analytic definition for $Z(\beta, V)$.

For conventional systems, Eq. (2) can often be approximated by a saddle point contribution at $\hat\beta$. For a closed system where E is fixed, the usual notion of a temperature is given by $\hat\beta^{-1}(\mathcal{E})$, $\mathcal{E} \equiv E/V$, and is related to E by the stationary condition: $E = -\frac{\partial \log Z}{\partial \beta}|_{\hat\beta}$. We shall refer to this as the "Boltzmann temperature".

We have considered elsewhere, in a quantum statistical treatment, analytic property of $Z(\beta, V)$ in β for various string theories.[1] We find that, generically, $Z(\beta, V)$ is analytic for $Re\,\beta > 0$ except at isolated points. For each string theory, because of the exponential growth in mass degeneracy, there is always an isolated rightmost singularity at $\beta = \beta_H$, i.e., the inverse Hagedorn temperature for that theory. There is a finite gap in their real parts between β_H and the next singularity to the left, and this gap is theory-dependent but calculable.

To be more specific, for an ideal gas of strings under the Maxwell-Boltzmann (MB) statistics, one has $Z(\beta, V) \simeq e^{\tilde{f}(\beta, V)}$, where $\tilde{f}(\beta, V)$ is the inverse Laplace transform of the single-particle density, $f(E, V)$. It can be shown that $\tilde{f}(\beta, V)$ is analytic for $Re\,\beta$ sufficiently large and its rightmost singularity is at $\beta = \beta_H$. Given $Z(\beta, V)$ as an analytic function of β, $\Omega(E, V)$ can be recovered throught Eq. (2), with $\beta_0 > \beta_H$. That is, *the totality of physics of microcanonical approach for free strings has been entirely encoded in the analyticity of $Z(\beta, V)$.*[1]

For strings, at low energies, Eq. (2) can be saturated by a saddle point at $\hat\beta(E)$, lying to the right of β_H. However, as the energy density $\mathcal{E}$ is raised, one reaches a point where either the saddle point moves to the left of β_H, or it gets close to β_H that the fluctuations about the saddle point become large. When this occurs, the saddle point approximation to Eq. (2) break down, and it defines the lower density scale $\mathcal{E}_H$ for the region of interest to heavy ion collisions spelled out earlier.

For $\mathcal{E} > \mathcal{E}_H$, whereas it is no longer meaningful to speak of a Boltzmann temperature, the statistical mechanics of free strings is still given unambiguously by Eq. (2). One can in fact push the contour in (2) to the left of the singular point, β_H, by a finite distance η, $\eta > 0$. As one moves past this point, one picks up an additional contribution involving the discontinuity across the cut. Denoting the discontinuity by $\Delta Z(\beta, V), \beta < \beta_H$, the large-$E$ behavior of $\Omega(E, V)$ is dominated by the singularity at β_H

$$\Omega(E, V) = - \int_{\beta_H - \eta}^{\beta_H} \frac{d\beta}{2\pi i} \Delta Z(\beta, V) e^{\beta E} + 0(e^{(\beta_H - \eta)E}), \qquad \eta > 0. \qquad (3)$$

Once $\Delta Z(\beta, V)$ is known, the dominant behavior of $\Omega(E, V)$ can be found. Therefore, the large-E limit of a free-string theory can best be approached by working first with the canonical quantity, $Z(\beta, V)$.[1,2] (Note that earlier related works of Hagedorn and co-workers[5] seem to have concentrated on the approach to the region $\mathcal{E} \sim \mathcal{E}_H$ from below. Our analysis, on the other hand, emphasizes on the region above $\mathcal{E}_H$.)

4. Critical Exponent γ and Classifications

We shall classify different types of effective string theories for confined QCD according by the critical exponent γ given in Eq. (1). In standard string theories, 2γ is equal to the number of "uncompactified" spatial dimensions. Therefore, the canonical value is $\gamma = 3/2$. However, since we are dealing with an effective string theory for QCD, we could not rule out the possibility that this critical index can take on an anomalous value. Indeed, as we suggest below, heavy ion collisions offer the unique possibility of measuring this critical exponent. This could in turn provide useful hint in our search for the realistic effective string theory for QCD.

It can be shown that, for $\gamma > 0$, $\tilde{f}(\beta)$ has a branch point at β_H but it is bounded. For $\gamma < 0$, $\tilde{f}(\beta)$ has a divergent algebraic branch point. Finally, when $\gamma = 0$, $\tilde{f}(\beta)$ has a logarithmic branch point at β_H. Model string theories have been constructed for both cases where $\gamma > 0$ and for $\gamma = 0$. Instead of providing an exhaustive analysis, we shall assume below that $\gamma \geq 0$. For $\gamma > 0$, $\tilde{f}(\beta)$ can be parametrized as $\tilde{f}(\beta) \sim g(\beta)(\beta - \beta_H)^\gamma + \lambda(\beta)$, where $\lambda(\beta)$ is regular at β_H. It follows that $Z(\beta)$ has the same type of singularity at β_H, i.e., $\Delta Z(\beta) \sim (\beta - \beta_H)^\gamma$. For $\gamma = 0$, one has $\tilde{f}(\beta) \sim -c \log(\beta - \beta_H) + \lambda(\beta)$. We shall assume below that $c = 1$, as suggested in the study of fundamental strings.[1-3]

The detailed statistical properties of a string gas turns out to be sensitive to the value of γ. This can best be brought out by studying the "inclusive distributions" for our ideal gas of strings. For instance, the single-string distribution with a definite energy ϵ under the MB statistics can be expressed as

$$\mathcal{D}(\epsilon; E) = \Omega(E)^{-1} f(\epsilon)[\Omega(E - \epsilon) + \delta(E - \epsilon)]. \tag{4}$$

The distribution $\mathcal{D}(\epsilon; E)$ is normalized so that, upon integration over ϵ, it yields the average number of strings, $\langle N \rangle$, in an ensemble with a total energy E, which can be more easily measured experimentally.

Exhaustive studies of this type can be found in Ref. 2, where we make use of the analytic properties of $\tilde{f}(\beta)$ and $Z(\beta)$. Here we shall mention several interesting features. For an ordinary gas of particles, $\epsilon \mathcal{D}(\epsilon, E)$ as a function of ϵ is typically $\epsilon^\alpha e^{-\beta \epsilon}$, i.e., the usual Boltzmann distribution. In the case of string gas, for $\gamma \geq 3/2$, we find that the distribution is peaked at the two ends [Fig. 2(a)] with a single energetic string soaking up most of the energy. This feature has been referred to in the past as the Frautschi-Carlitz picture. In particular, one finds that $\langle N \rangle$ becomes energy-independent at large E. On the other hand, for $\gamma = 0$, we have a scale-invariant energy distribution, with strings of all energies contributing equally to the total energy, *i.e.*, a plateau distribution, Fig. 2(b). One finds that $\langle N \rangle \sim \log E$, due to the $1/\epsilon$ tail of the distribution at large energies. Experimental detection of this unique signature would be most interesting. (For $0 < \gamma < 3/2$, see Ref. 2.)

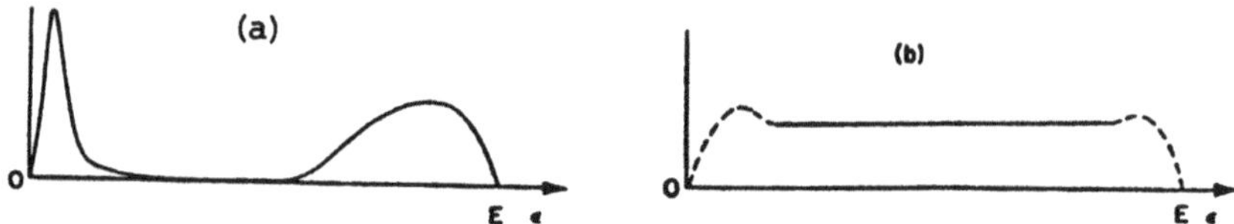

Fig. 2. Schematic plots of $\epsilon \mathcal{D}(\epsilon, E)$ for a string gas with (a): $\gamma \geq 3/2$, and (b): $\gamma = 0$.

5. Comments

While the free string gas already exhibits novel properties, it is of great interest to study the interacting gas. This could lead further insights for describing the onset of deconfinement, a feature which is absent in an ideal gas setting. Finally, I have emphasized here on how lessons learned from modern string studies could be used to study the physics of quarks and gluons at confinement scales. It is of course also true that insights from non-perturbative QCD studies, both theoretical and experimental, can also provide new hints for works on fundamental strings. Increasing future collaborative efforts in this direction could prove beneficial for both communities.

6. References

1. Nivedita Deo, Sanjay Jain and Chung-I Tan, *Phys. Lett.* **B 220**, 125 (1989).

2. Nivedita Deo, Sanjay Jain and Chung-I Tan, *Phys. Rev.* **D40**, 2626 (1989); *ibid.* **D45**, 3641 (1992).

3. J. Atick and E. Witten, *Nucl. Phys.* **B310**, 291 (1988); R. Brandenberger and C. Vafa, *ibid.* **B316**, 391 (1988); M. J. Bowick and S. Giddings, *ibid.* **B325**, 631 (1989); P. Salomonson and B. S. Skagerstam, *ibid.* **B268**, 349 (1986); N. Turok, *Physica* **A158**, 516 (1989); Nivedita Deo, Sanjay Jain and Chung-I Tan, "The Ideal Gas of Strings", in Proceedings of the International Colloquium on Modern Quantum Field Theory, Bombay, 1990 (World Scientific) and references therein.

4. A. Capella, U. Sukhatme, C-I Tan, and Tran T. V., "Dual Parton Model", Orsay preprint, LPTHE 92-38, and references therein.

5. H. Satz, *Fortschr. Phys.* **33** (1985) 4, 259-268, (in honor of R. Hagedorn).

COLLECTIVE PION EMISSION AND QCD VACUUM STRUCTURE

ANDRE KRZYWICKI [1]

Laboratoire de Physique Théorique et Hautes Energies, Bât. 211
Université de Paris-Sud, F-91405 Orsay, France[2]

ABSTRACT

Soft-pion emission in heavy-ion collisions is discussed using the semi-classical
approximation and an effective Lagrangian describing the excitations of the
physical QCD vacuum.

1. Introduction

I shall report about a work done in collaboration with Jean-Paul Blaizot [1]. I shall
not enter into technical details, which can be found in the original publication, trying
to give you a general idea about our motivations and results. Related considerations
can be found in the papers by Anselm and Ryskin [2] and by Bjorken [3].

There exist two extreme approximations which enable one to find relatively easily
some observable consequences of a given quantum field theory. The first consists
in treating the interaction as a perturbation. In QCD the perturbative calculations
have grown into a flourishing industry. The second approximation is the semi-classical
one. In QCD its use appears to be a bad strategy, at least when one seeks results
relevant for phenomenology. Indeed, the colour confinement and hadronization are
intrinsically quantum phenomena.

We nevertheless propose to employ the semi-classical approximation, but we suggest to combine it with the use of an effective Lagrangian [4] describing the observable
excitations of the QCD vacuum. The scope of this approach is limited to soft-pion
emission, but the advantage is that one deals directly with hadrons and the confinement problem is never explicitly met.

Our work can be relevant for the physics of heavy-ion collisions. At high energy
the rapidity densities of produced particles (mainly pions) can be as large as 10^3 and
the use of the semi-classical approximation is perhaps justified.

---

[1] E-mail: krz@qcd.circe.fr
[2] Laboratoire associé au C.N.R.S.

328

2. Discussion

In first approximation, the effective Lagrangian is that of the non-linear sigma model

$$L = \frac{f_\pi^2}{2}[(\partial\sigma)^2 + (\partial\vec{\pi})^2] \tag{1}$$

where $f_\pi = 93$ MeV is the pion decay constant and the fields satify the constraint

$$\sigma^2 + \vec{\pi}^2 = 1 \tag{2}$$

The equations of motion are equivalent to the following current conservation equations:

$$\begin{aligned}
\partial \cdot \vec{V} &= 0 \\
\partial \cdot \vec{A} &= 0
\end{aligned} \tag{3}$$

where

$$\begin{aligned}
\vec{V}_\mu &= \vec{\pi} \times \partial_\mu \vec{\pi} \\
\vec{A}_\mu &= \vec{\pi}\partial_\mu\sigma - \sigma\partial_\mu\vec{\pi}
\end{aligned} \tag{4}$$

are the Noether iso-vector and iso-axial currents associated with the global $SU(2) \times SU(2)$ symmetry of the theory.

Our work is inspired by Heisenberg's old papers [5] on the semi-classical description of multiparticle production. In particular, we adopt his idealized boundary condition: We assume that at time $t = 0$, the whole energy of the collision is localized within an infinitesimally thin slab with infinite transverse extent (instead of a pancake shaped region). The symmetry of the problem then implies that the pion field depends only on the invariant $s = t^2 - x^2$, where x is the longitudinal coordinate.

Thus, the theory becomes effectively $1 + 1$ dimensional, the currents (4) take the form $f(s)x_\mu$, $\mu, \nu = 0, 1$, and the current conservation implies that the function $f(s) = s^{-1}$ up to a multiplicative constant. Consequently

$$\vec{\pi} \times \vec{\pi}' = \frac{\vec{a}}{s} \tag{5}$$

and

$$\vec{\pi}\sigma' - \sigma\vec{\pi}' = \frac{\vec{b}}{s} \tag{6}$$

where $\vec{a}$ and $\vec{b}$ are integration constants characterizing the strength and direction in iso-space of the currents (4) in the state produced during the collision process. It is a strong assumption intrinsic to our approach that such classical currents do emerge soon after the beginning of this process.

After some algebra one finds that the fields obey the following linear equations :

$$(s\sigma')' + \frac{\kappa^2}{s}\sigma = 0 \tag{7}$$

and

$$(s\vec{\pi}')' + \frac{\kappa^2}{s}\vec{\pi} = 0 \tag{8}$$

where $\kappa = \sqrt{a^2 + b^2}$. Decomposing $\vec{\pi}$ along the axes of the triad $\vec{a}$, $\vec{b}$ and $\vec{c} = \vec{a} \times \vec{b}$ one finds the solution

$$\begin{aligned}
\pi_a &= 0 \\
\pi_b &= -\sin[\kappa \log \frac{s}{s_0}] \\
\pi_c &= \frac{a}{\kappa} \cos[\kappa \log \frac{s}{s_0}]
\end{aligned} \tag{9}$$

where s_0 is a scale parameter. For $s < s_0$ the field oscillates dramatically and the model cannot be trusted, since the gradient expansion of the effective Lagrangian does not make much sense.

It is instructive to compare equation (8) with the Klein-Gordon equation written assuming that the field depends on s only. One then finds $m_\pi^2/4$ in the place of κ^2/s, which suggests the following estimate for pion's proper formation time

$$\tau \approx \frac{2\kappa}{m_\pi} \tag{10}$$

The formation time cannot be smaller than the Compton wave-length of the particle. Thus, the self-consistency of the model requires that κ is of the order of unity, in any case not much smaller. For physical reasons, we conjecture that s_0 is of the order of 1 fm^2, or so.

This completes the presentation of the model. Let me now discuss a sample of results that can be obtained using the solution (9):

The energy-momentum/unit $\perp$ area

The energy-momentum tensor corresponding to the $1 + 1$ dimensional motion is

$$T_{\mu\nu} = \frac{2f_\pi^2\kappa^2}{s}(2u_\mu u_\nu - g_{\mu\nu}) \quad \mu, \nu = 0, 1 \tag{11}$$

Here, $u_\mu = x_\mu/\sqrt{s}$ is the covariant velocity of a covolume element with coordinate x_μ. Remarkably enough, (11) describes a relativistic fluid. Considered in $1 + 3$

dimensions, this fluid is however not in thermal equilibrium : there is no transverse pressure whatsoever.

The invariant energy density

One reads from (11) that the invariant energy density ϵ is

$$\epsilon = \frac{2f_\pi^2 \kappa^2}{s},\tag{12}$$

The energy density in the initial dense pion system is obtained from the above by setting $s = s_0$ (κ will be related to the soft-pion spectrum in a moment). Equation (12) can be used to obtain an alternative estimate of the formation time. Assuming that a pion is produced when ϵ becomes comparable to the pion mass divided by the cube of pion's Compton wave-length one gets

$$\tau \approx \frac{\sqrt{2}f_\pi \kappa}{m_\pi^2}\tag{13}$$

Within a factor of 2 this is the same as (10).

The soft-pion spectrum/unit $\perp$ area

For massless pions, the one-particle inclusive spectrum obtained from (9) reads

$$dN = \frac{f_\pi^2 \kappa}{\tanh(\pi\kappa)}\frac{dk}{\omega}\tag{14}$$

where $\omega =| k |$. For massive pions one should replace $dk/\omega \to dy$, y denoting the rapidity. It is amusing to note that the soft-pion emission is not accompanied by production of lepton pairs. The electro-magnetic current is proportional to the 3rd component of the isovector current and is a pure gradient. Thus the coupling to a photon with momentum k^μ is proportional to k^μ (this also follows from symmetry arguments: with our idealized boundary conditions, there is no other vector disponible). Thus, in Feynman's gauge, the amplitude for production of a pair, where the electron (positron) has momentum p (q), is proportional to $\overline{w}(p)\gamma_\mu k^\mu v(q)$, with $k = p + q$, and is zero by virtue of the Dirac equation.

The neutral-to-charged ratio

If there is no privileged direction in isospace, all orientations of the triad $[\vec{a}, \vec{b}, \vec{c}]$ are equally probable. Thus, for given κ, all the orientations of the pion field are also equiprobable. Let neutral pions constitute the fraction r of all soft pions produced in a collision. Then, the random variable r is distributed according to the law

$$dP(r) = \frac{1}{2\sqrt{r}}dr\tag{15}$$

Large fluctuations of the neutral/charged ratio have also been predicted in refs. 2 and 3.

3. Conclusion

The work reviewed in this talk is an exploratory one. We made an effort to simplify the problem as much as possible in order to get analytic results and achieve a maximum of clarity. One can improve the discussion in many respects, for example by introducing more realistic boundary conditions, by including higher-derivative terms in the effective Lagrangian, by considering $SU(3) \times SU(3)$ etc.

The weak point of the whole approach is that, as yet, we are unable to predict the cross-sections. As already mentioned, the self-consistency of the whole picture requires that κ is not too small. This means that the currents should be strong enough. But what is the probability of creating these rather strong classical iso-vector and iso-axial currents we are talking about? An answer to this question requires a better understanding of the dynamics operating at the very early stage of a heavy-ion collision, at times less than 1 fm/c say. Here, some new ideas are needed.

The approach could also receive a boost if some of the signatures of the coherent pion production were observed, like the large fluctuations of the neutral-to-charged ratio (maybe the so-called Centauro events observed in cosmic ray interactions can be interpreted that way). If the corresponding cross-sections are not too small, then the coherent emission of soft pions could perhaps become a new interesting playground for non-perturbative QCD.

References

1. J.-P. Blaizot and A. Krzywicki, *Phys. Rev.* **D46** in press

2. A.A. Anselm and M.G. Ryskin, *Phys. Lett.* **B266** (1991) 482.

3. J.D. Bjorken, preprints SLAC-PUB-5545 (May 1991) and SLAC-PUB-5673 (October 1991).

4. Classical references are: S. Weinberg, *Physica* **A96** (1979) 327; J. Gasser and H. Leutwyler, *Ann. Phys.* **158** (1984) 142, *Nucl. Phys.* **B250** (1985) 465 and 517. Explicit derivation of the effective Lagrangian starting with QCD has been discussed, for example, in: R.I. Nepomechie, *Ann. Phys. (NY)* **158** (1984) 67; J. Balog, *Phys.Lett.* **B149** (1984) 197; A.A. Andrianov, *Phys. Lett.* **B157** (1985) 425; P. Simic, *Phys.Rev.* **D34**(1986) 1903; J. Zuk, *Z. Phys.* **C29** (1985) 303; D. Espriu, E. de Rafael and J. Taron *Nucl. Phys.* **B345** (1990) 22.

5. W. Heisenberg, *Z. Phys.* **133** (1952) 65, and references therein.

TOP AND HIGGS MASSES FROM DYNAMICAL ELECTROWEAK SYMMETRY BREAKING

DAVID E. KAHANA
Center for Nuclear Research
Department of Physics
Kent State University
Kent, OH 44240 (USA)

ABSTRACT

The Standard model of electroweak interactions, with the gauge and Higgs bosons appearing as composites, is derived from a Nambu–Jona-Lasinio–type four-fermion interaction, assumed to be valid above a high scale μ. Simple relationships are found for the composite boson to top quark mass ratios and for the weak angle: $m_W^2 = 3/8\, m_t^2$, $m_H^2 = 4\, m_t^2$ and $\sin^2\theta_W = 3/8$ as in grand unified theories. Assuming three generations and a 'desert' hypothesis, these relationships are evolved with the full renormalisation group down to present experimental energies, yielding predictions for the top quark and Higgs-boson masses, near 155 GeV for the former and near 140 GeV for the latter. In this fashion, fermion-antifermion condensates can be shown to yield a top mass consistent with that indicated from electroweak loop corrections for LEP data.

1. Introduction

Recently, in very interesting works, Nambu[1], as well as Bardeen, Hill and Lindner[2], and others[3], have suggested replacing the Higgs mechanism[4] in the Standard Model[5] with a dynamical symmetry breaking approach generated by four-fermion interactions of the top quark. In fact, their model for the Higgs sector is that of Nambu and Jona-Lasinio (NJL)[6], and one may recover the usual Higgs meson as a composite of (t) and $(\bar{t})$. Several authors, notably Eguchi[7], Terazawa, Chikashige and Akama[8], and recently Bando, Kugo and Yamawaki[9], Suzuki[10] and others[11,12], have also treated vector mesons as composites within the NJL framework; perhaps the earliest work in this direction is by Bjorken[13], for QED. Here we also attempt to generate the entire electro–weak interaction from a specific, baryon number conserving, current–current choice of four–fermion Lagrangian. The W, Z and Higgs appear as coherent composites of all quarks and leptons, not just the heaviest quark. There are clear problems with such an approach which will become evident, but there are also possible rewards.

The effective four fermion theory used here is assumed to be the correct interaction at some high scale μ, and presumably is the low energy limit of some more basic theory. Subsequently, we evaluate the single fermion loops of the four-fermion theory to obtain an effective action for composites using a cutoff Λ, which is necessary because the theory contains the usual ultraviolet divergences of perturbation theory. This action is leading order in the $1/N$ expansion of the model. After fine-tuning of the vector and scalar couplings, quadratic and higher divergences are eliminated from the theory leaving only logarithmic divergences which can be handled by standard renormalisation techniques, and the action is rendered finite to leading order in $1/N$. The cutoff Λ must be retained, however, since if it is not the theory apparently becomes trivial. We find that the meson sector of the theory induced at leading order in $1/N$ by the NJL interaction can under certain restrictive conditions be cast into the form of the $SU_L(2) \times U_R(1)$ Standard model. Simple rational values are then obtained for the weak angle and meson to top quark mass ratios. It should be pointed out that these values are essentially independent of the cutoff Λ for large Λ, being ratios of logarithmically divergent quantities. The simplicity of the values for the weak angle and mass relationships suggests that they are, as in grand unified theories, un-renormalised values correct at a high mass scale μ, and this is indeed how we treat them. Below the scale μ the effects of the extended $SU(3) \times SU(2) \times U(1)$ interactions are included through the usual renormalisation group equations. This parallels the method of reference[2]; they differ in performing a matching at the scale Λ, where the boson wave function renormalisations vanish, through compositeness conditions.

As stated above, the tree level (bare) masses and mixing angles for the composite particles are specified by the 4–fermion theory, for example, $\sin^2\theta_W$ is found to be a simple rational number, reminiscent of grand unified models while the W and Z masses are similarly related to the top quark mass[8]. Several authors, in particular Marciano[14] and Nambu[1] have pointed out that one mechanism probably generates mass for both the top quark and the massive vector bosons, since the W, Z and top all seem to lie at more or less the same scale. Since we show that the standard electro–weak theory is recovered at the tree level, the electro–weak parameters may be taken as boundary conditions[2] for the standard model renormalisation group equations[14] at μ, and evolved downward to yield a determination of μ and a prediction of the ratios $\frac{m_t}{M_W}$ and $\frac{M_H}{m_t}$ at present experimental energies.

2. The Model

The NJL model we employ is defined by:

$$L = \bar\psi i\partial\!\!\!/\psi - \tfrac{1}{2}\left[(\bar\psi G_S\psi)^2 - (\bar\psi G_s\tau\gamma_5\psi)^2\right] - \tfrac{1}{2}G_B^2(\bar\psi\gamma_\mu Y\psi)^2 - \tfrac{1}{2}G_W^2(\bar\psi\gamma_\mu\tau P_L\psi)^2. \tag{2.1}$$

Here $\psi = [\psi_i]$ is a column vector: $i = \{t_L, t_R, b_L, b_R, e_L, e_R, \nu_L\}$, and this pattern is of course duplicated for each generation of fermions that is present. G_B and G_W are universal couplings for the isoscalar and isovector interactions respectively, while Y and τ are defined in the usual fashion. $P_L = \tfrac{1}{2}(1 - \gamma_5)$ is the projector on left-handed fermions. Breaking of weak isospin symmetry is introduced via the scalar coupling G_S, which is taken to be a diagonal matrix; we could of course incorporate Cabibbo-like mixings between the generations by allowing G_S also to have off-diagonal elements, but this is an added complication which we omit for now. We are also employing an Euclidean metric with $\gamma_\mu{}^2 = -1$. The choice made in equation

334

(2.1) would seem to be unique if the standard model for the quark–boson interactions is to be recovered, and could be derived for instance by the introduction of very massive vector bosons $\mathbf{W}^k_\mu$, $\mathbf{B}_\mu$ at some very high mass scale. The usual W and Z then arise from the lagrangian in Eq.(2.1), as much lighter composite images. The couplings G_S, G_B, G_W have dimensions $[L]$ as required by the four fermion coupling in $d = 4$, and are later taken to be inversely proportional to the cutoff scale Λ. As in reference[16] we construct the meson lagrangian using the methods of Gross and Neveu[17]. Auxiliary fields $F \equiv [\sigma, \pi^i, B_\mu, W^k_\mu]$ are introduced into the partition function:

$$Z(J) = \int \mathrm{D}[F] \int \mathrm{D}\psi\,\mathrm{D}\bar\psi\,\exp\left\{-\int \mathrm{d}^4x\,(L + L_{cl}(F)) + J[\bar\psi\psi]\right\},$$
$$L_{cl} = \tfrac{1}{2}(\sigma^2 + (\pi^i)^2) + \tfrac{1}{2}(B_\mu^2 + (W_\mu^k)^2) + JF \tag{2.2}$$

Shifting the fields according to:

$$\sigma = \sigma' + \bar\psi G_S \psi + J_S/G_S, \qquad B_\mu = B'_\mu + G_B \bar\psi\gamma_\mu Y\psi + J_B/G_B,$$
$$\pi = \pi' + i\bar\psi G_S\gamma_5\psi + J_\pi/G_S, \quad W^k_\mu = W'^k_\mu + G_W \bar\psi\gamma_\mu\tau^k P_L\psi + J_W/G_W \tag{2.3}$$

then cancels the four fermion interactions in favour of a coupling between the auxiliary fields and the fermions. The partition function is then essentially that for the coupled meson–fermion theory but lacking boson kinetic terms. We note that at this level gauge invariance is broken only by the quadratic classical terms L_{cl} for the auxiliary fields. The lagrangian for the shifted fields (in which primes may be dropped) then reads:

$$L' = \bar\psi\Delta\psi + \tfrac{1}{2}(\sigma^2 + (\pi^i)^2) + \tfrac{1}{2}B_\mu^2 + \tfrac{1}{2}(W_\mu^k)^2,$$
$$\Delta = i\gamma_\mu D_\mu + G_S(\sigma + i\pi \cdot \tau\gamma_5) \tag{2.4}$$

where D_μ is the covariant derivative:

$$D_\mu = \partial_\mu - iG_B B_\mu Y - iG_W W_\mu \cdot \tau P_L \tag{2.5}$$

The composite bosons, their propagators and their interactions are obtained by examining the one fermion loop effective action:

$$I_{eff} = I_{class} - (\mathrm{Tr}\,\ln\Delta - \mathrm{Tr}\,\ln\Delta_0)$$
$$= I_{class} - \mathrm{Tr}\,\ln\left[1 + \frac{1}{i\slashed{\partial} + M}(\tilde\Omega_S + \Omega_B + \Omega_W)\right] \tag{2.6}$$

in which:

$$\Omega_S = \Sigma + i\Pi \cdot \tau\gamma_5 = M + (\tilde\Sigma + i\Pi \cdot \tau\gamma_5), \qquad \tilde\Omega_S = \tilde\Sigma + i\Pi \cdot \tau\gamma_5$$
$$\Omega_B = G_B \slashed{B}Y$$
$$\Omega_W = G_W \slashed{W} \cdot \tau P_L \tag{2.7}$$

We have redefined scalar fields with more normal dimensions (L^{-1}) by absorbing the coupling G_S:

$$\Sigma = G_S\sigma, \qquad \Pi^k = G_S\pi^k \tag{2.8}$$

and introduced a vacuum expectation value through $\Sigma = M + \tilde{\Sigma}$. The expansion of the action in 2.6 uses the scalar meson condensate $\langle \sigma \rangle = \sigma_0$, $G_S \sigma_0 = M$; then $\Delta_0 = i\not{\partial} + M$ is the Dirac operator for massive quarks propagating in the condensed vacuum.

The model so far is a standard fermion–boson theory, manifestly gauge invariant in the fermion–boson sector, but lacking kinetic terms for both scalar and vector bosons. The classical 'hard' mass terms in Eq.(2.6) seem to violate gauge invariance at this point, but we will deal with this matter later.

3. The Boson Action

The masses and kinetic energies of all mesons may be generated naturally from Eq.(2.6). For example in the purely scalar sector, eliminating linear terms in the fluctuation of σ around the condensate leads to the (NJL) gap equation:

$$\text{Tr}\left[\frac{1}{i\not{\partial} + M}G_S\right] = \sigma_0 \tag{3.1}$$

and the second order action in Ω_S contains the kinetic energies and masses for the composite scalar:

$$\begin{aligned}
I_S &= \tfrac{1}{2}\text{Tr}\left[\frac{1}{i\not{\partial} + M}\tilde{\Sigma}\frac{1}{i\not{\partial} + M}\tilde{\Sigma}\right] + \tfrac{1}{2}\tilde{\sigma}^2 \\
&= \tfrac{1}{2}\text{Tr}\left\{\left[\frac{\partial^2}{(-\partial^2 + M^2)^2} + \frac{M^2}{(-\partial^2 + M^2)^2}\right]\tilde{\Sigma}^2 + \tfrac{1}{2}\frac{\partial^2}{(-\partial^2 + M^2)^2}(\partial_\mu\tilde{\Sigma})^2 + \cdots\right\}.
\end{aligned} \tag{3.2}$$

Third and fourth order terms would complete the rest of the standard Higgs lagrangian. In equation (3.2) we have dropped terms with more than two derivatives of the fields, and we will follow this procedure throughout. Performing the trace in 3.1 yields

$$\frac{1}{4\pi^2}\sum_i G_i^2 \int^\Lambda \frac{k^2 dk^2}{(k^2 + m_i^2)^2} = 1. \tag{3.3}$$

Combining (3.3) with (3.2) then cancels the 'hard' (proportional to Λ^2) mass terms in 3.2 for the scalar field, leaving only masses which vanish in the limit of no dynamical symmetry breaking, that is, when $\sigma_0 = 0$. Assuming that the top quark dominates, $m_t^2 >> m_i^2$, (3.3) becomes a condition on the top quark coupling[1,2]:

$$G_t^2 = \frac{4\pi^2}{n_c\left(\Lambda^2 - m_t^2\ln\left(\frac{\Lambda^2}{m_t^2} + 1\right)\right)}, \tag{3.4}$$

with the colour degeneracy $n_c = 3$. This fine tuning eliminates the Λ^2 singularities from the (NJL) theory. A similar fine tuning, or consistency condition, will need to be imposed in the vector boson sector to restore conservation of the $SU(2) \times U(1)$ currents. The kinetic terms for the Higgs may be put into normal form by wavefunction renormalisation:

$$\sigma' = Z_S^{1/2}\sigma; \qquad \sigma_0' \equiv v = Z_S^{1/2}\sigma_0,$$

$$Z_S^{1/2} = \tfrac{1}{2}\text{Tr}\left[\left(\frac{1}{-\partial^2 + M^2}\right)^2 G_S^2\right]. \tag{3.5}$$

A dimensionless scalar coupling $g_S = G_S Z_S^{-1/2}$ then replaces G_S in the action: except in the gap equation G_S no longer appears. Higher loops require only standard renormalisation of g_S; only logarithmic divergences are left after fine tuning of G_S. This procedure is not without ambiguity[2]; we take the approach in this paper that neither the couplings nor the masses can be unambiguously calculated. This will not however prevent us from making at least suggestive predictions, as the mass and coupling constant *ratios* should be less uncertain. Assuming a conventional normalisation for the scalar kinetic terms we find the standard NJL expression for the mass of the Higgs scalar:

$$m_S^2(\mu) = \frac{\text{tr}\left[\frac{2M^2}{-\partial^2 + M^2} G_S^2\right]}{Z_S}$$
$$m_H(\mu) = M_S(\mu) \sim 2m_t(\mu) \tag{3.6}$$

for $m_t(\mu) \gg m_i(\mu)$, $i \neq t$ and g_S^2 small. We assume that the renormalised parameters of the action are found at some scale μ, as yet undetermined.

It is now a straightforward, if somewhat tedious exercise to derive the vector boson action up to second order in derivatives. The algebra is simplified if one wishes to obtain only the terms of leading order in the limit that Λ becomes large (recall that the renormalisation factor $Z_S \sim \ln(\Lambda/\mu)$, is diverging logarithmically as $\Lambda \to \infty$). The surviving terms in this limit are:

$$\begin{aligned}
I = {}& \tfrac{1}{3}\text{tr}\left\{\left(\frac{1}{k^2 + M^2}\right)^2 \tfrac{1}{4}\left[Y F_{\mu\nu}(B) + P_L F_{\mu\nu}(W)\right]^2\right\} \\
& - \tfrac{1}{4}\text{tr}\left\{\left(\frac{1}{k^2 + M^2}\right)(Y B_\mu + P_L W_\mu)^2\right\} \\
& - \tfrac{1}{4}\text{tr}\left\{\left(\frac{1}{k^2 + M^2}\right)^2 (Y B_\mu + P_L W_\mu) M^2 (Y B_\mu + P_L W_\mu)\right\} \\
& + \tfrac{1}{4}\text{tr}\left\{\left(\frac{1}{k^2 + M^2}\right)^2 [\tfrac{1}{4}[\gamma_\mu W_\mu, M]^2 - [\gamma_\mu B_\mu(y_r - y_l) - \tfrac{1}{2}\{\gamma_\mu W_\mu, M\}]^2]\right\},
\end{aligned} \tag{3.7}$$

where $W_\mu = W_\mu \cdot \tau$. In (3.7) integrals over four dimensional k and x are understood. The symbol $F_{\mu\nu}(W)$ is the full, nonabelian field strength tensor. Apart from those mass terms which are not proportional to M^2, these terms are just the gauge field sector of the standard model with spontaneous symmetry breaking by the Higgs mechanism. An additional fine tuning, similar to that in the scalar gap equation (3.1), will be necessary to guarantee that the photon mass vanishes.

4. Quadratic Vector Boson Action

All terms quadratic in boson fields arise from second order in the expansion of the logarithm and may be written:

$$I^{(2)} = \sum_{a,b} \tfrac{1}{2}\text{Tr} \ln\left[\frac{1}{i\not{\partial} + M}\Omega_a \frac{1}{i\not{\partial} + M}\Omega_b\right], \tag{4.1}$$

with the term $I^{(2)}_{SS}$ already displayed. Not unexpectedly we set:

$$G_B \alpha = G_W \beta = E, \tag{4.2}$$

and

$$\begin{aligned} B_\mu &= \alpha A_\mu + \beta Z_\mu \\ W^3_\mu &= \beta A_\mu - \alpha Z_\mu. \end{aligned} \tag{4.3}$$

We present only the final results for the fields $W_\pm$, A, and Z. These are:

$$I^{(m)}_{AA} = \left[\tfrac{1}{2} + G^2_W \beta^2 \sum_i 2Q_i{}^2 \mathrm{tr}\left[\frac{\tfrac{1}{2}\partial^2 - m_i{}^2}{(-\partial^2 + m_i{}^2)^2}\right]\right] A_\sigma{}^2 \tag{4.4}$$

$$I^{(\mathrm{K.E.})}_{AA} = G^2_W \beta^2 \sum_i 2Q_i{}^2 \mathrm{tr}\left[\tfrac{1}{6}\frac{(\partial^2)^2}{(-\partial^2 + m_i{}^2)^4} - \tfrac{1}{4}\frac{m_i{}^2\partial^2}{(-\partial^2 + m_i{}^2)^4} + \tfrac{1}{4}\frac{m_i{}^2}{(-\partial^2 + m_i{}^2)^3}\right] A_{\sigma\lambda}{}^2 \tag{4.5}$$

$$I^{(m)}_{ZA} = \sum_i \tfrac{1}{4} C_i \mathrm{tr}\left[\frac{\tfrac{1}{2}\partial^2 - m_i{}^2}{(-\partial^2 + m_i{}^2)^2}\right] Z_\sigma A_\sigma \tag{4.6}$$

$$I^{(\mathrm{K.E.})}_{ZA} = \sum_i \tfrac{1}{4} C_i \mathrm{tr}\left[\tfrac{1}{6}\frac{(\partial^2)^2}{(-\partial^2 + m_i{}^2)^4} - \tfrac{1}{4}\frac{m_i{}^2\partial^2}{(-\partial^2 + m_i{}^2)^4} + \tfrac{1}{4}\frac{m_i{}^2}{(-\partial^2 + m_i{}^2)^3}\right] Z_{\sigma\lambda}A_{\sigma\lambda} \tag{4.7}$$

$$I^{(m)}_{ZZ} = \left[\tfrac{1}{2} + \sum_i \left\{\tfrac{1}{8}R_i\mathrm{tr}\left[\frac{\partial^2}{((-\partial^2 + m_i{}^2)^2)}\right] - \tfrac{1}{2}T_i\mathrm{tr}\left[\frac{m_i{}^2}{(-\partial^2 + m_i{}^2)^2}\right]\right\}\right] Z_\sigma{}^2 \tag{4.8}$$

$$I^{(\mathrm{K.E.})}_{ZZ} = \sum_i \mathrm{tr}\left[\left(\tfrac{1}{6}\frac{(\partial^2)^2}{(-\partial^2 + m_i{}^2)^4} + \tfrac{1}{4}\frac{m_i{}^2}{(-\partial^2 + m_i{}^2)^3}\right) R_i - \tfrac{1}{8}\frac{m_i{}^2\partial^2}{(-\partial^2 + m_i{}^2)^4}T_i\right] Z_{\sigma\lambda}{}^2 \tag{4.9}$$

$$\begin{aligned} I^{(m)}_{WW} = \Bigg[1 + G^2_W \sum_g &\Bigg[\tfrac{3}{2}\mathrm{tr}\left[\frac{\partial^2}{(-\partial^2 + m^2_{t_g})(-\partial^2 + m^2_{b_g})}\right] \\ &+ \tfrac{1}{2}\mathrm{tr}\left[\frac{\partial^2}{(-\partial^2 + m^2_{e_g})(-\partial^2 + m^2_{\nu_g})}\right]\Bigg]\Bigg] W^+_\mu W^-_\mu \end{aligned} \tag{4.10}$$

$$\begin{aligned} I^{(\mathrm{K.E.})}_{WW} = \tfrac{1}{6}G^2_W \sum_g &\Bigg[3\mathrm{tr}\left[\frac{(\partial^2)^2}{(-\partial^2 + m^2_{t_g})^2(-\partial^2 + m^2_{b_g})^2}\right] \\ &+ \mathrm{tr}\left[\frac{(\partial^2)^2}{(-\partial^2 + m^2_{e_g})^2(-\partial^2 + m^2_{\nu_g})^2}\right]\Bigg] W^+_{\sigma\lambda} W^-_{\sigma\lambda} \end{aligned} \tag{4.11}$$

The sums in the W–W terms are over generations only, the remaining traces are over Dirac matrices and momentum, and the anti-symmetric symbols $A_{\sigma\lambda}$ are just the linear terms in the covariant field tensors:

$$A_{\sigma\lambda} = (\partial_\sigma A_\lambda - \partial_\lambda A_\sigma).$$

The symbols Q_i, C_i, R_i, T_i, are defined by:

$$Q_i = \frac{y_{Li}}{2} + \frac{\tau_{3i}}{2} = \frac{y_{Ri}}{2}$$
$$C_i = 2\alpha\beta G_W G_B(y_{Li}^2 + y_{Ri}^2) - 2\alpha\beta G_W^2 \tau_{3i}\tau_{3i} + 2G_B G_W(\beta^2 - \alpha^2)\tau_{3i}y_{Li}$$
$$R_i = G_B^2\beta^2(y_{Li}^2 + y_{Ri}^2) - 2G_B G_W \alpha\beta y_{Li}\tau_{3i} + G_W^2\alpha^2\tau_{3i}\tau_{3i}$$
$$T_i = G_B^2\beta^2 y_{Ri}y_{Li} - G_B G_W \alpha\beta y_{Ri}\tau_{3i} \tag{4.12}$$

Performing sums over all the fermions we find also

$$\sum_i C_i = 16\, n_g G_W^2\, \tfrac{\beta}{\alpha}(\tfrac{8}{3}\beta^2 - 1)$$
$$\sum_i R_i = 8\, n_g G_W^2 \tag{4.13}$$

where n_g is the number of generations. We note a central result which emerges from a consideration of the quadratically divergent terms in Eq.(4.1–9):

$$I_{ZA}^{(m)} \sim \Lambda^2 G_W^2(\tfrac{5}{3}\beta^2 - \alpha^2) \tag{4.14}$$

The condition:

$$\sum_i C_i = 0. \tag{4.15}$$

in fact removes all divergent terms from this off-diagonal action, independent of regularisation. Remarkably, this same condition, in effect $\beta^2 \equiv \sin^2\theta_W = \tfrac{3}{8}$ as in the SU(5) GUT[18] model, allows one to simultaneously remove the Λ^2 divergent terms for the Z, W, and A fields, provided a simple consistency condition, in effect a second gap equation, is satisfied. The latter gap equation ensures the vanishing of the photon mass and is:

$$1 = n_g \frac{G_W^2\beta^2}{\pi^4}\sum_i Q_i^2 \int d^4k \frac{1/2k^2 + m_i^2}{(k^2 + m_i^2)^2} = 1 \tag{4.16}$$

with n_g the number of fermion generations. To this order one then obtains predictions for the W and Z masses at tree level (in the composite fields) in terms of the top quark mass. The dimensionless couplings g, and g' of the $SU(2)$ and $U(1)$ sectors may also be identified after wave function renormalisation similar to that in the scalar sector:

$$\frac{g}{2} = \frac{G_W}{\sqrt{Z_W}}, \qquad \frac{g'}{2} = \frac{G_B}{\sqrt{Z_B}} \tag{4.17}$$

We now finally quote results for the vector boson masses at tree level;

$$\mathcal{M}_Z^2 = \sum_i \text{tr}\left[\frac{(\tfrac{1}{4}R_i - \tfrac{1}{2}T_i)m_i^2}{(-\partial^2 + m_i^2)^2}\right] = \frac{G_W^2}{4\alpha^2}\sum_i \text{tr}\frac{m_i^2}{(-\partial^2 + m_i^2)^2}$$
$$Z_Z = \sum_i \text{tr}\left[\frac{1}{3}\frac{(\partial^2)^2}{(-\partial^2 + m_i^2)^4}\right]\tfrac{1}{4}R_i \tag{4.18}$$

We find for $m_t \gg m_i$ and small electro–weak couplings

$$M_Z^2(\mu) = \frac{\mathcal{M}_Z^2}{Z_Z} = \tfrac{3}{5}m_t^2(\mu) \tag{4.19}$$

and similarly

$$M_W^2(\mu) = \frac{1}{4} Z_W^{-1} \sum_i \mathrm{tr} \frac{m_i^2}{(-\partial^2 + m_i^2)^2} \simeq \frac{3}{8} m_i^2(\mu).$$

(4.20)

From 4.19 and 4.20 one has

$$M_W^2 / M_Z^2 = \alpha^2 = \cos^2 \theta_W(\mu) = 5/8$$

(4.21)

consistent with the value found from the vanishing of I_{ZA}. This value for the weak mixing angle is that predicted by grand unified theories[18], but the scale μ at which the present calculation will be taken is not necessarily a point of equality of couplings. In any case, the basic results of this work contained in Eq.(4.18–4.21) arose from $\log \Lambda$ terms in the action, are finite after wave function renormalisation, and virtually independent of the regularisation procedure.

5. Alternative Regularisation

The results of the last section could have been obtained by calculating the effective inverse propagators for the composite boson fields[19]. The NJL theory is not renormaliseable; some form of cutoff must be retained to make sense of the theory. Nevertheless, the second order fermion loops can be handled in a more or less conventional manner including wave function and charge renormalisation. For example, the unrenormalised scalar propagator, given here for a single fermion field, mass m, is:

$$\begin{aligned}
\Delta^{-1}(q^2) &= \frac{\delta(I_S + I_{class})}{\delta\sigma(q)\delta\sigma(-q)} \\
&= 1 + \frac{G_S^2}{(2\pi)^4} \int d^4k \, \mathrm{tr} \left\{ \left(\frac{1}{\rlap{/}k + m} \right) \left(\frac{1}{\rlap{/}k + \rlap{/}q + m} \right) \right\} \\
&= 1 + G_S^2 \pi(q^2)
\end{aligned}$$

(5.1)

The quadratic and logarithmic divergences can be made explicit by expanding the vacuum polarisation $\pi(q^2)$ to the lowest two orders in q^2:

$$\Delta^{-1}(q^2) = 1 + G_S^2 \pi(0) + G_S^2 q^2 \frac{\partial \pi}{\partial q^2}(0) + G_S^2 \hat{\pi}(q^2)$$

(5.2)

with $\hat{\pi}(q^2)$ finite in the limit $\Lambda \to \infty$. Direct use of the NJL scalar gap equation (3.1) then results in

$$\Delta^{-1}(q^2) = 2m^2 \frac{4G_S^2}{(2\pi)^4} \int \frac{d^4k}{(k^2 + m^2)^2} + \frac{q^2}{2} \frac{4G_S^2}{(2\pi)^4} \int d^4k \frac{k^2}{[k^2 + m^2]^3} + G_S^2 \hat{\pi}(q^2)$$

(5.3)

and a renormalised propagator may be defined by:

$$\begin{aligned}
\Delta_R^{-1}(q^2) &= \frac{1}{Z_S} \Delta_R^{-1}(q^2) \\
&= m_S^2 + q^2 + g_S^2 \pi_R(q^2)
\end{aligned}$$

(5.4)

with a renormalised coupling constant $g_S = G_S Z_S^{-1/2}$ and a renormalised mass parameter.

$$m_S^2 = \mathcal{M}_S^2 / Z_S = 4m^2 - g_S^2. \tag{5.5}$$

where $\mathcal{M}_S^2 = 1 + G_S^2 \pi(0) = Z_S m_S^2$ as in Eqn. 3.6. Evaluating the Higgs mass M_H^2 on the mass shell would induce a further small change of the order g_S^2. The results are finite, well determined, independent of the subtraction procedure and independent of any regularisation used to render momentum integrals finite. The dimensional 4-fermion coupling constant now only appears in the gap equation, being replaced elsewhere by the dimensionless coupling $g_S^2 = G_S^2 / Z_S$. The latter can in principle be determined by physical processes.

One can follow the same prescription for the photon, Z and W propagators. In the diagonalised version of the present calculation, with necessarily $\beta^2 = 3/8$, a single gap–like equation serves to fix the mass terms in the three vector boson propagators. Writing, for example, the unrenormalised photon function as

$$\Delta_{\rho\sigma}^{-1}(q^2) = \delta_{\rho\sigma} + \frac{E^2}{4} \pi_{\rho\sigma}(q^2) \tag{5.6}$$

the gap equation takes the form:

$$\Delta_{\rho\sigma}^{-1}(0) = 1 + E^2/4\pi_{\rho\sigma}^{(0)} = 0 \tag{5.7}$$

and simply states that the photon mass vanishes[13], or had we defined the renormalisation for the symmetric 4-fermion theory, i.e. with vanishing quark masses, the vector boson mass vanishes.

The results of the last section, Eqn. 4.19, then imply for example that the renormalised photon and Z inverse propagators have the form:

$$(\Delta_R^{-1}(q^2))_{\rho\sigma} = (q^2 \delta_{\rho\sigma} - q_\rho q_\sigma) + \frac{e^2}{4} \pi_{\rho\sigma}^R(q^2)$$
$$(\Delta_{RZ}^{-1}(q^2))_{\rho\sigma} = (q^2 \delta_{\rho\sigma} - q_\rho q_\sigma) + M_Z^2(\mu) + g'^2 \pi_{Z\rho\sigma}^R(q^2) \tag{5.8a}$$

with Eqn.(5.7) implying

$$(\Delta_{RZ}^{-1}(0))_{\rho\sigma} = \Delta_{\rho\sigma}^{-1}(0) + \mathcal{M}_Z^2(\mu)$$
$$= \mathcal{M}_Z^2(\mu) \tag{5.8b}$$

Only the leading weak–coupling independent terms have been kept in the Z (and W) masses. By direct calculation using the naive cutoff, one can verify that $\pi_{\rho\sigma}^R(q^2)$ is indeed to order Λ^{-2} what one expects from QED. In particular as $q \to 0$

$$\pi_{\rho\sigma}^R(q^2) \to -(q^2 \delta_{\rho\sigma} - q_\rho q_\sigma)\frac{q^2}{15} \tag{5.9}$$

Again the renormalised vector boson inverse propagators are independent of regularisation and finite functions of the dimensionless couplings, Eqn. 4.17, which are in principle determined by reference to physical processes. In particular, the electronic charge has been defined by

$$e = \frac{E}{2Z_A^{1/2}},\tag{5.10}$$

with $Z_A = Z_W = Z_Z$ for $\beta^2 = 3/8$ as given in Eqn. 4.18 to highest order in the electro-weak couplings.

To carry out the above program and in particular to establish the existence of a vector gap equation it is useful to employ some regularisation scheme for divergent integrals: That due to 't Hooft[20] is particularly appropriate to the present situation with the photon propagator for example given by

$$(\Delta_{\rho\sigma}^{-1}(q)) = \delta_{\rho\sigma} + \frac{G_W^2\beta^2}{(2\pi)^4}\sum_{abi}(4Q_i^2)C_aC_b\,\mathrm{tr}\int^\Lambda \frac{d^4k}{(2\pi)^4}\gamma_\rho\frac{1}{\not{k}+m_a}\gamma_\sigma\frac{1}{\not{k}+\not{q}+m_i}\tag{5.11}$$

where for each fermion i, a set of regulating fermions, masses $m_a(m_0 = m_i)$, are introduced satisfying

$$\sum C_a = \sum C_a m_a = \sum C_a m_a^2 = 0.\tag{5.12}$$

These fermions are coupled off–diagonally to the vector fields and eventually one allows $m_a \to \infty$, $a \geq 1$. Since in any case some cut–off must be retained to render the non–renormaliseable NJL theory meaningful, it is most direct to take $\Lambda < m_a$, $a \leq 1$. The non–gauge invariant coupling allows us to obtain a non–trivial vector gap equation, and for the photon and Z one obtains Eqs. (5.7) and (5.8b) in a form unaltered by regularisation. This is not strictly a renormalisation but an explicit choice of theory, a fine tuning of the dimensional vector couplings to yield zero boson mass in the absence of chiral symmetry breaking.

6. Renormalisation Group Evolution

The above results suggest the 4–fermion theory can be cast into a form equivalent to the Standard electro–weak theory for the composite bosons at tree level and including one fermion loop. We now assume the simple parameter relations we have obtained for the composites are achieved at some mass scale μ, near the grand unified scale and and hence in a region where the strong and electro–weak interactons are both weak. Thus the mass ratios we impose may be evolving rather slowly at such a scale. In detailed calculation one finds that only the Higgs self coupling $\lambda(\mu) = 2g_i^2(\mu)$ is changing at all rapidly, and we test our results by allowing a wide variation about the simple prediction $\frac{m_H^2}{m_t^2}(\mu) = 4$.

We imagine the strong interactions to be simply married to the electro–weak theory, and the relationships in Eqns. (4.19–4.21) evolved down to the scale $M_W = 80\,\mathrm{GeV}$. We will choose the scale μ such that $\sin^2\theta_W$ evolves down to the experimental value: 0.232[21]. This choice of the intermediate scale μ at which the standard model emerges is not well-defined, and the abrupt joining of the standard model to the theory above μ is probably unrealistic. However, we choose to do just

342

this, hopefully compensating for the vagueness by fitting μ, through the evolution of $\sin^2\theta_W$. To leading order for three generations of fermions we have[15]:

$$\sin^2\theta_W(M_W) = \frac{\alpha(M_W)}{\alpha_2(M_W)}$$
$$= \frac{3}{8}\left[1 - \frac{109}{18}\frac{\alpha(M_W)}{\pi}\ln\frac{\mu}{M_W}\right] + \cdots \tag{6.1}$$

There results $\mu = 7.5 \times 10^{12}\text{GeV}$ ($\mu = 3.7 \times 10^{14}\text{GeV}$) for $\sin^2\theta_W(M_W) = .232(.210)$, the former being the most recently accepted value of the weak angle[21,2]. The lower value of μ is somewhat below the usually accepted grand unified scale but still very high. Using this information it is now possible to evolve the ratio $\frac{m_t^2}{M_W^2}$. At the one loop level only the Higgs–top coupling g_t^2 is affected by the strong interaction. The relevant expressions are:

$$\frac{2\kappa_t}{\alpha_2} = \frac{4g_t^2}{g_2^2} = \frac{m_t^2}{M_W^2} \quad , \quad \kappa_t = \frac{g_t^2}{2\pi} \tag{6.2}$$

and

$$\kappa_t = \frac{2\alpha_3^{8/7}}{C + 9\alpha_3^{1/7}} \tag{6.3}$$

The latter is one of a general family of solutions for the fermion coupling renormalizaton group equation found by (KSZ)[22] but ignores additional electro–weak corrections at the scale μ[14]. The arbitrary parameter C may be determined by first evolving α_3 to μ and then using (6.2) and (6.3). One needs also, the evolution equations for $\alpha_2(\mu)$ and $\alpha_3(\mu)$ which we again take at the one-loop level[14,15,21]. From $\frac{m_t^2}{M_W^2} = 8/3$ there results

$$m_t(M_W) = 2.04 M_W \approx 160\text{GeV}. \tag{6.4}$$

A variation in the starting evolution scale by two orders of magnitude from 7.5×10^{12} GeV to 7.5×10^{14} GeV produces a small change in m_t from 163 GeV to 156 GeV, while an increase in the ratio $\frac{m_t}{M_W}(\mu)$ by 10% induces a shift from 163 GeV to 160 GeV.

The Higgs mass at the tree level is in NJL theories placed very near $M_H^2 = 4m_t^2$. Assuming a weakly coupled Higgs evolved by the usual scalar self-coupling relation

$$\lambda(\mu) = \frac{\lambda(M_W)}{1 - \frac{3\lambda}{2\pi^2}\ln\left(\frac{\mu}{M_W}\right)}, \tag{6.5}$$

one finds $M_H(M_W) = 137(142)\text{GeV}$ at $\mu = 7.5 \times 10^{12}(7.5 \times 10^{14}$ GeV. Given the uncertainty in matching the 4–fermion theory to the standard model perhaps a more drastic test of the stability of the prediction of the Higgs mass is called for. Arbitrarily taking $M_H^2 = 8m_t^2$ at 7.5×10^{12} GeV, moves the Higgs mass to 153 GeV, a small change easily understood in terms of the evolution in Eqn. 6.5 which produces rather compressed values for M_H at M_W from divergent values at μ. One might expect the Higgs coupling to be subject to strong cross-coupling with the top; we find however all other one loop corrections to Eqn. 6.5, including from the Z and W, virtually cancel at μ. Nevertheless a more careful treatment of this evolution should be performed.

Thus both top and Higgs masses are rather stably determined in this picture. It is almost certainly true that the barely bound Higgs at μ will evolve to a more deeply bound composite at $M_W{}^2$, an essential result of this dynamical symmetry breaking model.

7. Conclusions

It seems possible to recover the standard electro–weak theory from a four–fermion theory at a scale $\mu \sim 10^{13}$–10^{15} GeV, perhaps somewhat below the usual grand unification scale. The quark–meson sector, we reiterate, is consistently defined by setting $(G_W)W_\mu = (g_W/2Z_W^{1/2})\bar{W}_\mu$. No deeper basis is given for the four–fermion theory, nor for the asymmetric coupling of quarks to the mass-generating scalar field. As we have indicated, this theory could perhaps arise from massive vector bosons at this upper scale, with the observed W, and Z arising as very light composite images of their higher mass analogues. A more natural reason for the gauge symmetry at the present experimental mass scale might then emerge. Alternatively the four–fermion theory with an unusual choice for coupling constant $G \propto \Lambda^{-1}$ might be taken more seriously[23]. For consistency we should include the strong interactions in a similar manner, i.e., with composite gluons.

The development for the 4–fermion theory we present here follows in many ways that outlined in[2], but of course, we have added a specific $SU(2) \times U(1)$ current-current interaction to the four–fermion theory. In both presentations the four–fermion theory is expected to be valid at some high scale; we have introduced an intermediate scale μ, $\Lambda \gg \mu \gg M_W$, at which the four–fermion action can be taken to yield the standard $SU(2) \times U(1)$ theory with very simple expressions for the lagrangian parameters.

A confirmation of the Higgs and top mass values found in this work would imply new physics near the grand unified scale but probably somewhat different in kind from the usual GUTS theories. Certainly there is no necessity for explicit proton decay here. It may well be difficult in this approach to include a fourth generation given the sum rule constraint on fermion masses in terms of the W mass. One would require heavy t', b', and from our present knowledge of the Z width a rather massive neutrino, in this family. Simply taking $m_t = m_{t'} = m_{b'}$ and evolving with the fourth generation equivalents of Eqns. 6.2 and 6.3 results in all these three particles having masses near 115 GeV, still an experimental possibility. The Higgs mass is virtually unaltered by this exercise.

8. Acknowledgements

The author is grateful to Bill Marciano for encouragement, and invaluable clarifying instruction, without which it would have been difficult to complete this work. In addition thanks are due Frank Paige for discussions on high mass scales, and to Sally Dawson for many clarifying discussions. The author also thanks B. Müller for the invitation to speak at the conference and thanks KSU and the NSF for providing support.

9. References

1. Y. Nambu, Fermi Institute Preprint 89-08 (1989).
2. W. Bardeen, C. T. Hill, M. Lindner, Fermi-PUB-89/127-T.
3. V. A. Miranski, M. Tanabashi, K. Yamawaki, *Mod. Phys. Lett.* **A4**, 1043 (1989).
4. P. W. Higgs, *Phys. Lett.* **12**, 132 (1964).
5. S. Weinberg, *Phys. Rev. Lett.* **19**, 1264 (1967);
   A. Salam, in *Elementary Particle Theory; Relativistic Groups, and Analyticity.*
   Edited by N. Svartholm, *Proceedings of the Nobel Symposium#8* (Almqvist, Wiksell, Stockholm 1968);
   S. L. Glashow, J. Iliopulos, L. Maiani, *Phys. Rev.* **D2**, 1285 (1970).
6. Y. Nambu and G. Jona-Lasinio, *Phys. Rev.* **122**, 345 (1961).
7. T. Eguchi, *Phys. Rev.* **D14**, 2755 (1976).
8. H. Terazawa, Y. Chikashige and K. Akama, *Phys. Rev.* **D15**, 480 (1977).
9. M. Bando, T. Kugo and K. Yamawaki, *Phys. Rep.* **164C**, 219 (1988).
10. M. Suzuki, *Phys. Rev.* **D37**, 210 (1988).
11. D. E. Kahana, U. Vogl, *Phys. Lett.* **B244**, 10 (1990)
12. R. D. Ball, Invited talk, given at "Skyrmions and Anomalies", Mogillany, Poland, Feb. 21–23, 1987, to be published in the conference proceedings;
    A. Dhar, R. Shankar, S. R. Wadia, *Phys. Rev.* **D31**, 3256 (1985).
    D. Ebert amd H. Reinhardt, *Nucl. Phys.* **B271**, 188 (1986).
13. J.D. Bjorken, *Ann. Phys.* **24**, 174 (1963).
14. W. Marciano, *Phys. Rev. Lett.* **62**, 2793 (1989);
    Phys. Rev. Lett. **41**, 219 (1990).
15. W. Marciano, A. Sirlin, *Phys. Rev.* **D22**, 2695 (1980);
    M. Goldhaber, W. Marciano, in *Comments in Nuclear and Particle Physics.*
16. D. E. Kahana, *Phys. Lett.* **B229**, 9 (1989).
17. D. Gross and J. Neveu, *Phys. Rev.* **D10**, 3235 (1974).
18. H. Georgi, S. Glashow, *Phys. Rev. Lett.* **32**, 438 (1974);
    H. Georgi, H. Quinn, and S. Weinberg, *Phys. Rev. Lett.*, **33**, 451 (1974).
19. G.S. Guralnik and K. Tamvakis, *Nucl. Phys.* **B148**, 283 (1979).
20. G. 't Hooft, *Nucl. Phys.* **B33**, 173 (1971).
21. S. Fanchiotti, and A. Sirlin, *Phys. Rev.* **D41**, 319 (1990).
22. J. Kubo, K. Sibold, and W. Zimmermann, *Phys. Lett.* **B220, 185 (1989)**.
23. K.G. Wilson, *Phys. Rev.* **D7**, 2911 (1973).

COLOR–SEXTET QUARK CONDENSATION IN QCD

KYUNGSIK KANG

Physics Department, Brown University, Box 1843,
Providence, RI 02912, USA

ALAN R. WHITE

High Energy Phsyics Division, Argonne National Laboratory
Argonne, IL 60439, USA

ABSTRACT

We present the minimal sextet–quark condensation model as an attractive alternative to the standard model. The model is constructed by a simple and natural extension beyond the standard model and has many new interesting features some of which can be tested readily at the existing colliders. A crucial test may be a short–lived axion– like η_6 of mass 30–60 GeV, which can produce high energy photons diffractively at hadron colliders and radiatively in e^+e^- colliders.

1. Introduction

The idea of technicolor[1] is to search for an alternative to spontaneous symmetry breaking (SSB) and to break the electroweak gauge symmetry **dynamically**, without the use of Higgs scalars, while preserving the successful mass relation of the weak gauge bosons. There are two possibilities to achieve dynamical symmetry breaking (DSB): either[1] one introduces a QCD–like technicolor group (TG) which confines at $\Lambda_{TC} \sim 1\,\mathrm{TeV}$, along with anomaly–free techniquark doublets ψ in a fundamental representation of TG but in singlets of the QCD group, or[2] one uses extra heavy quarks belonging to some higher dimensional representation of the familiar QCD group $SU(3)$. In the first approach, the global chiral symmetry $SU(2)_L \times SU(2)_R$ in the technicolor sector is broken down to $SU(2)$ by techniquark condensate $\langle \bar{\psi}_L \psi_R \rangle \sim (\Lambda_{TC})^3$, causing a triplet of massless composite technipions which are absorbed by the weak gauge bosons to become their longitudinal components. It is then possible to obtain $M_W = M_Z \cos\theta_W = \frac{g}{2}\Lambda_{TC}$. To generate the quark and lepton masses within a dynamical context, one needs to embed TG in a larger extended technicolor group (ETG)[3] which is broken down to TG at a scale $\Lambda_{ETC} > \Lambda_{TC}$ and gives mass to the ordinary quarks and leptons radiatively through the heavy ETG gauge boson couplings to fermions and technifermions. This gives typically a fermion mass $m_f \sim \Lambda_{TC}^3/\Lambda_{ETC}^2$. However the ETG gauge boson couplings cause unwanted flavor–changing neutral current (FCNC) interactions in the ordinary quark sector due to fermion–fermion couplings. The $K_L - K_S$ mass difference, a $\Delta S = 2$ process, leads to $\Lambda_{ETC} > 500$ TeV, which in turn predicts a typical fermion mass m_f to be a few MeV at the most. A way to overcome the FCNC difficulty is to force the technicolor coupling constant to run very slowly[4], i.e., to "walk" by considering the technicolor theory near a fixed point of the beta function that describes the evolution of the technicolor coupling constant. The assumption is that the beta functions for the technicolor coupling is small above Λ_{TC} scale and the anomalous dimension of the fermion mass operator can be large all the way up to Λ_{ETC}. In such cases, the techniquark condensate is given by $\langle \bar{\psi}_L \psi_R \rangle \sim \Lambda_{TC}^3 (\Lambda_{\mathrm{ETC}}/\Lambda_{TC})^\gamma$ where γ is the anomalous dimension of the bilinear operator $\bar{\psi}_L \psi_R$. Thus it is possible to achieve less spectacular enhancements of the techniquark condensate for a walking technicolor theory in which asymptotic freedom is nearly saturated. Of course care must be given to maintain the stability of the fixed point and asymptotic freedom even when higher–loop corrections are taken into account.

In the second approach of DSB, new quarks belonging to a higher color representation of the ordinary color $SU(3)$ group of QCD play the role of the techniquarks of the technicolor model, provided that they condense at the scale required for the dynamical electroweak symmetry breaking. In fact it is well–known that the ordinary quarks can form condensates at a scale $\leq 1\mathrm{GeV}$, which then break the global chiral symmetry and produce pions as the Goldstone bosons. This scale is obviously far too small to give the right mass relation for the weak gauge bosons. Recently several authors[5] suggested heavy top quark condensation as a possibility of DSB because of the experimental lower bound $m_t > 89$ GeV. Detailed investigation of the minimal top quark condensation model with the four fermion interaction $(\bar{\psi}_L t_R)(\bar{t}_R \psi_L)$, ψ_L being the third generation doublet (t_L, b_L), predicts the top quark mass $m_t = 220 \sim 230$ GeV at the scale of new physics $\Lambda \simeq 10^{15} \sim 10^{19}$ GeV. The result is dangerously close to the indirect experimental upper bound of m_t based on the analysis of radiative corrections to the ρ–parameter.

With the extra quarks belonging to higher representations $R > 3$ under color $SU(3)$, it is possible to achieve much larger enhancement of the condensate because the strength of the quark–antiquark binding potential is proportional to $C_2(R)\alpha_S(\mu_R)$ where $C_2(R)$ is the quadratic Casimir invariant of the representation $\mathbf{R}$ and is bigger than $C_2(3) = 4/3$ for $R > 3$. In that case, the fermion mass operator in exotic quark sectors can condense at a scale much larger than 1 GeV and perhaps as large as the desired electroweak symmetry breaking scale. This type of DSB has a number of attractive features. First of all, the chiral breaking scale of QCD provides the electroweak symmetry breaking scale, which can lead to many new phenomena due to mixing of electroweak and strong interactions. Secondly, it is possible to construct[2] minimal DSB model with a flavor doublet of sextet quarks that has a fixed point in the two–loop beta function and yet preserves asymptotic freedom. The zero seems to survive to very high–order loop corrections and therefore the minimal sextet–quark condensation model (MSQCM) has the required property that one assumes for a walking technicolor model. In addition MSQCM can be constructed by extending naturally the gauge and fermion sectors of the standard model under color $SU(3)$ and electroweak $SU(2) \times U(1)$ by adding a flavor doublet of sextet quarks. While the model is one of the simplest extensions of the standard model, the spectrum of the new particles is rather unambiguous and can be tested in part at the existing colliders. We initially[6] suggested, based on the total cross-sections along with the forward elastic $p\bar{p}$ scattering amplitude at CERN and Fermilab Tevatron collider energies, a possible signal of color–sextet symmetry breaking, i.e., diffractive production of an axion–like particle η_6 originating from the sextet–quark condensations. In particular we pointed out that mini–Centauro and Geminion events in very-high–energy cosmic ray experiments[7] could actually be identified as the hadronic and two–photon decay products of the η_6 respectively and led us to suggest to look for the two–photon decay mode of η_6 with mass 30± 10 GeV in hadron diffractive process. It now appears[8] that the most distinctive of the Geminion events lead to a higher mass of 60-70 GeV for the η_6. Also it was realized that a massive η_6 would eventually be observable as a rare radiative decay of Z^0 at LEP. We emphasize that the existence of η_6 is a uniquely distinctive sign of sextet–quark condensation in that there is no comparable state in other models. While there are η–like states in technicolor models[9], they have either much heavier mass because of the technicolor gauge anomaly or carry color.

In the next section, we describe briefly the structure of the MSQCM. Section 3 contains possible evidences for the sextet–quark condensates. We will see that the η_6 may provide a new source for radiative decays of the Z^0.

2. The Minimal Sextet–Quark Condensation Model (MSQCM)

The MSQCM is a straight–forward modification of the standard $SU(3) \times SU(2) \times U(1)$ model by adding a flavor doublet of color sextet quarks $Q_6 = (U, D)$ with $(6^*, 2, 1/3)$ to the usual fermions of the standard model but with no elementary Higgs scalars. There are two additional doublets of new leptons that are required by anomaly cancellation. The Lagrangian is given by

$$\mathcal{L} = \mathcal{L}_{\text{gauge}} + \mathcal{L}_{\text{fermion}} + \mathcal{L}_{4f} \tag{1}$$

where $\mathcal{L}_{\text{gauge}}$ is the Lagrangian for the color $SU(3)$ and electroweak $SU(2)$ and $U(1)$ gauge fields, $\mathcal{L}_{\text{fermion}}$ is the kinetic terms of the ordinary and additional fermions constructed by making use of the appropriate $SU(3) \times SU(2) \times U(1)$ covariant derivatives, and $\mathcal{L}_{4f}$ represents all relevant four–fermion interaction terms of scale dimension 6 that are flavor–chiral and gauge invariant.

With the usual six flavors of ordinary quarks and two flavors of sextet quarks, asymptotic freedom is still preserved in MSQCM but there is a zero in the QCD beta function

$$\mu \frac{\partial g}{\partial \mu} = -\frac{g^3}{16\pi^2} \left[b_0 + b_1 \left(\frac{g^2}{16\pi^2} \right) + b_2 \left(\frac{g^2}{16\pi^2} \right)^2 + \cdots \right] = \beta(g) \tag{2}$$

where

$$b_0 = \frac{11}{3} C(A) - \frac{4}{3} \sum_R I_R N_R = 11 - \frac{2}{3}(N_3 + 5N_6) = \frac{1}{3}$$

$$b_1 = +\frac{34}{3} C(A)^2 - \frac{20}{3} C(A) \sum_R I_R N_R - 4 \sum_R C(R) I_R N_R \tag{3}$$

$$= 102 - \frac{38}{3} N_3 - \frac{250}{3} N_6 = -140\frac{2}{3}$$

and b_2 is given[10], in the momentum subtraction scheme, approximately by $b_2 \simeq 0.01(b_1/b_0)^2$. Here $C(A)$ and $C(R)$ are the quadratic Casimir operators of the adjoint and $\mathbf{R}-$ representations respectively and I_R and N_R denote the second index and multiplicity of the representation $\mathbf{R}$. Notice that the stability of the fixed point is guaranteed if $b_0 b_2/b_1^2 < \frac{1}{4}$, which is well satisfied by the MSQCM. From the smallness of the third term in Eq. (2) for $\alpha_S = \frac{g^2}{4\pi} = -4\pi b_1/b_0 \simeq \frac{1}{33}$ and of the expansion parameter $g^2/16\pi^2 = \alpha_S/4\pi \simeq 1/412$ it is clear that the zero should survive to very high order in the perturbation expansion of $\beta(g)$. In other words, the MSQCM has the desired property of the walking technicolor.

There is a $U(2) \times U(2)$ chiral flavor symmetry of the sextet–color quark sector in MSQCM. Normal QCD chiral dynamics may cause chiral–symmetry breakdown in the sextet quark sector at a mass scale $F_{\pi_6} \sim 250$ GeV, much higher than the usual scale F_π of the chiral symmetry breaking in the color–triplet quark sector. This can be understood from an elementary picture in which the strength of the effective quark–antiquark binding potential is propotional to $C_2(R)\alpha_S(\mu_R)$ and chiral symmetry breaking occurs at some common critical value of the flux string tension so that $C_2(6)\alpha_S(F_{\pi_6}) \sim C_2(3)\alpha_S(F_\pi)$. As the QCD coupling $\alpha_S(\mu_6)$ evolves (actually "walks" slowly), a $\bar{Q}_6 Q_6$ condensates forms at a scale $F_{\pi_6} = |\langle \bar{Q}_6 Q_6 \rangle|^{1/3} \sim 250$ GeV triggering the chiral–symmetry break–down spontaneously and producing four massless Goldstone bosons, an isotriplet π_6 and an isosinglet η_6. Then through the coupling terms of the weak gauge boson fields of the standard model to the sextet–quark currents in $\mathcal{L}_{\text{fermion}}$, the isotriplet π_6 are absorbed by the gauge bosons to become the longitudinal components of $W^\pm$ and Z^0 with the mass relation $M_W = M_Z \cos\theta_w \sim g F_{\pi_6}$. In analogy with the familiar color–triplet quark sector, π_6 and η_6 couple longitudinally to the isotriplet and isosinglet (under the unbroken $SU(2)$) axial–vector currents respectively. Actually the isosinglet axial–vector current would contain a small admixture[11] of color–triplet quarks so that it is free of the QCD $U(1)$–anomaly and transforms orthogonal under the $U(1)$ flavor symmetry to another combination which is made mostly of the isosinglet axial–vector current of the color–triplet quarks with a small color–sextet component and couples to the QCD $U(1)-$ anomaly. We note that η_6 is effectively an axion since it is the Goldstone boson associated with a $U(1)$ axial chiral symmetry of the color–sextet quarks. As pointed out above, in the MSQCM, α_S evolves very slowly above the electroweak scale and interactions involving small instantons, integrated over a large momentum range, can produce[12] a large anomalous dimension for sextet quark operators and in particular a much larger η_6 mass than a conventional axion mass of $\mathcal{O}(100\text{keV})$.

The existence of the η_6 provides a unique signal of sextet quark symmetry breaking, which is unparalleled by other models, because one generally does not expect any new strong interaction of the Higgs sector, elementary or dynamical, and the technicolor sector to manifest themselves until way above the electroweak scale. The η_6 of the MSQCM allows the early appearance of such a new interaction. In elementary Higgs models, the analogous Higgs scalar to the η_6 do not give either the anomalous two–photon decay or the off–shell non–local coupling that can allow an additional source for photon production. Of course, there will be in MSQCM new QCD baryons and mesons of pseudoscalar, vector, axial–vector etc. states that are made of sextet quarks, but they are all expected to be heavy, as the color–sextet quark masses are expected to be larger than F_{π_6}. In fact, an estimate based on the nontrivial solution of the ladder Schwinger–Dyson equation for the Q_6 propagator in the quenched planar approximation gives[13] the color–sextet quark mass to be about 350 GeV for t–quark mass range of 77–160 GeV. In short the existence of a particle with the properties of the η_6 is a very special signal of the existence of a color–sextet quark sector that can be tested at the existing collider energies.

3. η_6 Production and Decay

Because of the interplay between QCD and electroweak interactions in the MSQCM, a host of new phenomena is expected[6] at the energy scale $> F_{\pi_6}$. In addition, there will be new baryons and mesons which are made of color–sextet quarks. They are all expected to be heavy, i.e., $M \geq M_{Q_6}$, some of which may already have been hinted by the high–energy cosmic exotics. The cross–section for perturbative QCD production[15] of Q_6 is, apart from color factors, the same as that of color–triplet quarks having the same mass M_{Q_6} and the hard cross–section is expected to be smaller than 1 Pb at the Fermilab Tevatron collider. Other possibly related new phenomena include the existence[14] of higher order composite operator contributions to hadron final states with at least six jets which may be observable at LHC and strong production[15] of W^+W^- and Z^0Z^0 pairs which also may have been indicated already by experiments. However the effects of the η_6, as an axion–like particle associated with the axial $U(1)$ symmetry of color–sextet quarks, may be much easier to detect[16]. We initially proposed[6] to look for a massiv η_6 via its two photon decay made in hadronic diffractive processes. We argued that the recent measurements of high energy elastic $p\bar{p}$ scattering amplitude and total cross sections at CERN and Fermilab collider energies might be due to diffractive production of the η_6 that has the properties of two particular classes of cosmic events known as "geminions" or "binocular" events and "mini-centauros". Diffractive production of the η_6 may be an explanation of the puzzle of small diffractive cross sections at collider energies.

As a light axion, the η_6 can have a major two–photon decay modes via the anomaly. Using the usual PCAC relation between the isosinglet axial–vector current and the η_6 field, the decay rate of the η_6 into two photons can be calculated from anomaly, giving

$$\Gamma(\eta_6 \to 2\gamma) \sim \left(m_{\eta_6}^3 \alpha_{em}^2 e_q^4 / 16\pi^3 F_{\eta_6}^2\right) N^2 \sim \left(\frac{100 keV}{m_{\eta_6}}\right)^2 \text{sec} \tag{4}$$

where e_Q is the charge of sextet quarks, $N = 6$ is the color number, and F_{η_6} is the decay constant. If $m_{\eta_6} \sim 60$ GeV, the lifetime of the η_6 from Eq. (4) is comparable with that of the π^0 for $F_{\eta_6} \sim F_{\pi_6} \sim 250 \text{GeV}$, i.e., $3 \times 10^{-17}\text{sec}$. The η_6 can be produced[17] radiatively from Z^0. The branching ratio for $Z^0 \to \eta_6 + \gamma$ is calculable from the triangle anomaly giving $\Gamma\left(Z^0 \to \eta_6\gamma\right)/\Gamma\left(Z^0 \to \mu^+\mu^-\right) \sim 10^{-5}$. This decay should eventually be observable at LEP. In this connection, it is interesting to speculate that the two events of $Z^0 \rightarrow \mu^+\mu^-\gamma\gamma$ with $M_{2\gamma} \sim 60\text{GeV}$ observed by the L_3 experiment[18] at LEP may be related with η_6 production.

A strongly interacting Q_6 sector has an absorption effect for e^+e^- pair production via the interference between the diffractive excitation of a photon into a $\bar{Q}_6 Q_6$ state that decays into e^+e^- and the electromagnetic production of e^+e^- pairs. The production of hadrons and 2γ via the η_6, together with direct production of the Z^0, will drastically modify the properties of high–energy photon–initiated air showers and the development of electromagnetic clusters within hadron initiated air showers. This may be an explanation of the muon–rich photon showers[19] and wide range of anomalous shower development seen in high–energy cosmic ray events.

Besides the CDF group which[20] has been attempting to search for the η_6 in the diffractive $p\bar{p}$ scattering, an interesting physics possibility at a photon linear collider has been proposed[21] to search for a host of new particles including the η_6 that have appreciable two–photon couplings. In the meantime, the UA4-2 group has repeated the original $UA(4)$ experiment. Diffractive production of the η_6 would result in characteristically distinct energy behaviors in total cross–section, which can readily be checked at existing collider energies. In any case, the existence of the η_6 provides a new hitherto unknown source for photon production from Z^0.

In short, the MSQCM has a number of attractive features. It is by far the most natural and the simplest extention of the standard model without elementary Higgs scalars and has a uniquely distinguished advantage in t it can be tested at the existing colliders.

4. Acknowledgements

This work is supported in part by the U. S. Department of Energy, Contract DE-AC02-76-ER03130.A31 (Report No. BROWN-HET-873) and Contract W-31-109-ENG-38 (Report No. ANL-HEP-CP-92-71).

5. References

1. S. Weinberg, *Phys. Rev.* **D13**, 974 (1976); **D19**, 1277 (1979); L. Susskind, *Phys. Rev.* **D20**, 2619 (1979).

2. W. J. Marciano, *Phys. Rev.* **D21**, 2425 (1980); E. Braaten, A. R. White and C. R. Willcox, *J. Mod. Phys.* **A1**, 693 (1986); K. Kang and A. R. White, *J. Mod. Phys.* **A2**, 409 (1987).

3. S. Dimopoulos and L. Susskind, *Nucl. Phys.* **B155**, 237 (1979); E. Eichten and K. Lane, *Phys. Lett.* **B90**, 125 (1980).

4. B. Holdom, *Phys. Rev.* **D24**, 1441 (1981).

5. V. A. Miranski, M. Tanabastin and K. Yamawaki, *Mod. Phys. Lett.* **A4**, 1043 (1989); W. J. Marciano, *Phys. Rev. Lett.* **62**, 2793 (1989); W. A. Vardeen, C. T. Hill and M. Lindner, *Phys. Rev.* **D41**, 1647 (1990).

6. K. Kang and A. R. White, *Phys. Rev.* **D42**, 2425 (1989).

7. S. Hasegawa, in VI International Symposium on Very High Energy Cosmic Ray Interactions, Taubes, France (1990).

8. T. Arizawa, Private Communication.

9. K. Lane, Snowarass Proceedings (1982).

10. O. V. Tarasov, A. A. Vladimirov and A. Yu Zharkov, *Phys. Lett.* **93B**, 429 (1980); A. R. White, ANL-HEP-PR-84-31.

11. T. E. Clark, C. N. Leung, S. T. Love and J. L.Rosner, *Phys. Lett.* **B177**, 413 (1986).

12. B. Holdom and M. E. Peskin, *Nucl. Phys.* **B208**, 397 (1982).

13. K. Fukazawa, T. Muta, J. Saito, I. Watanabe and M. Yonegawa, *Prog. Theor. Phys.* **85**, 111 (1991).

14. R. E.Chivukula, M. Golden and E. H. Simmons, *Phys. Lett.* **B207**, 453 (1991); E. H. Simmons, *Phys. Lett.* **B259**, 125 (1991).

15. A. R. White, *Mod. Phys. Lett.* **A2**, 397 (1987).

16. K. Kang and A. R. White, Proc. Beyond the Standard Model II, Univ. of Oklahoma, OK (1990); Proc. Joint International Lepton-Photon Symp. & Europhysics Conf. on H.E. Phys., Geneva, Switzerland (1991).

17. T. Hatsuda and M. Umezawa, *Phys. Lett.* **B254**, 493 (1991).

18. J. Wenninger - L3 Collaboration ,in Rencontre de Moriond (1992).

19. G. Yodh, *Nucl. Phys. B* (Proc. Suppl) **12**, 277 (1990).

20. N. D. Giokaris et al, *Nucl. Phys. B* (Proc. Suppl.) **25B**, 40 (1992).

21. D. L. Borden, D. Bauer and D. O. Caldwell, SLAC-Pub-5715.

CONCLUDING REMARKS: VACUUM STRUCTURE AND MATTER GENESIS

R. VINH MAU

*Division de Physique Théorique, Institut de Physique Nucléaire 91406 Orsay Cedex
and LPTPE, Université Pierre et Marie Curie, 4 Place Jussieu 75252 Paris Cedex 05 (France)*

ABSTRACT

The possibility of building a theory of the origin of matter based on the complex
structure of the vacuum in the electroweak standard model and the chiral anomaly
is reviewed.

I was asked by Berndt Müller and Herb Fried to give the summary talk
of this Workshop but the topics covered here are so diverse that I find very hard
to summarize them in a coherent way. Instead, I would like to address an issue
which, I think, bears some relevance to the present meeting, namely the possibility
of building a theory of the origin of matter based on the properties of the vacuum
(not of QCD but of the electroweak (EW) standard model.)

During this week, the structure of the QCD vacuum has been scrutinized in
many aspects. The least we can say today is that this structure is very complex.
Actually, the complexity of the vacuum structure is not only proper to QCD but is
a common property of non abelian gauge theories[1]. In particular, the electroweak
(EW) standard model possesses also a vacuum with a complex structure, and, as
will be argued later, it is the combination of this complex vacuum structure along
with the chiral anomaly which leads to fermion number non conservation. This
means that the standard model can by itself generate the baryons and electrons
that make up the matter in the Universe. This feature, in turn, can provide an
explanation of the predominance of matter over antimatter in the Universe, the so
called baryon asymmetry in the Universe (BAU).

It has been long thought, for lack of better reasons, that the BAU is a boundary
condition set at the beginning of time. Besides the fact that this assumption is an
unaesthetic one, it is now believed that if the Universe is baryon asymmetric to day
it was almost baryon symmetric at the earliest time[2].

The necessary conditions for BAU

If BAU is not a boundary condition then it requires of course baryon number
non conservation. However, this is not sufficient to generate a net baryon asymmetry. As pointed out by Sakharov[3], one needs two more conditions:

- the theory must have C and CP violation, otherwise equal numbers of
baryons and antibaryons are produced, giving no net increase

- there must be a departure from thermal equilibrium at some epoch during
the evolution of the Universe, otherwise the dominant dynamics drives the system
to equal mixtures of baryons and antibaryons.

Before elaborating on the mechanism for the EW standard model to generate BAU, let me mention that the grand unified theories (GUTs) were thought to be a natural theory for BAU since they meet all three of these conditions. GUTs violate baryon number and CP. The BAU results from the out of equilibrium decay of superheavy bosons at temperatures below the GUT scale ($\simeq 10^{14} - 10^{15}$ GeV). One can easily produce the observed ratio $\eta = $ baryon number/entropy $\sim 10^{-11}$.

However, the proton has not yet been observed to decay. Moreover, there is another serious problem due to the large temperature range between the GUT scale ($10^{14} - 10^{15}$ GeV) and the EW scale ($10^2 - 10^3$ GeV). As it will be shown later, baryon number proceeding through electroweak anomaly will wash out any asymmetry produced by GUTs unless the baryon number minus the lepton number is not conserved. Another advantage in favor of the EW baryogenesis is that the physics at $\sim 10^2$ GeV is under better control than that at 10^{15} GeV.

Baryon Number Violation in the EW Standard Model

For sake of simplicity, let us consider the EW standard model with an SU(2) gauge group and a Higgs field. This model possesses a non trivial vacuum structure, i.e., in addition to the usual perturbative vacuum configuration $A_i = \vec{\sigma}.\vec{A_i} = 0$ and $\phi = \phi_0$, there is an infinite number of pure gauge configurations

$$\begin{cases} A_i & = & \frac{2i}{g}\partial_i U U^+ \qquad U \in SU(2) \\ \phi & = & T(U)\phi_0 \end{cases} \tag{1}$$

characterized by an integer topological number (Chern-Simons number)

$$N_{CS} = \frac{\epsilon_{ijk}}{64\pi^2} \int d^3 x g^2 Tr(A_i \partial_j A_k - \frac{2ig}{3} A_i A_j A_k) \tag{2}$$

all with zero energy.

The classical potential energy E of the fields can be schematically represented as in Fig. 1.

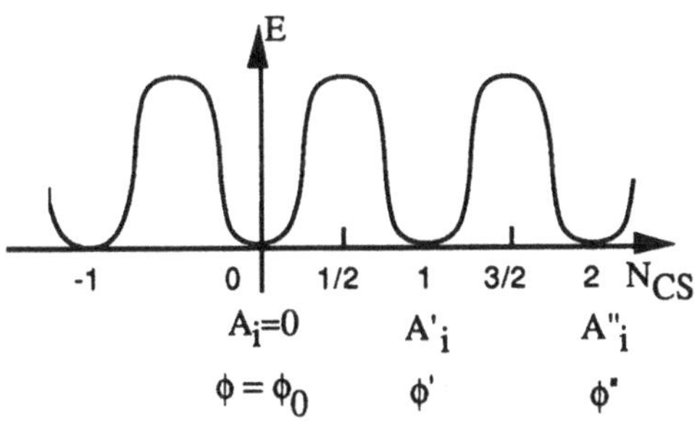

Fig. 1

In order to change from a vacuum configuration with one integer value of N_{CS} to that with another integer value, it is necessary to pass through a non vacuum, i.e.,

354

finite energy field configurations. Between the vacua, it has been shown[4] that there exist unstable, time-independent solutions of the field equations called *sphalerons* with $N_{CS} = \frac{1}{2}, \frac{3}{2}, \cdots$ Also, at these saddle points, the height of the barrier has been found to be $4\pi\frac{M_w}{g^2} \sim 8 - 14$ TeV (M_w : W boson mass).

Associated with the twisting of the gauge field from one vacuum state to another one is the violation of chiral fermion number through the chiral anomaly[5]. At the semi classical level, baryon number, lepton numbers are conserved in the standard model. However, as pointed out by t'Hooft[6], the quantum corrections invalidate these conservation laws at the absolute level. Due to the chiral anomaly the currents only obey partial conservation laws:

$$\partial_\mu L^\mu = \frac{1}{64\pi^2}\epsilon_{\mu\nu\alpha\beta}\left(\frac{1}{2}g^2 \ Tr \ F^{\mu\nu}F^{\alpha\beta}\right) \tag{3}$$

$$\partial_\mu B^\mu = n_f\partial_\mu L^\mu$$

where $F_{\mu\nu}$ is the strength tensor for the SU(2) gauge fields, g is the weak coupling constant and n_f the number of quark families. It can be easily shown that the R.H.S. of eq. (3) is proportional to $\partial_\mu K^\mu$ where

$$K_\mu = \frac{g^2}{64\pi^2}\epsilon_{\mu\nu\alpha\beta}Tr(A_\nu\partial_\alpha A_\beta - \frac{2ig}{3}A_\nu\partial_\alpha A_\beta) \tag{4}$$

The time component of K_μ is simply the Chern Simons number density of eq. (2). Integrating eq. (3) over 4-dimensional space between 3-dimensional hyperplanes $t = t_1$ and $t = t_2$, yields

$$\Delta L = [N_{CS}(t_2) - N_{CS}(t_1)] \tag{5}$$

$$\Delta B = n_f\Delta L$$

Thus, non conservation of baryon number B and lepton number L is related to a change in the Chern-Simons number N_{CS} of the SU(2) gauge vacuum.

Under normal conditions (i.e. $T = 0$ and/or low densities) the change from one vacuum configuration to another can be achieved by quantum tunneling. The transition rates are suppressed by a semi-classical factor[6]

$$e^{-4\pi/\alpha_w} \sim 10^{-150}$$

Therefore at zero temperature, the baryon number non conservation is negligible. This can be easily understood in terms of the very large barrier (~ 10 TeV) separating vacua of definite baryon number due to the sphaleron.

At finite temperature, thermal excitations over the barrier are, however, expected to increase the transition rate, leading to no suppression of the baryon number violation[7]. At intermediate temperatures, the rate for penetrating the sphaleron barrier is essentially given by the Boltzmann factor associated with forming a sphaleron,

$$\Gamma \sim (\alpha_w T)^4 \ e^{-E_{sp}(T)/T} \tag{6}$$

where the energy of the sphaleron at temperature T is given by

$$E_{sp}(T) = B\frac{M_w(T)}{\alpha_w}, \qquad M_w(T) : \text{W boson mass at temperature T}$$

At temperatures well above the electroweak phase transition, chiral symmetry is restored. The Higgs field has zero expectation value, and hence all particle masses vanish. The variation of the W boson mass with temperature is shown in Fig. 2.

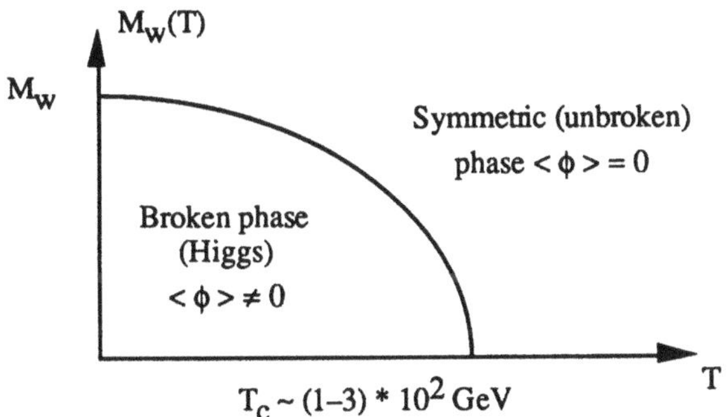

Fig. 2

Thus, for temperatures $T \gg T_c \sim 10^2$ GeV, the transition rate becomes

$$\Gamma \sim (\alpha_w T)^4$$

This rate is very large, and any pre-existing baryon and lepton asymmetry would be wiped out by the time of the electroweak phase transition. In particular, the BAU generated from GUTs at $T \sim 10^{15}$ GeV disappears at that time unless $B - L$ is not conserved.

The Electroweak Phase Transition

At any fixed temperature, the Higgs field ϕ adopts the value at which the free energy density is minimum. Since the EW phase transition temperature ($\sim 10^2$ GeV) is far above the quark-hadron phase transition temperature ($\sim 10^2$ MeV) the particles and antiparticles are quarks and leptons. Also, in this range of temperature the net quark and lepton number densities are small and therefore their chemical potentials are negligible. Then, in the simple approximation that the bosons and the fermions form a plasma of free quarks, leptons, gluons, W bosons etc..., the free energy density or the effective potential is given by

$$F(\phi,T) = V(\phi) \mp (k_B T)^4 \frac{1}{2\pi^2} \sum_{i=1}^{117} \int_0^\infty dk \; k^2 \log[1 \pm \exp(-(k^2 + \frac{\lambda_i^2 \phi^2}{(k_B T)^2})^{1/2})] \qquad (7)$$

where i refers to the particle species, λ_i is the coupling constant of particle i to the Higgs field (so that $\lambda_i \phi$ is the particle i mass) and $V(\phi)$ is the Higgs potential at zero temperature. The shapes of $F(\phi,T)$ are shown in Fig. 3.

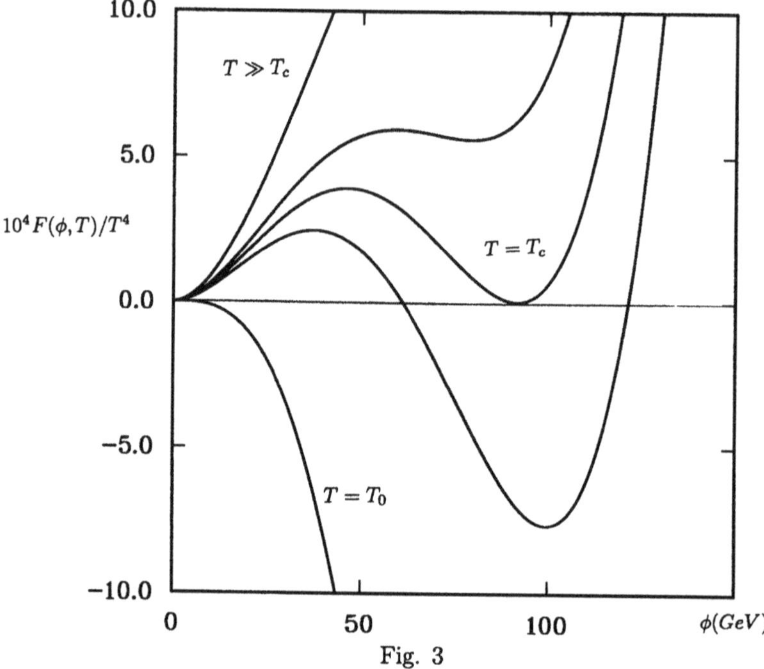

Fig. 3

For $T \gg T_c$, chiral symmetry is restored, the only minimum of the free energy occurs at $\phi = 0$ corresponding to the unbroken phase (the false vacuum). As the Universe cools down, a second minimum at a value $\phi \neq 0$, appears. This second minimum becomes degenerate with the unbroken phase minimum at a critical temperature T_c where a 1st order phase transition begins to take place. The phase transition proceeds via nucleation of bubbles of the broken phase $\phi \neq 0$ in the unbroken phase $\phi = 0$. The broken phase corresponds to the true vacuum. If a bubble of the broken phase appears with a too small radius it collapses, but if its radius is larger than some critical value, it expands. At some temperature lower than that corresponding to the absolute instability of the unbroken phase, all bubbles coalesce and produce a large bubble filling up the whole space.

The Bubble Nucleation

The dynamics of the bubble formation was studied in some details by Linde[8], and he proposed the following expression for the bubble formation rate per unit volume:

$$\Gamma_{\mathrm{bub}}(T) = T \left(\frac{S_3}{2\pi T} \right)^{3/2} \left[\frac{\det'(-\nabla^2 + \frac{\partial^2 F}{\partial \phi^2}(\phi,T))}{\det(-\nabla^2 + \frac{\partial^2 F}{\partial \phi^2}(\phi,T))} \right]^{-1/2} e^{-\beta S_3} \tag{8}$$

where det means that the zero modes have been removed. The 3-dimensional action S_3 is defined by

$$S_3 = \int d^3x \left[\frac{1}{2}(\vec{\nabla}\phi)^2 + F(\phi, T) \right] \tag{9}$$

The so called bounce solution is to be used in the above action

$$\frac{d^2\phi}{dr^2} + \frac{2}{r}\frac{d\phi}{dr} = \frac{\partial F}{\partial \phi}(\phi, T) \tag{10}$$

This solution satisfies the boundary conditions $\phi \to 0$ as $r \to \infty$, and $\frac{d\phi}{dr} \to 0$ as $r \to 0$. The prefactor appearing in Γ_{bub} is very hard to calculate but by dimensional analyzis it must be roughly proportional to T^4 so that

$$\Gamma_{\text{bub}}(T) \sim T^4 \left(\frac{S_3}{2\pi T} \right)^{3/2} e^{-\beta S_3} \tag{11}$$

Fig. 4 illustrates the evolution of the bounce action S_3 with temperature.

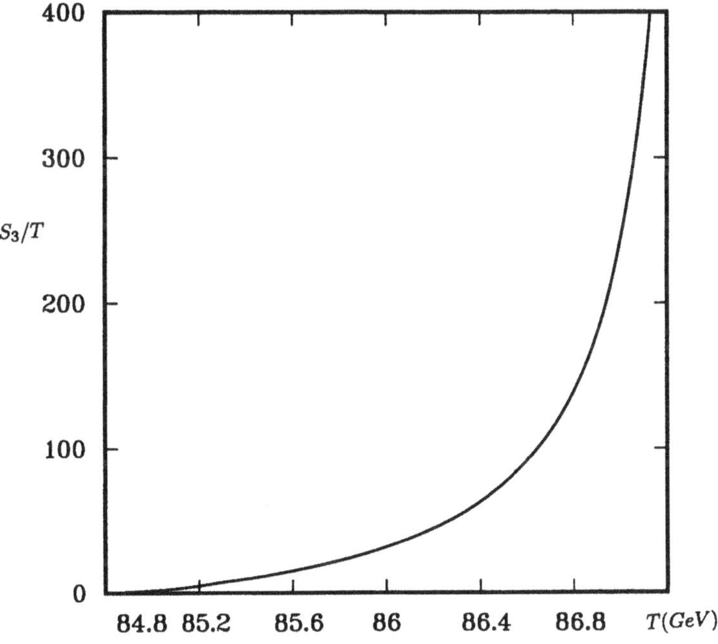

Fig. 4

It was obtained[9] by solving numerically eq. (10) with a Higgs meson mass $M_H \sim 50$ GeV and a top quark mass $M_t = 100$ GeV. Right after the phase transition, the high potential barrier between the unbroken and broken phases produces a very large action but as the Universe cools the barrier drops, and the action decreases exponentially.

The Bubble Expansion

The bubble nucleation rate given by eq. (8) or eq. (11) is the formation rate of bubbles that are large enough to expand. For a spherically symmetric expanding bubble centered at the origin, the Higgs field $\phi(\vec{r},t)$ satisfy the evolution equation

$$-\vec{\nabla}^2\phi + \frac{\partial^2\phi}{\partial t^2} + \eta\frac{\partial\phi}{\partial t} = -\frac{\partial F}{\partial\phi} \tag{12}$$

This equation is obtained by assuming that the total free energy (including the kinetic energy) of an expanding bubble is transfered during its propagation to the hot plasma of the unbroken phase. This gives rise to the damping term $\eta\frac{\partial\phi}{\partial t}$, and the parameter η can be estimated by assuming that the moving bubble front is equivalent to a beam of Higgs particles interacting with the particles in the plasma[10],[11]. Equation (12) can be solved numerically with boundary conditions corresponding to perturbations around the critical profiles of Fig. 3. The solutions of eq. (12) are shown in Fig. 5[9].

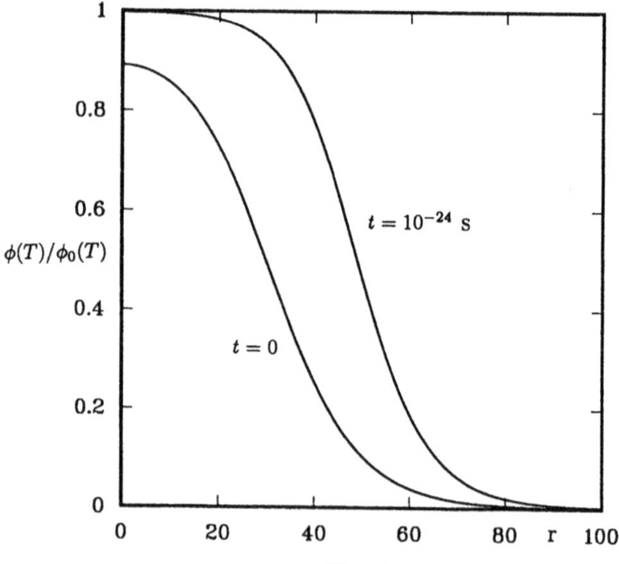

Fig. 5

The profile of the Higgs field at $T = 86.7$ GeV for $t = 0$ and $t = 10^{-24}$ s.

The Baryon Production

For the minimal EW standard model with one Higgs doublet, the CP violating contributions to the free energy density come from the relative phases in the quark mass matrix and turn out to be too small. With two Higgs doublets and CP violation in the Higgs sector, fermion loops approximation (e.g. in Fig. 6) gives rise contributions of the following form[12],[13]

$$F_{CP}(\phi,T) = \frac{g^2}{32\pi^2}\frac{1}{M^2 T^2}\theta Tr(\epsilon_{\mu\nu\alpha\beta}F^{\mu\nu}F^{\alpha\beta}) \tag{13}$$

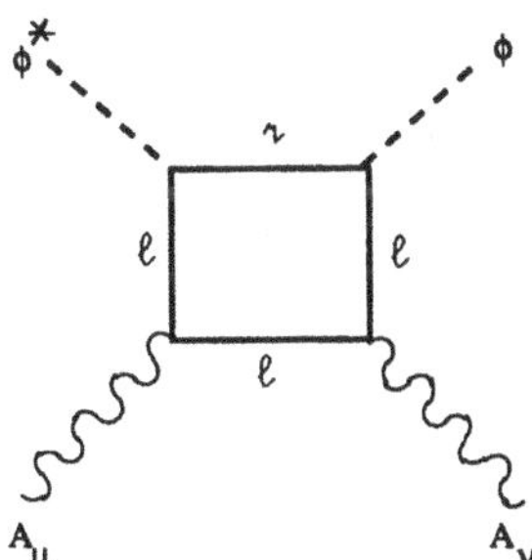

Fig. 6

where θ is the relative phase of the Higgs fields, the coefficient $\frac{1}{M^2}$ depends on the Higgs boson and the top quark masses and the VEV of the Higgs field at $T = 0$. Using the anomaly equation (3) and integrating by parts, one can related F_{CP} to the baryon number density ρ_B and the variation of the Higgs field. In the adiabatic limit where the Higgs field changes slowly, dropping the   spatial gradient

$$F_{CP} = -\frac{g^2}{3M^2 T^2}\frac{\partial\theta}{\partial t}\rho_B$$

The total free energy density can be written as

$$F_{\text{tot}} = F(\phi,T) + \frac{39}{4}\frac{1}{T^2}\rho_B^2 - \frac{1}{3M^2 T^2}\frac{\partial\theta}{\partial t}\rho_B \tag{14}$$

where the 1st two terms correspond to the CP conserving contributions. At thermal equilibrium $\frac{\partial F_{tot}}{\partial \rho_B} = 0$, which yields $\rho_B^{eq} = \frac{2}{39}\frac{1}{3M^2}\frac{\partial}{\partial t}\theta$.
The regime of interest is that near thermal equilibrium where the baryon number density satisfies the relaxation equation

$$\frac{d\rho_B}{dt} = \Gamma(T)\left[\rho_B(t) - \rho_B^{eq}(t)\right] \tag{15}$$

where $\Gamma(T)$ is the rate of sphaleron formation in the bubble.

360

This has the solution

$$\rho_B(t) = \frac{2g^2}{117M^2} \int_{-\infty}^{t} \left(T^2 \frac{\partial \theta}{\partial t'} \Gamma(t') e^{-\int_{t'}^{t} \Gamma(t'')dt''} \right) dt' \tag{16}$$

This solution exhibits the good properties that outside the bubble $\rho_B(t)$ is small because the sphaleron rate there is large enough to wash out any net baryon production, but inside the bubble $\rho_B(t)$ is not suppressed since the sphaleron rate there is exponentially small. Eq. (16) also shows that the baryons are essentially produced in the region where $\frac{\partial \theta}{\partial t}$ is the most significant i.e. near the bubble wall.

After the phase transition, the baryon number violation rate due to the sphaleron formation may be large compared to the expansion rate of the Universe. In this case, any asymmetry produced during the transition will be quickly wiped out. The requirement that the transition be low enough that this not to occur places constraints on the models. In the minimal standard model with one Higgs doublet, the Higgs mass must be smaller than a critical value $M_{crit} \sim 45$ MeV[14]. This is to be compared with the recent experimental lower bound from LEP which is ~ 57 GeV. For models with extended Higgs sector M_{crit} may be larger but depends on the details of the interaction, generally $M_{crit} \sim M_w$.

Conclusions

In this lecture, arguments are put forward to voice that there exists a possibility of explaining the baryon asymmetry of the Universe from the anomalous electroweak phase transition at $T \sim 100$ GeV. The CP violation in the minimum standard model with only one Higgs doublet gives insufficient baryon number violation to account for the value $\eta \sim 10^{-11}$. The multi-Higgs models are more promising since CP violating phases are larger.

Although the idea of BAU from anomalous EW standard model is very simple and clear once it is proposed, the detailed calculations are more foggy. The dynamics of the EW phase transition should be studied more carefully. For example, the order of the phase transition is not yet established, the prefactor appearing in the bubble nucleation rate has been only estimated but not really calculated, the nature of the bubble propagation, relativistic versus non relativistic, is still under discussion. On the other hand, the calculation of the CP violating contribution to the free energy is far of being complete. All these uncertainties, of course, afflict the prediction of the asymmetry parameter η. According to Linde[15] "the literature contains contradictory claims and statements on almost every important question".

Nevertheless, I believe that the concomittant presence of a rather well formulated theory with many still unsolved problems renders the EW theory of matter genesis even more interesting, besides the fact that it concerns one of the most challenging issues in Theoretical Physics and Cosmology.

Acknowledgements

Parts of this talk are based on work in progress with W.N. Cottingham and D. Kalafatis. I would like to acknowledge gratefully their collaboration.

References

[1] R. Jackiw and C. Rebbi, *Phys. Rev. Lett.* **37** (1976) 172
C.G. Callan, R.F. Dashen and D.J. Gross, *Phys. Lett.* **63B** (1976) 334.

[2] See e.g. E.W. Kolb and M.S. Turner, The Early Universe, Addison-Wesley Publ. Co (1990) 157.

[3] A.D. Sakharov, Pisma zh ETP 5 (1967) 32.

[4] N.S. Manton, *Phys. Rev.* **D28** (1983) 2019
F.R. Klinkhamer and N.S. Manton, *Phys. Rev.* **D30** (1984) 2212.

[5] S. Adler, *Phys. Rev.* **177** (1969) 2426
J.S. Bell and R. Jackiw, *Nuovo Cimento* **51** (1969) 47
W.A. Bardeen, *Phys. Rev.* **184** (1969) 1841.

[6] G. 't Hooft, *Phys. Rev. Lett.* **37** (1976) 8; *Phys. Rev.* **D14** (1976) 3432.

[7] V.A. Kuzmin, V.A. Rubakov and M.E. Shaposhnikov, *Phys. Lett.* **155B** (1985) 36.

[8] A.D. Linde, *Nucl. Phys.* **B216** (1983) 421.

[9] D. Kalafatis, private communication.

[10] N. Turok, Princeton preprint PUPT-91-1273.

[11] W.N. Cottingham, D. Kalafatis and R. Vinh Mau, In prepartion.

[12] M.E. Shaposhnikov, *JETP Lett.* **44** (1986) 465; *Nucl. Phys.* **B287** (1987) 757; Nucl. Phys. **B299** (1988) 797.

[13] L. Mc Lerran et al., *Phys. Lett.* **256B** (1991) 451.

[14] A.I. Bochkarev et al., *Phys. Lett.* **244B** (1990) 275.

[15] A. Linde et al., SLAC Preprint SU-ITP 92-6 (1992).

Authors

Participants

Per A. Amundsen
Rogaland University Center
P.O. Box 2557, Ullandhaug
Stavanger, N-4004
Norway

Véronique Bernard
PHTH, CRN
23, rue du Loess, BP 20 G
Strasbourg, F-67037
France

Tamás S. Biró
Institut für Theor. Phys.
Justus-Liebig-Univ. Giessen
Giessen, D-6300
Germany

Jacques Bossart
Institut für Theor. Physik
ETH Zürich-Hönggerberg
Zürich, CH-8093
Switzerland

Conrad Burden
Dept. of Physics
Australian National University
Canberra, ACT 2601
Australia

Peter Carruthers
Dept. of Physics
Univ. of Arizona
Tucson, AZ 85721

Hans Günter Dosch
Institut für. Theor. Physik
Universität Heidelberg
Philosophenweg 16
Heidelberg, D-6900, Germany

Emilio Elizalde
Dept. ECM, Fac. of Physics
University of Barcelona
Diagonal 647
Barcelona, E-08028, Spain

Marshall Baker
Dept. of Physics
Univeristy of Washington
Seattle, WA 98195

Christoph Best
Institut für Theor. Physik
J. W. Goethe-Universität
Frankfurt Am Main, D-6000
Germany

Jean-Paul Blaizot
Service de Physique Theorique
C.E.N. Saclay
Gif-sur-Yvette, F-91191
France

Detlev Bückers
Institut für Theor. Physik
Univ. Tübingen
Auf der Morgenstelle 14
Tübingen, D-7400, Germany

Alfons Capella
LPTHE
Universite Paris-Sud
ORSAY, F-91405
France

Mohamed Chabab
Lab. de Phys.-Math.
Univ. de Montpellier
Montpellier Cedex, 34095
France

Alessandro Drago
Dept. of Physics
Univ. Ferrara
Via Paradiso, 12
Ferrara, 44100, Italy

Stefano Forte
INFN, Sezione di Torino
via P. Giuria 1
Torino, I-10125
Italy

Herbert M. Fried
Department of Physics
Brown University
Providence, RI 02912

V. N. Gribov
Landau Institute
GSP-1 117940
Moscow, 117 334
Russia

Henning Heiselberg
Niels-Bohr Institute
Blegdamsvej 17
Copenhagen, DK-2100
Denmark

Jiří Hošek
Nuclear Physics Institute
2506 Rez near Prague
Czechoslovakia

David E. Kahana
Center for Nuclear Research
Kent State University
Kent, OH 44240

Kyungsik Kang
Dept. of Physics
Brown University
Providence, RI 02912

André Krzywicki
Univ. de Paris XI
LPTHE, Bat. 211
Orsay Cedex, F-91405
France

Michel LeBellac
INLN, Parc Valrose
Université de Nice
Nice Cedex 2, F-06108
France

Pawel Górnicki
MPI für Kernphysik
Saupfercheckweg
Heidelberg, D-6900
Germany

R. Haymaker
Dept. of Physics & Astronomy
Louisiana State University
Baton Rouge, LA 70803

Kent Hornbostel
Dept. of Physics
Ohio State University
Columbus, OH 43210

Sidney H. Kahana
Physics Department
Brookhaven National Laboratory
Upton, New York 11973

Dimitri Kalafatis
Div. de Physique Théorique
IPN/Orsay
1, rue Georges Clemenceau
Orsay Cedex, F-91406, France

Siegfried Krewald
KFA Jülich
Institut für Kernphysik
Jülich, D-5170
Germany

Gideon Lana
School of Physics and Astron.
Tel Aviv University
Tel Aviv, 69978
Israel

Peter Levai
Central Res. Inst. for Physics
P.O. Box 49
Budapest 114, H-1525
Hungary

Harald Markum
Inst. f. Kernphysik
Tech. Univ. Wien
Vienna, A-1040
Austria

Berndt Müller
Dept. of Physics
Duke University
Durham, NC 27708-0305

Robi Peschanski
Service Physique Théor.
CEN Saclay
Gif-sur-Yvette Cedex, F-91191
France

Janos Polonyi
CTP 6-314
MIT
Cambridge, MA 02139

Johann Rafelski
Dept. of Physics, Bldg. 81
University of Arizona
Tucson, AZ 85721

Patrick Reynaud
Institut Non Linéaire de Nice
UMR C.N.R.S.
Nice Cedex 2, F-06108
France

Georges Ripka
Service Physique Théor.
CES
Gif-sur-Yvette Cedex, F-91191
France

Serge Rudaz
School of Physics & Astron.
University of Minnesota
Minneapolis, MN 55455

André Martin
TH
CERN
Geneva 23, CH-1211
Switzerland

Poul Olesen
Niels Bohr Institute
Blegdamsvej 17
Copenhagen, DK-2100
Denmark

Bernard Pire
CPT
Ecole Polytechnique
Palaiseau Cedex, F-91128
France

Jean Potvin
Dept. of Science & Math.
Parks College, St. Louis Univ.
Cahokia, IL 62206

Hugo Reinhardt
Institut für Theor. Physik
Univ. of Tübingen
Tübingen, D-7400
Germany

Jean-Marc Richard
Inst. des Science Nucléaires
Univ. Joseph-Fourier
Grenoble Cedex, F-38026
France

Craig Roberts
Physics 203
Argonne National Lab.
Argonne, IL 60439-4843

Ivo Sachs
ETH Zürich
Institut für Theor. Physik
Zürich, CH-8093
Switzerland

368

Martin Schaden
Institut für Theor. Physik
Tech. Univ. München
Garching, D-8046
Germany

Alec J. Schramm
Dept. of Physics
Duke University
Durham, NC 27708-0305

J. Ely Shrauner
Physics Department
Washington University
St. Louis, MO

Madeleine Soyeur
Lab. Nationale Saturne
CEN de Saclay
Gif-sur-Yvette Cedex, F-9119
France

Katsumi Tanaka
Dept. of Physics
Ohio State University
Columbus, OH 43210

Vicente Vento
Dept. Fisica Teorica
Univ. of Valencia
Burjasot, Valencia, E-46100
Spain

Ken Yee
Physics Theory #510A
Brookhaven National Lab
Upton, NY 11973

Dominique Schiff
LPTHE, Bat. 211
Université Paris XI
Orsay Cedex, F-91405
France

Dieter Schütte
Institut für Theor. Physik
Universität Bonn
Bonn, D-5300
Germany

Edvard Shuryak
Department of Physics
SUNY
Stony Brook, NY 11794

Chung-I Tan
Dept. of Physics
Brown University
Providence, RI 02912

Peter Tandy
Dept. of Physics
Kent State University
Kent, OH 44242

R. VinhMau
Div. Physique Théorique
L.P.N.
Orsay Cedex, F-94106
France

Craig Roberts

David Kahana

Edvard Shuryak

Robert Vinh Mau

William Cipolla (Peter Carruthers in back)

V. N. Gribov

Hans Günther Dosch

Eduardo Elizalde

Hugo Reinhardt

André Martin

Kent Hornbostel

Janos Polonyi

Top: The American University of Paris

Left: Richard Haymaker

Bottom: Poul Olesen

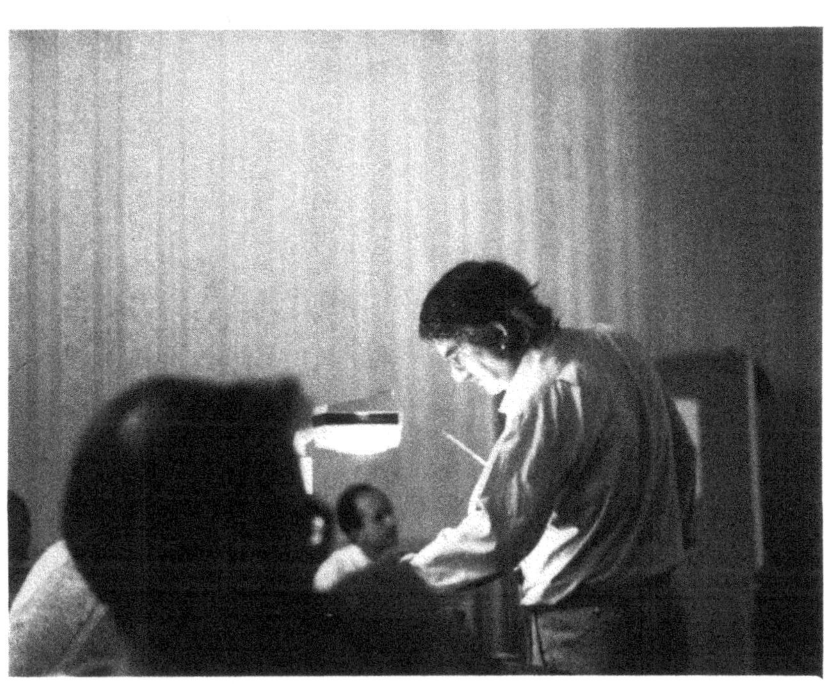

www.ingramcontent.com/pod-product-compliance
Lightning Source LLC
Chambersburg PA
CBHW060748240726
48664CB00009BA/1661